Salicylic Acid Signalling in Plants

Salicylic Acid Signalling in Plants

Special Issue Editors

Tibor Janda
Magda Pál
Gabriella Szalai

MDPI • Basel • Beijing • Wuhan • Barcelona • Belgrade • Manchester • Tokyo • Cluj • Tianjin

MDPI

Special Issue Editors
Tibor Janda
Agricultural Institute,
Centre for Agricultural Research
Hungary

Magda Pál
Agricultural Institute,
Centre for Agricultural Research
Hungary

Gabriella Szalai
Agricultural Institute, Centre for Agricultural Research
Hungary

Editorial Office
MDPI
St. Alban-Anlage 66
4052 Basel, Switzerland

This is a reprint of articles from the Special Issue published online in the open access journal *International Journal of Molecular Sciences* (ISSN 1422-0067) (available at: https://www.mdpi.com/journal/ijms/special_issues/salicylic_acid_signalling).

For citation purposes, cite each article independently as indicated on the article page online and as indicated below:

LastName, A.A.; LastName, B.B.; LastName, C.C. Article Title. *Journal Name* **Year**, *Article Number*, Page Range.

ISBN 978-3-03928-981-3 (Hbk)
ISBN 978-3-03928-982-0 (PDF)

Cover image courtesy of Szalai Gabriella.

Contents

About the Special Issue Editors

Tibor Janda is the Head of the Department of Plant Physiology, Agricultural Institute, Centre for Agricultural Research in Martonvásár, Hungary. He received his M.S. degree from the József Attila University in Szeged as a molecular biologist and biotechnologist in 1992. He is a member of several scientific committees and editorial boards. His research interests include the effects of low temperature and photoinhibition on photosynthesis; physiology and biochemistry of abiotic stress tolerance in plants, especially in cereals; the effects of salicylic acid and related compounds on stress tolerance of crop plants; the role of antioxidants in stress tolerance of plants; the role of light in the development of stress tolerance in plants; and cross-talk between signaling and acclimation processes in plants. He has published over 200 peer-reviewed articles and book chapters, and he has received more than 3000 independent citations

Magda Pál is a Senior Research Associate of Department of Plant Physiology, Agricultural Institute, Centre for Agricultural Research in Martonvásár, Hungary. She received her M.S. degree from the University of Debrecen as a molecular biologist and biotechnologist in 2001. She is a member of several scientific committees and editorial boards. Her research interests include physiology and biochemistry of abiotic stress tolerance in plants, especially in cereals; cadmium-induced plant responses and cadmium tolerance at metabolite and gene expression levels; the effects of salicylic acid and polyamines on stress tolerance of crop plants; and the relationship between polyamines and other plant hormones or protective compounds. She has published over 100 peer-reviewed articles and book chapters, and she has received more than 700 independent citations.

Gabriella Szalai is scientific consultant with the Department of Plant Physiology, Agricultural Institute, Centre for Agricultural Research in Martonvásár, Hungary. She received her M.S. degree from the József Attila University in Szeged as a molecular biologist and biotechnologist in 1993. She is a member of several scientific committees and editorial boards. Her research interests include the physiology and biochemistry of abiotic stress tolerance in plants, especially in cereals; the effects of salicylic acid and related compounds on stress tolerance of crop plants; the role of antioxidants in the stress tolerance of plants; the role of light in the development of stress tolerance in plants; and cross-talk between signaling and acclimation processes in plants. She has published approx. 200 peer-reviewed articles and book chapters, and she has greceivedot more than 3500 independent citations.

International Journal of
Molecular Sciences

MDPI

Editorial

Salicylic Acid Signalling in Plants

Tibor Janda *, Gabriella Szalai and Magda Pál

Department of Plant Physiology, Agricultural Institute, Centre for Agricultural Research, H-2462 Martonvásár, Hungary; szalai.gabriella@agrar.mta.hu (G.S.); pal.magda@agrar.mta.hu (M.P.)
* Correspondence: janda.tibor@agrar.mta.hu; Tel.: +36-22-569509

Received: 7 April 2020; Accepted: 8 April 2020; Published: 10 April 2020

Abstract: Ten articles published in the "Special Issue: Salicylic Acid Signalling in Plants" are summarized, in order to get a global picture about the mode of action of salicylic acid in plants, and about its interaction with other stress-signalling routes. Its ecological aspects and possible practical use are also discussed.

Keywords: acclimation; biosynthesis; cross talk; hormones; phenolics; plant growth and development; plant stress; salicylic acid; signalling

---

1. Introduction

Salicylic acid (SA) is ubiquitously distributed in the whole plant kingdom. The basal level of SA differs widely among species. It is generally present either in the free fraction or in the form of glycosylated, methylated, glucose-ester, or amino acid conjugates. In plants, SA can be synthesized via two distinct and compartmentalized enzymatic pathways, both requiring the primary metabolite chorismate. L-phenylalanine, derived from chorismate, can be converted into SA via the precursors free benzoic acid, benzoyl glucose, or *ortho*-hydroxy-cinnamic acid, depending on the plant species. Chorismate can also be converted into SA via isochorismate in the chloroplast. Several physiological processes in which SA may play a role have been reported, including seed germination, growth regulation, flower induction, thermogenesis, and especially, the regulation of plant responses under biotic or abiotic stress conditions. SA may be involved in different signalling processes. For example, various hormones involved in plant defence mechanisms crosstalk with SA, and both negative and positive interactions have been reported. SA signalling also leads to the reprogramming of gene expression and protein synthesis. It may affect the antioxidative metabolism, and it modulates cellular redox homeostasis. However, in spite of the extensive work on SA-related processes, the exact mode of action is poorly understood.

2. What Is Known So Far?

The effects and the role of SA in plant physiological processes have been widely studied for a long time. However, there are still a lot of open question in this field. This is also indicated by the fact that there are four review papers published on this topic, and they represent a wide range of approaches. These include a "classical" approach related to the role that SA plays in the presence of a stress factor—in this case, the heavy metal cadmium (Cd) [1]. Another review paper deals with another "classical" question: what is the mechanism of the SA perception, and which compounds are able to bind SA and transfer the message to further signalling routes [2]? The exact mechanisms related to SA are still poorly understood. This is also indicated by the fact that a relatively new approach, the involvement of SA in endoplasmic reticulum stress, has also been reviewed [3]. Finally, the last review addresses a possible practical use of SA, and its ecological consequences are also discussed [4].

To the best of our knowledge, heavy metal stress was one of the first processes where the protective role of SA in the case of an abiotic stressor was demonstrated. The treatment of cucumber or tobacco plants with SA induced tolerance against copper toxicity [5]. Similarly, SA treatment also ameliorated the effects Pb^{2+} or Hg^{2+} stresses in rice [6]. The pre-treatment of barley seedlings with SA reduced the oxidative damage caused by Cd, leading to increased biomass [7]. Cd is one of the most important pollutants, and may cause severe stress to plants, and through the food chain to humans, too. In order to reduce Cd injury, plants have evolved various cross-linked strategies, such as binding it to the cell wall [8], chelation with phytochelatins [9], and involvement of the antioxidant system [10].

The fact that SA signalling may also be a part of Cd tolerance is indicated by the fact that usually, exposure to Cd induces the endogenous SA level better than any other abiotic stressors [11]. Guo and co-workers recently reviewed the possible methods of application of exogenous SA to protect plants from Cd toxicity; they also reviewed the possible mechanisms that are induced by SA, and which play a role in the defence processes [1]. It is very important to emphasise that besides application mode, the effects of exogenous SA depend on various other factors, such as the concentrations of Cd and SA, the plant species, and the plant's developmental stage. Based on various tested systems, Guo and co-workers concluded that the levels of SA in pre-soaking treatments or when used as foliar spray are generally higher than those in hydroponic treatments. This indicates that the uptake of SA through the root system is better regulated than via the leaves or the seeds [1]. The mode of action of SA during Cd stress is still poorly understood, and results are often controversial. In spite of the high number of successful applications of exogenous SA, results using mutant or transgenic *Arabidopsis* plants indicate that a high SA level may also be harmful [12]. However, using the *sid2* mutant supports the view that SA is necessary for efficient defence processes [13]. These results support the view that the SA level must also be fine-tuned in order to achieve the most effective protective effect. The defence against Cd is usually complex in plants, and it seems that SA affects various elements of it. SA has a dose-dependent effect on plant growth; however, while SA is also linked with other plant growth regulators, this effect may also be secondary, due to induction other protective mechanisms [14]. As a direct protective mode of action, SA may also increase the accumulation of Cd in the cell wall and prevent its translocation into other cell organelles. The possible effect of SA on the cell wall components has also been demonstrated on biotic stresses, where the cell wall is indeed the first line of the defence system [15]. As in the case of many others stressors, the induction of the antioxidant system seems a general mechanism to protect plants against the oxidative damage induced by excess levels of heavy metals. For example, in wheat, although a direct relationship was not found between the initial SA levels and the degree of Cd tolerance, the increased SA level in the root during cadmium stress was related to the enhancement of the internal glutathione cycle, thus inducing the antioxidant and metal detoxification systems [16]. However, these results cannot be generalized in many species, and most probably the novel role of SA in Cd toxicity will likely continue to be unveiled [1].

The study of SA-associated signalling pathways began almost simultaneously with the discovery of the role of SA in defence against pathogens. One of the most critical question was, and still is: what is (are) the main receptor(s) of the SA signal, and which proteins bind the stress-induced SA? NPR1 (nonexpresser of pathogenesis-related protein 1), as one of the salicylic acid binding proteins (SABPs), is a key transcriptional regulator of SA signalling. However, the many controversial data on the mode of action of SA indicates that SA signalling is not a simple "yes" or "no" question. Pokotylo and co-workers recently reviewed the possible groups of SABPs [2], listing a high number of different candidates. These include the well-known factors, such as catalase [17,18] or NPR proteins [19,20], as well as some others that are less frequently connected with SA signalling. Besides the characterization of the possible SABPs, it is also an important finding that although some SABPs are well-synchronised at the transcriptomic level, genes responsible for different SABPs can be both positively and negatively expressed in response to elicitation. Clarification of the interactions between the different regulating factors, and a better understanding the molecular mechanisms of interaction between SA and its binding proteins are among the main problems in the future.

The endoplasmic reticulum (ER) has multiple cellular functions, including protein synthesis. Unfavourable environmental conditions may disturb the ER homeostasis, causing ER stress [21]. This is accompanied by the accumulation of unfolded or misfolded proteins triggering the unfolded protein response. SA is a multifaceted compound, and it has been proposed that SA may also have role during ER stress. Poór and co-workers recently summarised the possible mode of action of SA in this process [3]. SA probably acts via regulating the redox homeostasis and inducing, directly or indirectly, specific transcription factors. Furthermore, the involvement of polyamines can also be assumed. Unfortunately, there are still a lot of open question in this field. Answering these questions about the role of SA and other phytohormones in ER stress and in the unfolded protein response may help to develop new strategies in agricultural research and practice, in order to protect plants against environmental stresses.

Present agriculture will probably have to face an important challenge soon: feeding the quickly increasing human population may require a second green revolution. Improvement of the stress tolerance of cultivated plants, and thus the increase of crop yields and nutritional values in environmentally friendly ways, is a crucial task in food production. Exogenous application of naturally occurring, biologically active compounds like SA can be an alternative approach to improving crop productivity under changing environmental conditions. The replacement of synthetic chemicals with natural plant secondary compounds could be an excellent option from economic and environmental points of view. With the application of these naturally occurring biologically active compounds, the acclimation processes can be intensified. However, there are still several open questions that must be answered before any of these compounds can be recommended in a responsible manner for practical use, especially under field conditions. The compounds' effects must be known both for the plants where they are used, and for the wider environment. A recent review by Filgueiras and co-workers provides a detailed overview about the induction pathways induced by SA at the primary (genetic, metabolomic, and physiologic changes in plants), secondary (effects on herbivores and pathogens attacking the plants), and tertiary levels (with consequences for herbivore populations) [4]. For example, in addition to the investigation of the primary molecular effects of SA, another potential approach for the control of herbivores in agricultural systems through defences related to SA is attraction of natural enemies of the insects. The release of volatile SA-related compounds may also contribute to the reduction of the herbivore populations [22].

3. What Are the New Directions?

In spite of the intense research on the mode of action and possible practical use of SA, there are still a lot of open questions. Six research articles have also been published in the present "Special Issue". These papers cover a wide range of aspects of SA-related mechanisms. They include a "classical" demonstration of the importance of SA in biotic stress [23] and the importance of the use of various mutants related to SA signalling [24,25], as well as focus on the crosstalk with other stress-related compounds [25,26]. There are also works investigating the effects of environmental factors on the SA-related signalling [26,27], and on the role of different SA analogue compounds [28].

The involvement of SA in biotic stress-related processes have been widely studied. However, Shi and co-workers were the first to demonstrate the responses of SA biosynthesis genes and molecular functions of SA in the response to anthracnose disease in tea plants [23]. This disease is induced by *Colletotrichum* fungi and may cause a 5–20% loss of tea yield. Data support the view that SA and its related signalling networks, including the induction of pathogenesis-related protein 1 (PR1) and connection with other plant hormones, play a pivotal role in the activation of tea immunity to anthracnose disease. They also provided a transcriptome dataset to profile gene expression and metabolic networks associated with tea plant immunity against anthracnose [23].

The fact that up to now, more than forty *Arabidopsis* mutants or transgenic lines with modified levels of SA or constitutively activated SA signalling pathways have been described also indicates the importance of SA signalling. The mutant collection presented by Pluhařová and co-workers provides a very valuable tool to better understand the mechanisms underlying trade-offs between growth and defence in plants. The authors present a novel research study, providing new insights clarifying a link between SA and plant behaviour under environmental stresses [24]. They showed a negative correlation between the SA content and the rosette size, but not the root growth. This is especially important, because hydroponically-added SA may often reduce root growth more than the shoots. Their results also draw attention to the importance of light intensity when data obtained under different growth conditions are compared.

SA signalling is not a simple linear route. SA may interact with several other stress-related compounds. The importance of polyamines in living cells has been known for several decades. However, their signalling role has only become apparent in recent years [29]. Tajti and co-workers recently demonstrated a possible crosstalk between the SA and polyamine signalling pathways [25]. They showed that the SA-deficient *Arabidopsis* mutant, *sid2* plants, could be characterised with a different polyamine metabolism. The *sid2* mutant plants also showed different responses to the exogenous polyamine treatments than the wild-type plants. Significant differences in the SA content and synthesis were also found between wild type and the SA-deficient, mutant *Arabidopsis* plants after polyamine treatments.

The efficiency of adaptation processes is influenced not only by genotype, but also by environmental factors. It has been shown that light during the acclimation period is required for the development of both the efficient freezing tolerance in frost-hardy cereals, and for the cold tolerance in the chilling-sensitive maize plants [30,31]. However, the interaction of light-mediated signalling and the development of cold tolerance processes is still poorly understood. A recent metabolomic research presented by Pál and co-workers demonstrated that different light conditions during the cold acclimation period differentially affected certain stress-related mechanisms in young maize plants, both in the root and in the leaves [26]. Chilling-induced accumulation of SA was light-dependent. The observed changes were also accompanied by hormonal and metabolic shifts.

SA signalling depends not only on the abiotic, but on the biotic environmental conditions too. It has been demonstrated that the inoculation of *Mentha x piperita* plants with different *Rhizobacteria* strains may increase the endogenous SA production [32]. Cappellari and co-workers have also now demonstrated that the inoculation of *M. piperita* with plant growth-promoting *Rhizobacteria* may modify the effects of exogenous application of SA or methyl jasmonate. Exogenous SA increased the total phenol content in this plant, and depending on the concentration used, certain *Rhizobacteria* could improve this effect. The exogenous SA also modified the production of the main monoterpene compounds [27]. These results suggest that a combination of SA or other plant growth regulator with certain plant growth-promoting *Rhizobacteria* may improve the production of secondary metabolites in aromatic plants.

It was suggested a long time ago that besides SA, some of its related compounds (synthetic ones or its natural precursors) may also have similar biological effects as SA itself [33]. Recently, Palmer and co-workers characterised several SA analogues, which were able to increase the strength of interactions among NPR3/4 in a yeast two-hybrid system. They induced NPR1 accumulation and the expression of PR1 in *Arabidopsis* plants, and were also able to inhibit the growth of plant pathogen bacteria, such as the citrus greening pathogen *Candidatus liberibacter* [28].

4. Conclusions

We hope that the reviews and studies published here in this special issue bring a closer understanding of the role of SA and an insight into its complex tasks, as well as a new direction for where there are still gaps and open questions that need to be explored, both at the metabolite and gene expression level, in the use of agriculturally important crop plants or mutant model plants, and not least of all, in both basic research and practical usage.

Author Contributions: Conceptualization, T.J.; original draft preparation, T.J.; writing—review and editing, T.J., G.S., and M.P.; funding acquisition, G.S. All authors have read and agreed to the published version of the manuscript.

Funding: This work was funded by a grant from National Research, Development and Innovation Office (grant No. K124430).

Acknowledgments: We thank all contributors to this "Special Issue".

Conflicts of Interest: The authors declare no conflict of interest.

Abbreviations

ER	Endoplasmic reticulum
NPR	Nonexpresser of pathogenesis-related
PR1	Pathogenesis-related protein 1
SA	Salicylic acid
SABP	Salicylic acid binding protein

References

1. Guo, B.; Liu, C.; Liang, Y.; Li, N.; Fu, Q. Salicylic acid signals plant defence against cadmium toxicity. *Int. J. Mol. Sci.* **2019**, *20*, 2960. [CrossRef] [PubMed]
2. Pokotylo, I.; Kravets, V.; Ruelland, E. Salicylic acid binding proteins (SABPs): The hidden forefront of salicylic acid signalling. *Int. J. Mol. Sci.* **2019**, *20*, 4377. [CrossRef] [PubMed]
3. Poór, P.; Czékus, Z.; Tari, I.; Ördög, A. The multifaceted roles of plant hormone salicylic acid in endoplasmic reticulum stress and unfolded protein response. *Int. J. Mol. Sci.* **2019**, *20*, 5842. [CrossRef] [PubMed]
4. Filgueiras, C.C.; Martins, A.D.; Pereira, R.V.; Willett, D.S. The ecology of salicylic acid signaling: Primary, secondary and tertiary effects with applications in agriculture. *Int. J. Mol. Sci.* **2019**, *20*, 5851. [CrossRef] [PubMed]
5. Strobel, N.E.; Kuc, J.A. Chemical and biological inducers of systemic resistance to pathogens protect cucumber and tobacco plants from damage caused by paraquat and cupric chloride. *Phytopathology* **1995**, *85*, 1306–1310. [CrossRef]
6. Mishra, A.; Choudhuri, M.A. Effects of salicylic acid on heavy metal-induced membrane deterioration mediated by lipoxygenase in rice. *Biol. Plant.* **1999**, *42*, 409–415. [CrossRef]
7. Metwally, A.; Finkemeier, I.; Georgi, M.; Dietz, K.-J. Salicylic acid alleviates the cadmium toxicity in barley seedlings. *Plant Physiol.* **2003**, *132*, 272–281. [CrossRef]
8. Wójcik, M.; Tukiendorf, A. Cadmium uptake, localization and detoxification in *Zea mays*. *Biol. Plant.* **2005**, *49*, 237–245. [CrossRef]
9. Pál, M.; Horváth, E.; Janda, T.; Páldi, E.; Szalai, G. The effect of cadmium stress on phytochelatin, thiol and polyamine content in maize. *Cereal Res. Commun.* **2006**, *34*, 65–68.
10. Schützendübel, A.; Schwanz, P.; Teichmann, T.; Gross, K.; Langenfeld-Heyser, R.; Godbold, D.L.; Polle, A. Cadmium-induced changes in antioxidative systems, hydrogen peroxide content, and differentiation in Scots pine roots. *Plant Physiol.* **2001**, *127*, 887–898. [CrossRef]
11. Pál, M.; Horváth, E.; Janda, T.; Páldi, E.; Szalai, G. Cadmium stimulates the accumulation of salicylic acid and its putative precursors in maize (*Zea mays*) plants. *Physiol. Plant.* **2005**, *125*, 356–364. [CrossRef]
12. Tao, S.; Sun, L.; Ma, C.; Li, L.; Li, G.; Hao, L. Reducing basal salicylic acid enhances *Arabidopsis* tolerance to lead or cadmium. *Plant Soil* **2013**, *372*, 309–318. [CrossRef]
13. Guo, B.; Liu, C.; Li, H.; Yi, K.; Ding, N.; Li, N.; Lin, Y.; Fu, Q. Endogenous salicylic acid is required for promoting cadmium tolerance of *Arabidopsis* by modulating glutathione metabolisms. *J. Hazard. Mater.* **2016**, *316*, 77–86. [CrossRef] [PubMed]
14. Gondor, O.K.; Pál, M.; Darkó, É.; Janda, T.; Szalai, G. Salicylic acid and sodium salicylate alleviate cadmium toxicity to different extents in maize (*Zea mays* L.). *PLoS ONE* **2016**, *11*, e0160157. [CrossRef] [PubMed]
15. Nishimura, M.T.; Stein, M.; Hou, B.H.; Vogel, J.P.; Edwards, H.; Somerville, S.C. Loss of a callose synthase results in salicylic acid-dependent disease resistance. *Science* **2003**, *301*, 969–972. [CrossRef] [PubMed]
16. Kovács, V.; Gondor, O.K.; Szalai, G.; Darkó, É.; Majláth, I.; Janda, T.; Pál, M. Synthesis and role of salicylic acid in wheat varieties with different levels of cadmium tolerance. *J. Hazard. Mater.* **2014**, *280*, 12–19. [CrossRef]
17. Chen, Z.; Klessig, D.F. Identification of a soluble salicylic acid-binding protein that may function in signal transduction in the plant disease-resistance response. *Proc. Natl. Acad. Sci. USA* **1991**, *88*, 8179–8183. [CrossRef]

18. Horváth, E.; Janda, T.; Szalai, G.; Páldi, E. In vitro salicylic acid inhibition of catalase activity in maize: Differences between the isozymes and a possible role in the induction of chilling tolerance. *Plant Sci.* **2002**, *163*, 1129–1135. [CrossRef]
19. Wu, Y.; Zhang, D.; Chu, J.Y.; Boyle, P.; Wang, Y.; Brindle, I.D.; De Luca, V.; Després, C. The *Arabidopsis* NPR1 protein is a receptor for the plant defense hormone salicylic acid. *Cell Rep.* **2012**, *1*, 639–647. [CrossRef]
20. Fu, Z.Q.; Yan, S.; Saleh, A.; Wang, W.; Ruble, J.; Oka, N.; Mohan, R.; Spoel, S.H.; Tada, Y.; Zheng, N.; et al. NPR3 and NPR4 are receptors for the immune signal salicylic acid in plants. *Nature* **2012**, *486*, 228–232. [CrossRef]
21. Liu, J.X.; Howell, S.H. Endoplasmic reticulum protein quality control and its relationship to environmental stress responses in plants. *Plant Cell* **2010**, *22*, 2930–2942. [CrossRef] [PubMed]
22. Filgueiras, C.C.; Willett, D.S.; Junior, A.M.; Pareja, M.; El Borai, F.; Dickson, D.W.; Stelinski, L.L.; Duncan, L.W. Stimulation of the salicylic acid pathway aboveground recruits entomopathogenic nematodes belowground. *PLoS ONE* **2016**, *11*, e0154712. [CrossRef] [PubMed]
23. Shi, Y.-L.; Sheng, Y.-Y.; Cai, Z.-Y.; Yang, R.; Li, Q.-S.; Li, X.-M.; Li, D.; Guo, X.-Y.; Lu, J.-L.; Ye, J.-H.; et al. Involvement of salicylic acid in anthracnose infection in tea plants revealed by transcriptome profiling. *Int. J. Mol. Sci.* **2019**, *20*, 2439. [CrossRef] [PubMed]
24. Pluhařová, K.; Leontovyčová, H.; Stoudková, V.; Pospíchalová, R.; Maršík, P.; Klouček, P.; Starodubtseva, A.; Iakovenko, O.; Krčková, Z.; Valentová, O.; et al. "Salicylic acid mutant collection" as a tool to explore the role of salicylic acid in regulation of plant growth under a changing environment. *Int. J. Mol. Sci.* **2019**, *20*, 6365. [CrossRef] [PubMed]
25. Tajti, J.; Hamow, K.Á.; Majláth, I.; Gierczik, K.; Németh, E.; Janda, T.; Pál, M. Polyamine-induced hormonal changes in *eds5* and *sid2* mutant *Arabidopsis* plants. *Int. J. Mol. Sci.* **2019**, *20*, 5746. [CrossRef]
26. Pál, M.; Janda, T.; Majláth, I.; Szalai, G. Involvement of salicylic acid and other phenolic compounds in light-dependent cold acclimation in maize. *Int. J. Mol. Sci.* **2020**, *21*, 1942. [CrossRef]
27. Cappellari, L.D.R.; Santoro, M.V.; Schmidt, A.; Gershenzon, J.; Banchio, E. Improving phenolic total content and monoterpene in *Mentha x piperita* by using salicylic acid or methyl jasmonate combined with *Rhizobacteria* inoculation. *Int. J. Mol. Sci.* **2019**, *21*, 50. [CrossRef]
28. Palmer, I.A.; Chen, H.; Chen, J.; Chang, M.; Li, M.; Liu, F.; Fu, Z.Q. Novel salicylic acid analogs induce a potent defense response in *Arabidopsis*. *Int. J. Mol. Sci.* **2019**, *20*, 3356. [CrossRef]
29. Pál, M.; Szalai, G.; Janda, T. Speculation: Polyamines are important in abiotic stress signaling. *Plant Sci.* **2015**, *237*, 16–23. [CrossRef]
30. Janda, T.; Majláth, I.; Szalai, G. Interaction of temperature and light in the development of freezing tolerance in plants. *J. Plant Growth Regul.* **2014**, *33*, 460–469. [CrossRef]
31. Szalai, G.; Majláth, I.; Pál, M.; Gondor, O.K.; Rudnóy, S.; Oláh, C.; Vanková, R.; Kalapos, B.; Janda, T. Janus-faced nature of light in the cold acclimation processes of maize. *Front. Plant Sci.* **2018**, *9*, 850. [CrossRef] [PubMed]
32. del Rosario Cappellari, L.; Santoro, M.V.; Schmidt, A.; Gershenzon, J.; Banchio, E. Induction of essential oil production in *Mentha x piperita* by plant growth promoting bacteria was correlated with an increase in jasmonate and salicylate levels and a higher density of glandular trichomes. *Plant Physiol. Biochem.* **2019**, *141*, 142–153. [CrossRef] [PubMed]
33. Janda, T.; Szalai, G.; Antunovics, Z.; Horváth, E.; Páldi, E. Effect of benzoic acid and aspirin on chilling tolerance and photosynthesis in young maize plants. *Maydica* **2000**, *45*, 29–33.

International Journal of
Molecular Sciences

MDPI

Review

Salicylic Acid Signals Plant Defence against Cadmium Toxicity

Bin Guo [1,2,*], Chen Liu [1], Yongchao Liang [2,3], Ningyu Li [1] and Qinglin Fu [1,*]

1 Institute of Environment, Resource, Soil and Fertilizer, Zhejiang Academy of Agricultural Sciences, Hangzhou 310021, China; liuchen@mail.zaas.ac.cn (C.L.); liningyu259@126.com (N.L.)
2 Key Laboratory of Environment Remediation and Ecological Health, Ministry of Education, Hangzhou 310058, China; ycliang@zju.edu.cn
3 College of Environmental and Resource Sciences, Zhejiang University, Hangzhou 310058, China
* Correspondence: guob@mail.zaas.ac.cn (B.G.); fuql161@aliyun.com (Q.F.); Tel.: +86-571-86404041 (B.G.); +86-571-86404360 (Q.F.)

Received: 20 May 2019; Accepted: 13 June 2019; Published: 18 June 2019

Abstract: Salicylic acid (SA), as an enigmatic signalling molecule in plants, has been intensively studied to elucidate its role in defence against biotic and abiotic stresses. This review focuses on recent research on the role of the SA signalling pathway in regulating cadmium (Cd) tolerance in plants under various SA exposure methods, including pre-soaking, hydroponic exposure, and spraying. Pretreatment with appropriate levels of SA showed a mitigating effect on Cd damage, whereas an excessive dose of exogenous SA aggravated the toxic effects of Cd. SA signalling mechanisms are mainly associated with modification of reactive oxygen species (ROS) levels in plant tissues. Then, ROS, as second messengers, regulate a series of physiological and genetic adaptive responses, including remodelling cell wall construction, balancing the uptake of Cd and other ions, refining the antioxidant defence system, and regulating photosynthesis, glutathione synthesis and senescence. These findings together elucidate the expanding role of SA in phytotoxicology.

Keywords: salicylic acid; cadmium; reactive oxygen species; antioxidant defence system; glutathione

1. Introduction

Cadmium (Cd) is one of the most toxic pollutants for all living organisms with a long biological half-life [1,2]. It exists in soil naturally by weathering of the parent rocks and accumulates through anthropogenic activities, such as deposition of airborne Cd from smelting and mining, wastewater irrigation, application of Cd-contaminated phosphate fertilizers and soil amendments [3–6]. Cd is a non-essential element for plants, and its uptake by plants has posed a serious health issue to humans through the food chain. For this reason, Cd phytotoxicity is a major subject in current research on plant biology [7].

Cadmium can be absorbed easily by plant roots, and transported to plant shoots, leading to various visible toxic symptoms, such as growth retardation, wilting, leaf chlorosis, and cell death [8,9]. The mechanisms of Cd toxicity include replacing and inactivating essential elements and destroying protein structure, thereby interfering with various physiological processes, such as photosynthesis [10], respiration [11], element assimilation [12], and cell division [13]. To defend against Cd injury, plants utilize many coordinated strategies, such as binding Cd to the cell wall [14], vacuolar retention through chelation via phytochelatins (PCs) [15] and upregulation of the antioxidant system [16]. Since all of the above mechanisms are closely cross-linked, it is important for plant survival that these metabolic reactions are adjusted accordingly via regulating signals.

Salicylic acid (SA), a simple phenolic molecule, has long been recognized as a potent phytohormone that regulates plant development and defence in higher plants [17]. The synthesis of SA occurs by two distinct and compartmentalized pathways. One pathway derives from phenylalanine and takes place in the cytoplasm. First, phenylalanine ammonia-lyase (PAL) converts phenylalanine into cinnamic acid; cinnamic acid is then decarboxylated to form benzoic acid and finally undergoes 2-hydroxylation to generate SA [18]. Mutation of PAL genes in *Arabidopsis* results in 50% decrease in pathogen-induced SA accumulation, suggesting that the PAL pathway indeed contributes to SA biosynthesis [19]. Another biosynthetic pathway is through isochorismate synthase that catalyses the conversion of chorismate into isochorismate. [20]. In this pathway, SA is generated in chloroplasts from chorismate by the synthesis of two isochorismate synthases, *ICS1* and *ICS2* [21]. Analysis of SA-deficient mutants, *sid2*, revealed that loss of ICS1 suppresses the pathogen-induced SA accumulation [20], whereas loss of both ICSes results in further reduction of SA concentration [21].

It has been well established that SA is a key signal regulating local and systemic plant defence responses against pathogens [20,22]. In this signalling pathway, SA-binding proteins, such as SABP2, NPR3, NPR4 and NPR1 with high affinity for SA, are considered to be SA receptors that induce the expression of pathogenesis-related (PR) proteins and trigger systemic acquired resistance (SAR) [23]. Moreover, SA has shown important roles in mediating plant responses to abiotic stresses [24], including drought [25], chilling [26], osmotic stress [27], thermogenesis [28] and heavy metal toxicity [29]. Recently, the beneficial roles of SA in enhancing plant Cd tolerance, which has been reported in a wide range of plant species, have drawn much attention. However, the negative effect of SA was also noted in castor bean seedlings in which pretreatment with SA aggravated Cd damage [30]. The effects of SA on Cd-stressed plants depend on many aspects, including the application mode, the concentrations of Cd and SA and endogenous SA in the tested plants as well as the different species and developmental stages of the plants. This review covered recent studies on (1) modes of SA treatment, (2) the diverse physiological roles of SA in reducing Cd toxicity, and (3) future prospects for research concerning SA in Cd-stressed plants.

2. Salicylic Acid (SA) Treatment Methods

Table 2 summarizes the recent findings in the literature on the roles of SA in Cd-stressed plants. Except for one study with castor bean and three experiments with SA-deficient mutants, most of the studies show that SA alleviates the effects of Cd damage. However, very few of these literatures are actually relevant for coming to a solution for Cd toxicity. Exogenous applications of SA are mainly performed via three modes: spraying SA onto leaves, presoaking seeds with SA solution, and hydroponic treatment of roots to SA. There were 46 international publications involved in 100 pairs of exogenous SA and Cd treatments and 5 publications involved in endogenous SA treatments. Figure 1 represented SA exposure concentrations plotted against Cd exposure concentrations. Among these studies, one in which Cd and SA treatments with 560 mg L^{-1} and 3000 μM, respectively, was excluded because the concentrations used were too high and may be misleading. Taken together, the following information can be inferred: (1) Concentrations used in Cd treatments range from 0.56 to 300 mg L^{-1} (mg kg^{-1}) with a mean of 31.6 mg L^{-1} (mg kg^{-1}) and a variable coefficient of 160%; (2) concentrations of SA spray treatments range from 10 to 2170 μM with a mean of 585 μM and a variable coefficient of 104%; (3) concentrations of SA used in seed presoaking treatments range from 100 to 1000 μM with a mean of 464 μM and a variable coefficient of 70%; and (4) concentrations of SA used in hydroponic exposure treatments range from 1 to 500 μM with a mean of 148 μM and a variable coefficient of 117%.

From these studies, it was determined that the levels of SA in spray or pre-soaking treatments are generally higher than those in hydroponic treatments, suggesting that the regulation of SA levels in plants through root uptake is more efficient than that in the other two modes of application. Furthermore, each mode has high variable coefficients of concentrations either for Cd or for SA treatments. It is interesting to note that partial application of SA to different organs (seeds, leaves or roots) helps plants to establish a systemic defence against Cd toxicity.

Table 1. Effect of salicylic acid (SA) on cadmium (Cd) tolerance in plants.

	SA Treatment	Cd Treatment	Timeline	Plant Species	Main Responses * Means Negative or No Effect	Reference
Spraying	600 µM, 10 days	22.5 mg L^{-1}	Simultaneous	Potato (*S. tuberosum* L.)	I, II, III, VII	[31]
	100 µM, 1 time	30, 60 and 120 mg kg^{-1} (pot)	Simultaneous	Peppermint (Mentha piperita)	I, II, III	[32]
	50 µM, 4 times in a 3-day interval	75, 150, and 300 mg kg^{-1}	Simultaneous	Oilseed rape (*Brassica napus*)	II, III, V, VI	[33]
	10, 50, 100, and 200 µM each day treated for 50 mL last 4 days	44.8 mg kg^{-1}	Pretreatment	Melon (*Cucumis melo* L.)	I, II, III	[34]
	500 µM, 1 time	40 mg kg^{-1}	Pretreatment	Soybean (*Glycine max L. cv. Liaoxing 1*)	I, II, III	[35]
	2170 µM 1 time	56 and 112 mg kg^{-1}	Simultaneous	Radish (*Raphanus sativus*)	I, *IV	[36]
	1000 µM for 10 mL, 45 times in a day interval	100 and 200 mg L^{-1}	Simultaneous	Indian mustard (*Brassica juncea*)	I, II, III, IV, V	[37]
Presoaking	500 µM for 24 h.	112 mg L^{-1} for 72 h.	Pretreatment	Mungbean (*Vigna radiata* L. Wilczek)	I, II	[38]
	250 or 1000 µM for 8 h	5.6 and 11.2 mg L^{-1} for 10 days	Pretreatment	Flax (*Linum usitatissimum* L.)	Lipids	[39]
	250 or 1000 µM for 8 h	5.6 and 11.2 mg L^{-1} for 10 days	Pretreatment	Flax (*Linum usitatissimum* L.)	I, II	[40]
	500 µM for 12 h	0.56, 1.12, and 5.60 mg L^{-1} for 7 days	Pretreatment	Kentucky bluegrass	I, II, III, *IV, V	[41]
	500 µM for 12 h	56 and 112 mg kg^{-1} for 56 days	Pretreatment	Wheat (*Triticum aestivum* L. cv. Giza 168)	I, II, III, *IV	[42]
	500 µM for 6 h	1.12, 1.68, and 2.80 mg L^{-1} for 14 days	Pretreatment	Maize (*Zea mays*)	VI	[43]
	250 or 1000 µM for 8 h	5.6 and 11.2 mg L^{-1} for 10 days	Pretreatment	Flax (cv. Viking)	I, II	[44]
	250 or 1000 µM for 8 h	5.6 and 11.2 mg L^{-1} for 10 days	Pretreatment	Flax (*Linum usitatissimum* L.)	I, *IV, VI	[45]
	100 µM for 12 h	5.6 and 11.2 mg L^{-1} for 6 days	Pretreatment	Legume (*Phaseolus aureus* and *Vicia sativa*)	I, II	[46]
	250 and 500 µM for 12 h	5.6 mg L^{-1} for 12 days	Pretreatment	Bean (*R. communis* cv. Zibi 5)	I, III, IV	[30]
	250 and 1000 µM for 12 h	5.6 and 11.2 mg L^{-1} for 12 days	Pretreatment	Flax (*Linum usitatissimum* L.)	I, II, III, IV, V	[47]
	500 µM for 20 h	11.2, 44.8 and 112 mg kg^{-1} for 30 days	Pretreatment	Wheat (*Triticum aestivum* L.)	I, II, III	[48]
	100 µM for 3 h	3, 5, and 7 mg kg^{-1} for 3 days	Pretreatment	Soybean (Balkan, L608)	II, III, *IV	[49]
	500 µM for 6 h	25, 50, and 100 mg kg^{-1}	Pretreatment	Hemp (*Cannabis sativa* L.)	I, II, III, *IV	[50]
	500 µM for 6 h	1.12, 1.68, and 2.80 mg L^{-1} for 14 days	Pretreatment	Maize (*Zea mays* L., hybrid Norma)	I, II, III, IV	[51]
	100 µM for 16 h	11.2 and 112 mg L^{-1} for 1 day	Pretreatment	Rice (cv: Longai)	I, II	[52]
	100 µM for 1, 3, 6 h	3 and 5 mg L^{-1} for 7 days	Pretreatment	Alfalfa (*Medicago sativa* L. cv. Evropa)	I, IV, V	[49]
	100 µM for 8 h	1.12, 11.2, and 112 mg L^{-1} for 1 day	Pretreatment	Rice (cv: Longai)	I, II, *IV	[53]
	500 µM for 6 h	2.8 mg L^{-1} for 12 days	Pretreatment	Barley (*Hordeum vulgare* cv Gerbel)	I, II, III, IV, V, VI, VII	[54]

Table 2. Effect of salicylic acid (SA) on cadmium (Cd) tolerance in plants.

	SA Treatment	Cd Treatment	Timeline	Plant Species	Main Responses * Means Negative or No Effect	Reference
Hydroponic application	10 µM for 15 days	16.8 mg L^{-1} for 15 days	Simultaneous	Rice (*Oryza sativa* L. Galileo))	I, II, III	[55]
	20 µM for 1 day	150 mg L^{-1} for 9 days	Pretreatment	*Nymphaea tetragona* Georgi	II, III, *IV,V	[56]
	50 µM for 7 days	1.12 mg L^{-1} for 7 days	Simultaneously	*Lemna minor*	II, III, IV, V	[57]
	50 µM for 1 day	11.2 mg L^{-1} for 8 h	Pretreatment	Wheat (*Triticum aestivum* L.)	I, *IV, Hormones	[33]
	500 µM for 24 h	56 mg L^{-1} for 1 day	Pretreatment	Maize (*Zea mays* L., hybrid Norma)	II, III, *IV, VI	[58]
	100, 200, 300 and 400 µM for 14 days	11.2 mg L^{-1} for 14 days	Simultaneous	Ryegrass (*Lolium perenne* L.)	I, II, III, *IV, V	[50]
	50 µM for 10 days	5.6 mg L^{-1} for 10 days	Simultaneous	Rice (*Oryza sativa* cv. HUR3022)	I, II, III	[59]
	100 µM for 14 days	22.4 mg L^{-1} for 14 days	Simultaneously	Peanut (Arachis hypogaea L.)	I, II, III, *IV, V	[60]
	250 and 500 µM for 10 mins	1.68 mg L^{-1} for 3 and 6 h	Post-treatment	Barley (*Hordeum vulgare* L.) cv. Slaven	I, II, Auxin	[61]
	200 µM for 14 days	11.2 mg L^{-1} for 14 days	Simultaneously	Ryegrass (*Lolium perenne* L.)	I, II, III, VI	[62]
	10, 50 and 100 µM for 7 days	2.24 mg L^{-1} for 3 days	Pretreatment	Bean (*Phaseolus vulgaris*)	I, II, III, IV, V	[63]
	60, 120, 250 and 500 mM	5.6 mg L^{-1} for 5 days	Pretreatment	Soybean (*Glycine max* L., A6445RG)	II, III, *IV, V, VI, VII (HO−1)	[64]
	1, 10, and 100 µM for 72 h	5.6 mg L^{-1} for 1 day	Pretreatment	Alfalfa (*Medicago sativa* L. cv Zhongmu No.1)	I, II, *IV, VI, VII (HO−1)	[65]
	3000 µM for 3 h	560 mg L^{-1} for 1 day	Pretreatment	Rice (*Oryza sativa L.*, cv. Taichung Native 1)	II, IV	[66]
	10 µM for 72 h	5.6 mg L^{-1} for 6 days	Pretreatment	Rice (*O. sativa* cv Jiahua 1)	I, II, IV	[67]
	10 µM for 72 h	5.6 mg L^{-1} for 6 days	Pretreatment	Rice (*O. sativa* cv Jiahua 1)	I, II	[68]
	10 µM for 24 h	5.6 mg L^{-1} for 6 days	Pretreatment	Rice (*O. sativa* cv Jiahua 1)	I, II, IV	[65]
	1, 10, and 100 µM	3 and 6 mg L^{-1} for 3 days	Simultaneous	Soybean (*Glycinemax* L. cv SG1)	*I, *IV, IV	[54]
	500 µM for 24 h	2.8 mg L^{-1} for 10 days	Pretreatment	Barley leaves (*Hordeum vulgare* cv Gerbel)	I, II, III, IV, V, VI, VII	[69]
	500 µM for 24 h	56 mg L^{-1} for 1 day	Pretreatment and simultaneously	Maize (*Zea mays* L., hybrid Norma)	*I, *II, *III, *VI	[70]
SA mutants	Up and down-regulating endogenesis SA	5.6 mg L^{-1} for 7 days	-	*NahG,snc1*	I, II, III, IV, VII	[71]
	Down-regulating endogenesis SA	0.56 mg L^{-1} for 12 days	-	*Sid2*	I, II, III, IV, V, VI, VII	[72]
	SA accumulation	16.8 mg L^{-1} for 28 days	-	*Lycium chinense*	II, III, IV, VII(*LcGSHS*)	[73]
	Up and down-regulating endogenesis SA	5.6, 11.2, and 16.8 mg L^{-1} for 7days	-	Accumulating mutant *snc1*, *npr1−1*, Reducing mutant *nahG*, *snc1/nahG*	*I, *II, *III	[74]
	Down-regulating endogenesis SA	56 mg L^{-1} for 5 days	-	*NahG*	*II, *III, *VII (CAT1)	[75]

I Growth, II antioxidant system, III photosynthesis, IV Cd uptake, V Ion uptake, VI phytochelatins, VII SA or Cd-induced genes.

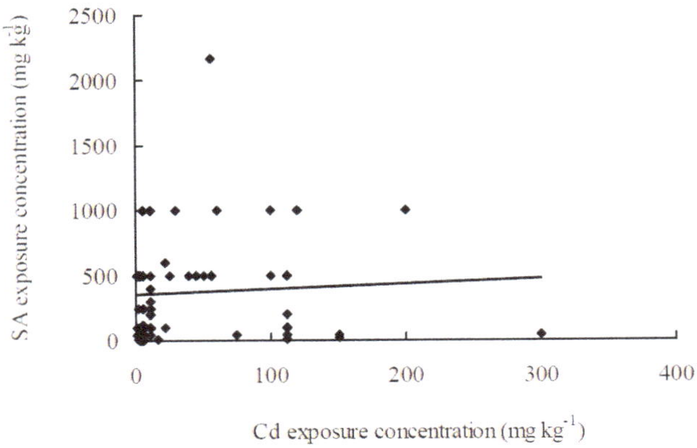

Figure 1. Representation of SA exposure concentrations plotted against Cd exposure concentrations, drawn from the data reported in Table 2. * Table 2 and Figure 1 are adapted from the Reference [76].

2.1. SA Spray

Seven studies reported that SA spray alleviated Cd damage to plants. The species included potato, peppermint, oilseed rape, melon, soybean, radish, and Indian mustard. For instance, foliar application of 600 µM SA onto potato over 10 consecutive days significantly decreased the concentrations of reactive oxygen species (ROS) in leaves and stimulated the antioxidant enzyme mechanism and the related genes (*StSABP2*, *StSOD* and *StAPX*) under 200 µM Cd stress [31]. However, the concentrations used in SA application in the literature ranged from 10 to 2170 µM, and the frequency of applications was completely different. Such large gaps make it difficult to provide guidance for implementing SA spray treatments.

2.2. Presoaking of Seed with SA

There are more studies (19 articles) conducted on seed presoaking than on SA spray treatments. The presoaking periods ranged from 3 to 24 h, and the presoaking levels ranged from 100 to 1000 µM. Moreover, the tested plants were mainly food crops, i.e., rice, maize, wheat, barley, and bean, and industrial crops, i.e., bluegrass, flax, alfalfa, hemp, and castor bean. Similarly, most of the presoaking reports suggested that pretreatment with SA alleviated the subsequent Cd-induced damage to plant growth and the antioxidant system. Notably, presoaking method is more practical for agricultural and/or economic use than spray treatment.

2.3. Hydroponic Application

Indeed, hydroponic treatment of SA is less practicable because Cd is not a problem in hydrographic environment. However, this type of studies is needed since SA or Cd treatment can be controlled more precisely and effectively, which is beneficial to the phytotoxicology research. There were 20 articles conducted on this application method. The SA concentrations ranged from 10 to 500 µM, and the treatment period ranged from 3 to 72 h. Furthermore, the timeline of SA application was different with pretreatment, simultaneous treatment and post-treatment methods. Again, hydroponic treatment of plant roots to SA mitigated Cd toxicity systemically in plants. However, more research work is needed on the transport and signalling pathways of SA through different organs under Cd stress.

2.4. SA Mutants

In the tested experiments, the exposure times of Cd and SA were relatively short, and the levels of Cd and SA were relatively high. Hence, plants received the SA signal rapidly as an instant response to the "acute" Cd stress. Furthermore, information in the literature differs in terms of plant species, treatments and concentrations. All of the above uncertain issues may lead to unpredictable results. For example, when the treatment concentration of SA exceeded the appropriate range, oxidative toxicity occurred [77]. Recently, studies (5 articles) focused on the role of endogenous SA in regulating Cd tolerance by comparing SA-deficient or SA-accumulating mutants with wild-type plants. Surprisingly, although *sid2* and *NahG* are both SA-deficient *Arabidopsis* lines, these two mutants exhibited contrasting responses. In *sid2* mutants, the mutation in the gene encoding isochorismate synthase (ICS) aggravated Cd toxicity compared to the wild-type plants [72]. In contrast, the *NahG* transgenic lines manifested higher Cd resistance than the wild-type plants [74,75].

3. Possible Roles of SA in Alleviating Cadmium (Cd) Toxicity

Taking a comprehensive view of SA roles in response to Cd toxicity, this review focuses on the recent advances in the physiological and molecular mechanisms of the following aspects: plant growth, Cd immobilization and distribution, element assimilation, photosynthesis, ROS and the antioxidant defence system, glutathione, and senescence.

3.1. Plant Growth

Cadmium exposure inhibits plant growth [78]. It also causes morphological changes in leaves and roots, such as leaf chlorosis and lignification of cell walls in root tissues [79]. The Cd-induced growth inhibition is mainly due to reduction of net photosynthetic rate [80], inactivation of enzymes involved in CO_2 fixation [81], disturbance of element metabolism [82], and induction of lipid peroxidation [83].

As a multifaceted phytohormone, SA mediates physiological and biochemical processes during all plant developmental stages, including seed germination, vegetative growth, seed production, and senescence [84,85]. *Arabidopsis* mutants with constitutively high levels of SA, such as *cpr5* [86], *acd6−1* [87], *agd2* [88] and *pi4kIIIβ1β2* [89] exhibit dwarf phenotype. In contrast, the SA-depleted *Arabidopsis* NahG transgenic plants have a longer vegetative stage and higher growth rate compared with wild-type plants [18]. The biochemical events involved in the regulation of cell division and growth by SA still need to be clarified [90], which may be crosslinked with AUX, ROS, Ca^{2+} and mitogen-activated protein kinase (MAPK) pathways [84].

In the presence of Cd, exogenous treatment with SA showed a positive effect on the growth of various plant species, such as flax [47], bluegrass [41], radish [36], and rice [67,68]. Soybean seedlings treated with 6 mg kg^{-1} Cd for 72 h showed retarded growth symptoms in roots, stems and leaves [91]. SA applied simultaneously at the levels of 1 μM, 10 μM and 100 μM significantly reversed these inhibitory effects. In barley, Cd exposure reduced the dry weight of shoots and roots by approximately 35%, whereas pretreatment with SA resulted in significant recovery of all the growth parameters [54].

Exogenous treatment with SA has a dose-dependent effect on plant growth, as observed in the *Arabidopsis* mutants with unnecessary or deficient SA levels. Presoaking treatment with 10 to 500 μM SA increased the germination of Kentucky bluegrass seeds, while the germination sharply declined under 1000 to 5000 μM SA treatments [41]. The Cd-induced inhibitory effects on ryegrass growth were significantly alleviated by low SA concentrations, but no effects were found for the high SA concentration [69]. Some plants, such as hemp [50], are vulnerable to high levels of SA but still benefit from SA treatment when they suffer from Cd toxicity. In castor bean seedlings, SA treatment significantly worsened plant growth in both the presence and absence of Cd [30].

Furthermore, inconsistent conclusions were drawn in experiments with *Arabidopsis* transgenic plants or mutants. The Cd-inhibited growth in *Arabidopsis* was aggravated by unnecessary SA in *snc1* mutants and mitigated by the depletion of SA in *nahG* transgenic lines [74]. It seems that endogenous SA negatively regulates *Arabidopsis* tolerance to Cd. However, the *sid2* mutants with SA-deficient phenotype showed a Cd-sensitive phenotype that manifested as having accentuated Cd-induced growth inhibition [72].

3.2. Cd Immobilization in the Cell Wall

The plant cell wall, as a complex composed of sugars, proteins and phenols, is the first barrier against Cd toxicity and limits Cd translocation into the cytosol [78]. The hemicellulose and pectin in the cell wall are major components for Cd deposition due to their negative charges [92]. Findings between hyperaccumulating and non-hyperaccumulating ecotypes of *Sedm alfredii* show that roots with higher cell-wall polysaccharides and activity of pectin methylesterase are more impermeable to Cd [93]. Using energy-dispersive X-ray microanalysis, Cd binding to the cell wall was directly visualized in *Thlaspi caerulescens* [14]. In addition, long-term Cd treatments may interfere synthesis and composition of cell wall, such as inducing lignification, which in turn impact Cd sequestration in a more complicated way [94]. Whereas for SA, the signalling pathway is also involved in regulation of cell wall components. Genetic research showed that *pmr4* mutants are resistant to pathogens due to mutation of a callose synthase, while blocking the SA signaling pathway restore the susceptibility, suggesting callose or callose synthase negatively regulates the SA pathway [95]. SA application altered the lignin and hemicellulose composition of cell wall in *Brachypodium distachyon* by increase in caffeic acid, ferulic acid and p-coumaric acid content [96].

Therefore, it is hypothesized that pretreatment with SA may increase Cd accumulation in the cell wall and prevent Cd translocation into other cell organelles. Until now, only a few experiments with peanut and ryegrass have shown that SA treatment reduced Cd influx by rearrangement of the cell wall composition [60,62,69]. It has been reported that either SA pretreatment or Cd treatment alone strengthened the root cell wall in rice [97]. However, it was also found that SA treatment with Cd application failed to synergistically affect the cell wall construction or the activity of enzymes related to lignin synthesis, such as polyphenol oxidase (PPO), peroxidase (POD) and PAL. This might be because the strong toxicity of Cd maximized the process of lignification, which concealed the SA signalling role in cell wall construction.

3.3. Cd Uptake and Translocation

The effect of SA on the construction of the cell wall is closely related to Cd uptake and translocation, but the results are still controversial. Some studies have indicated that the treatment of the plant with SA could decrease Cd assimilation and root-to-shoot translocation. For instance, SA application substantially reduced Cd uptake and ameliorated Cd-induced growth inhibition in radish roots [36]. Pretreatment of flax with SA significantly decreased the Cd levels in different organs of the plant [47]. This was also reported in other plant species, i.e., Kentucky bluegrass [41], ryegrass [62,69], alfalfa [65], oilseed rape [33], and wheat [98]. The suppression of Cd uptake and translocation might be due to SA-induced reduction in the uptake, inhibition or activation of element translocators that dislocate Cd into vacuoles [50,91].

However, the role of SA signalling in preventing Cd transport between plant organs is not always physiological. SA and Cd applied simultaneously increased Cd assimilation in soybean [91]. In SA-pretreated barley, tissue Cd contents were unaltered both in vacuoplasts and mesophyll cells at the organ and the whole-plant level [54]. An interesting examination of Cd translocation was conducted using a split-root system with rice [67]. In this system, an appreciable amount of Cd was transferred from Cd-treated root parts to non-Cd-treated root parts. However, pretreatment with 10 μM SA of the whole rice roots did not restrict but promoted this transport process. As for SA mutants, in *sid2* and *NahG*, SA deficiency did not influence Cd assimilation either in shoots or in roots, indicating that SA might not mediate an avoidance mechanism in plants [72,75].

3.4. Element Uptake

Cadmium uptake by plants involves the competition of nutrients by using the same membrane transporters, hence interfering with ionic homeostasis [54,78]. Expression studies with the Fe transporter cloned from *Arabidopsis*, *IRT1*, facilitate the Cd influx across the root-cell plasma membrane [99].

Expression of the wheat cDNA *LCT1* in *S. cerevisiae* mediates both Ca and Cd transport into the cytosol of plant cells [100].

The beneficial role of SA could be attributed to its maintenance of the optimal nutrition status of plants. For instance, the Cd-induced disturbance in ion uptake, including K, Fe, Ca, Mg, Mn, and Zn uptake, was alleviated by SA treatment, in flax [47], ryegrass [69], rice [101], peanut [60], Kentucky bluegrass [41] and oilseed rape [33]. This could be explained by the alterations of plasma membrane properties by SA, increasing the activity of H^+-ATPase [102], which facilitates the assimilation of nutrients under Cd toxicity.

The effect of SA on element uptake is especially relevant for K, which plays a key role in regulating H^+-ATPase in the plasma membrane. As early as 1981, the inhibitory function of SA was observed for K absorption in oat roots [103]. However, under Cd stress, a positive correlation between SA and K was noted in soybean seedlings [91] and the SA mutant, *sid2* [72], and SA further interfered with the uptake of Fe and Mg in roots and shoots. The interaction between SA and Cd is also involved in the assimilation of S, the key element of sulfhydryl groups that chelate Cd ions into less vulnerable organelles in plant cells. In the absence of Cd, SA treatment increased the S content of barley roots [54]. In contrast, the S levels were much higher in the *sid2* mutant leaves than in the wild-type plants under Cd-free conditions [72].

3.5. Photosynthesis

The mineral nutrient stress induced by Cd results in severe alterations in photosynthesis in terms of chloroplast structure, chlorophyll concentration, and activities of carboxylating enzymes [51]. Meanwhile, SA may act as an important photosynthesis regulator by the influence of RuBisCO activity, contribution to light acclimation and redox homeostasis, and the function of the stomatal switch [84].

Pretreatment with SA prevented Cd-induced chlorophyll destruction in maize [51], soybean [64], oilseed rape [33] and flax [47]. Conversely, depletion of SA further lowered the chlorophyll concentrations in Cd-treated *sid2* mutants [72]. The Cd-inhibited activity of RuBisCo and carbonic anhydrase was recovered by exposure to 0.1 mM SA in peppermint [32]. Moreover, SA application increased the carotenoid synthesis of soybean seedlings, whereas it decreased the flavonoid content under Cd stress [35]. The alleviating role of SA could be due to the restored K contents in leaves, strengthening stomatal closure [72], which is synchronized with the net photosynthetic rate, transpiration, capability for CO_2 fixation and inhibition of the activities of chlorophyll-degrading enzymes [35,50,62].

In addition, non-stomatal factors signalled by SA also play roles in the maintenance of photosynthetic capacity under Cd stress. The application of SA to Cd-treated barley leaf slices significantly slowed the decreasing trends in the photosynthetic yield of photosystem II (PSII) [54]. Spraying 0.1 mM SA onto melon leaves induced an increase in Fv/Fm in Cd-stressed plants, indicating that SA improved PSII efficiency [34]. Similarly, treatment with SA in rice prevented the unnecessary energy transference from PSII to PSI induced by Cd toxicity [55]. In this study, SA increased the cyclic electron transport around PSI in thylakoid membranes and protected the Mn-cluster of the oxygen-evolving complex from Cd damage.

The photosynthetic response to SA is both dose- and species-dependent. Pretreatment with 250 and 500 μM SA in castor bean leaves failed to affect chlorophyll levels but aggravated the negative effect on photosynthesis induced by Cd [30], which might have been associated with an increase in stomatal limitation. The reduced SA level in *nahG* plants resulted in the maintenance of photosynthetic efficiency by low photoinhibition under Cd stress [71]. The response to endogenous SA in Cd tolerance might be associated with the regulation of photosynthetic electron transport, starch degradation and PSII structures at the transcriptional level.

3.6. Reactive Oxygen Species (ROS) and Antioxidant Defence System

The production of ROS in plant cell is an unavoidable consequence of oxygen metabolism, especially during respiration and photosynthesis [104]. The major mode of Cd toxicity in plants is to induce ROS production and results in oxidative injuries in plants [30]. Although Cd does not participate directly in cellular redox reactions (such as Fenton reaction), it can indirectly elevate ROS accumulation in the cellular environment by the disturbance of electron transport, destroying the structure of antioxidant enzymes and interfering with antioxidant molecule synthesis. Cd indirectly modulates the activity of the plasma membrane NADPH oxidase, increasing the formation of $O_2^{\bullet-}$ and H_2O_2, which has been found in tobacco [105], rice [106], and lupine roots [107]. Meanwhile, the enhanced demand for glutathione (GSH) for Cd chelation causes rapid loss in antioxidative defence [16].

It is well known that SA signals plant resistance through modulation of ROS metabolism, especially H_2O_2. The mode of SA action involves binding directly to CAT and ascorbate peroxidase (APX), two major H_2O_2-scavenging enzymes [108], inhibiting their activities in plants. This finding was also confirmed in *sid2* mutants, SA-deficient *Arabidopsis* plants, whose leaf CAT activity is higher than that in wild-type plants [72]. Using the DAB staining method, the SA-induced accumulation of H_2O_2 was visualized in rice leaves [66]. SA may regulate H_2O_2 accumulation through a self-amplifying feedback loop in *Arabidopsis*. In this process, SA acts as an electron donor of CAT that effectively slows down the peroxidation cycle, hence sharply decreasing the efficiency of H_2O_2 elimination [108]. Another important mechanism may occur in mitochondria. SA blocks electron flow from the substrate dehydrogenises to the ubiquinone pool and triggers H_2O_2 generation [109].

Both Cd stress and SA accumulation can elevate H_2O_2 production in cells. Therefore, SA treatment may aggravate the oxidative stress induced by Cd toxicity. Surprisingly, a large body of studies found that SA treatment alleviates Cd-induced oxidative stress in plants. In fact, H_2O_2 has a dual role in plant biology as both a toxic byproduct and a key regulator against many abiotic stresses in plants [110]. The increase in H_2O_2 status stimulated by SA pretreatment acts as a crucial message to "set up" the antioxidant system and then induces plants to resist subsequent Cd stress. The beneficial effects of SA were found in most of the studies in which SA treatment was performed, either by spraying, presoaking, or by hydroponic incubation, in advance of the application of Cd stress (Table 2). For example, pretreatment with 10 μM SA for 72 h initially caused H_2O_2 accumulation in vitro in rice roots. Correspondingly, the levels of GSH, non-protein thiols (NPT), and ascorbic acid (AsA) and the activities of CAT, superoxide dismutase (SOD) and POD were elevated compared to those of non-SA treated roots during the subsequent Cd exposure period [67,68]. The expression level of selected genes (*StSABP2*, *StSOD* and *StAPX*) was enhanced in SA-treated potato plants under Cd stress [31]. Pretreatment with SA improved the antioxidant capacity and tolerance of Cd-induced oxidative stress in rice leaves, which was similar to the effects of H_2O_2 pretreatment [66]. Moreover, the role of SA in mitigating Cd damage was also confirmed in many other species, such as flax [39], mustard [37], oilseed rape [33], ryegrass [62,69], melon [34], etc.

Although a number of reports on the benefits of SA signalling exist in the literature, some questions have not yet been fully answered. SA treatment alleviated Cd toxicity in barley but lowered the activities of antioxidant enzymes [54]. SA could act as an oxidant at high levels due to its capability to increase H_2O_2 generation and oxidation of proteins in plant leaves [77]. In contrast, it was reported that SA may act as a direct scavenger to reduce excessive H_2O_2 injury in the pea during Cd stress [111]. In addition, SA may directly signal the expression of defence-related enzymes, such as haem oxygenase−1 (HO−1), to alleviate Cd-triggered oxidative toxicity by reestablishing redox homeostasis [64]. Moreover, the crosstalk and response pathways between SA and other phytohormones should also be noted. SA combined with NO treatment synergistically counteracted Cd-induced oxidative damage in peanut and ryegrass [60,62]. Under Cd stress, pretreatment of wheat seedlings with SA significantly promoted the synthesis of dehydrins, the abscisic acid (ABA)-signalled proteins that can neutralize and bind unnecessary H_2O_2 [98].

The results of the SA mutants studies are still contradictory. *snc1* mutants with high intrinsic levels of SA possess high POD activity, which can generate a large amount of ROS and manifest the Cd-sensitive phenotype [74]. In contrast, SA deficiency in NahG transgenic lines mitigates the oxidative stress induced by Cd toxicity [74,75]. In contrast, *sid2* mutants with an SA-deficient phenotype aggravated Cd-induced oxidative damage compared with their wild-type plants.

3.7. Glutathione and Chelation

Glutathione (γ-Glu-Cys-Gly, GSH) is one of the most important reducing equivalents in plants, protecting plants against Cd-induced oxidative damage. Furthermore, it is also a key molecular compound or a basic component of phytochelatins (PCs) involved in Cd chelation and thereby confines Cd to less sensitive organelles, such as vacuoles [15,112].

A series of genetic reports using single mutants or transformants have shown direct evidence that endogenous SA signalling is linked to GSH biosynthesis. The catalase-deficient *Arabidopsis* mutant, *cat 2*, induces SA levels and a wide range of SA-dependent responses alongside the upregulation of GSH [113]. In contrast, attenuation of GSH levels was correlated with decreased SA contents in *cat2 atrbohF* compared with *cat2* [114]. It has been shown that the GSH concentrations were much lower in Cd-stressed *sid2* leaves than in wild-type plants [72]. Low GSH levels were also found in other SA-deficient mutants, such as *npr1−1* and *mpk4−1* [115,116]. Furthermore, the exogenous application of SA has been shown to enhance S assimilation in barley roots [54] and elevate GSH content in some plant species, such as peppermint [32], flax [45], peanut [60] and rice [68].

The regulatory role of SA in GSH biosynthesis may be related to serine acetyltransferase (SAT) transcription, the precursor gene that catalyses cysteine formation. Increased free SA levels, both by genetics and by exogenous application, lead to an increased specific activity of SAT and GSH in *Arabidopsis* [117]. Under Cd stress, although depletion of SA in *sid2* mutants significantly enhanced the uptake of S, a key element for GSH construction, down-regulated transcription of SAT-c and SAT-p in *sid2* blocked the process of GSH biosynthesis and resulted in lower GSH levels compared with the wild-type plants [72]. Glutathione synthetase (GSHS) is a rate-limiting enzyme that catalyses the second step of GSH synthesis in plants. The expression of the Cd-induced *LcGSHS* transcript, a GSHS gene isolated from *L. chinense*, is controlled by the endogenous SA-dependent pathway and results in greater GSH accumulation and Cd tolerance in transgenic *Arabidopsis* [73]. Furthermore, SA has also been shown to mediate the synthesis of glutathione reductase (GR1), the pivotal enzyme for regenerating and maintaining GSH in the reduced state [118]. During Cd exposure, SA deficiency significantly decreased GR1 transcription in *sid2* mutants and resulted in lower GSH levels and GR activity compared with the wild-type plants [72]. However, another SA-deficient transgenic line, *NahG*, manifested high GSH accumulation and a high GSH/GSSG ratio in the presence of Cd [74,75]. These inconsistent results suggest that the role of SA in regulating GSH synthesis under Cd stress requires further investigation.

3.8. Senescence

Cd toxicity accelerates the ageing process in plant cells referred to as senescence. The Cd-induced morphological changes associated with senescence were shown in pine roots [16]. Cd exposure accelerated the senescence process of *Arabidopsis*, as indicated by an increase in SAG12 expression, a typical senescence marker gene [119].

Although senescence is a negative physiological process, it is one of the most important stages that plants undergo to maintain organ homeostasis and to escape from unfavourable conditions. Under biotic stress, SA is well known to induce senescence of infected tissues to build up a physical barrier against the spread of pathogens. In SA-deficient *Arabidopsis* plants, the transcript of SAG12 was considerably reduced or undetectable [88]. Whether treatment with Cd and SA manifests antagonistic or synergistic effects on plant senescence is still unclear. Generally, Cd distributes unevenly in field environment. Therefore, some parts of plant root may suffer Cd toxicity severely but some parts may not. An interesting hypothesis is that SA might accelerate Cd-induced senescence of stressed root parts, and then benefit the whole plant to elude Cd damage by adjusting the root growth direction towards the non-Cd contaminated environment. To test this hypothesis, a split-root experiment was conducted in which half of the roots were exposed to Cd while the other half were not exposed [67]. They found that Cd treatment caused senescence of the stressed part of the roots and stimulated the growth of the non-stressed part of the roots. However, SA pretreatment had no effect on the senescence of the Cd-exposed roots and did not lower the Cd uptake, which might be because the low level of exogenous SA used was insufficient to trigger senescence. Nonetheless, this hypothesis can be tested by using *Arabidopsis* mutants, such as *snc−1* with constitutively high concentrations of endogenous SA, since this mutant manifests a senescence phenotype and lower uptake of Cd compared to the wild-type plants [74].

4. Future Insights and Conclusions

Research over the past 20 years has strongly indicated that SA is a very promising molecule for the reduction of Cd toxicity in plants. Here, we reviewed reports describing the promoting role of SA in Cd resistance under various treatment methods, including pre-soaking, hydroponic exposure, and spraying. Figure 2 proposed the possible roles of SA in alleviating Cd toxicity to plants. However, there still remains a contradiction between the effects of SA at low and high doses. Furthermore, several unsolved and central questions concerning homeostasis, gene expression, and crosstalk with other phytohormones are still not fully understood.

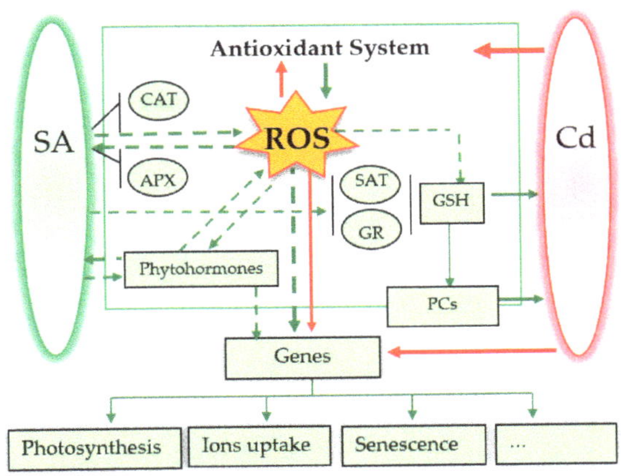

Figure 2. Possible roles of SA in alleviating Cd toxicity to plants. The dotted arrows mean possible signalling pathways. Red and green arrows indicate damage and positive effects, respectively. * Figure 2 is adapted from the reference [76].

4.1. SA Homeostasis

It is well known that plants can increase their intracellular concentration of SA to combat various environmental stresses, including Cd toxicity [120]. Exposure of Cd for 7 days increased the SA levels of leaves in young maize seedlings [121]. Compared with SA-deficient mutants, Cd stress significantly increased the SA level in wild-type *Arabidopsis*, indicating that Cd-induced SA synthesis occurs through the NahG- or SID2- pathways [72,74,75]. Exogenous SA treatment highlights the expression of *OsWRKY45* and increases the endogenous concentrations of SA in Cd-stressed rice and Lemna minor plants [57,66]. Under biotic stress, the mechanism of SA transport in plant cells has been illuminated. During pathogen attack, the hydroxyl group of SA is conjugated by glucose, resulting in formation of the SA glucoside (SAG). Then, SAG is actively transported from the cytosol into the vacuole, where it functions as an inactive storage form that can release free SA [122]. However, the balance of SA homeostasis in vivo under Cd stress still needs to be further monitored.

4.2. SA-Related Gene Expression

The Cd toxicity and SA signal may share the same origins in regard to gene expression. Characterization and identification of the SA receptor during Cd stress are highly anticipated.

4.2.1. Nonexpressor of Pathogenesis-Related (NPR) Protein

Nonexpressor of PR (NPR1 and NPR3/4), the canonical signal transducer of SA, regulates many gene expressions in seed germination, flowering, and senescence processes [123]. It has been uncovered that both NPR1 and NPR3/4 are SA receptors in regulating SA-mediated plant immunity [124]. Further genetic analysis indicated that NPR1 is a transcriptional activator while NPR3/4 are repressors functioning independent of NPR1 [125]. However, few studies have focused on the issue of whether SA-induced NPR is involved in Cd tolerance in plants. Surprisingly, it has been found that an increase in plant biomass was coupled with SA accumulation in *npr1−1* (a SA-deficient mutant) after 12-h-Cd exposure, suggesting that the regulation of Cd tolerance is not related to the NPR1 signalling pathway [74].

4.2.2. Mitogen-Activated Protein Kinase (MAPK)

Another important example of SA-regulated gene expression involves mitogen-activated protein kinase cascades (MAPK), which acts as a negative regulator in plant growth and development in response to endogenous and environmental cues [126]. SA accumulation in *mpk4* mutants result in a severely dwarf phenotype [127]. Cd exposure quickly enhanced the kinase activity of MPK6, while the *mpk6* mutation enhanced Cd tolerance by alleviating oxidative stress [128]. However, until now, no studies have been conducted to evaluate the role of SA in signalling Cd resistance through MAPK regulation.

4.2.3. ATP-Binding Cassette (ABC) Transporters

ATP-binding cassette (ABC) transporters belong to a large family that utilizes the energy of ATP binding and hydrolysis to transport elements across cellular membranes. In particular, they are involved in sequestering Cd or PC-Cd complexes into vacuoles to alleviate Cd toxicity [129]. An ABC transporter from soybean was identified that was strongly and rapidly induced by SA treatment [130]. Therefore, it is of interest to explore whether ABC transporters are involved in SA signalling and SA-induced Cd tolerance.

4.3. Crosstalk with Other Phytohormones

In addition to crosstalk with ROS transduction, the coordination between signalling of SA and other phytohormones is also an important aspect to consider [131]. SA pretreatment mitigated the Cd-induced disturbance in the levels of indoleacetic acid, cytokinins and ABA in wheat seedlings [98], and alleviated Cd toxicity in barley root tips by inhibiting auxin-mediated ROS generation [61]. It was also reported that SA combined with NO counteracted the negative effects of Cd on ryegrass plants [62]. However, the direct or indirect influence of SA signalling on the balance of plant hormones needs to be determined.

In conclusion, investigations of SA signalling roles shed new light on the approaches to enhancing Cd tolerance with phytohormones. In addition to classical methods of exogenous application, these studies can now be complemented by the creation of a new generation of SA-excessive or SA-deficient mutants. Furthermore, with the characteristics of low cost and high efficiency, SA application shows promising use in helping plants defend against Cd toxicity. Seed presoaking and spraying with SA are pragmatic approaches for this purpose. In the meantime, novel roles of SA in Cd toxicity will likely continue to be unveiled.

Funding: Supported by Department of Science and Technology of Zhejiang Province, China (2010C33158, 2015C03020), Fundamental Research Funds for the Central Universities (2019FZJD007), State Key Laboratory Breeding Base for Zhejiang Sustainable Pest and Disease Control (2010DS700124-ZZ1802), Natural Science Foundation of China (No. 41001184), and National Key R&D Program of China (2018YFD0800202).

Conflicts of Interest: The authors declare no conflict of interest.

Abbreviations

AsA	ascorbic acid
ABA	abscisic acid
ABC transporters	ATP-binding cassette transporters
APX	ascorbate peroxidase;
Cd	cadmium;
CAT	catalase;
GR	glutathione reductase
GSH	glutathione
GSHS	glutathione synthetase
HO−1	haem oxygenase−1
MAPK	mitogen-activated protein kinase
NPR1	Nonexpressor of PR1
NPT	non-protein thiols
PAL	phenylalanine ammonia-lyase
PCs	phytochelatins
POD	peroxidase
PPO	polyphenol oxidase
PSII	photosystem II
ROS	reactive oxygen species
SA	salicylic acid
SAT	serine acetyltransferase
SOD	superoxide dismutase

References

1. Vig, K.; Megharaj, M.; Sethunathan, N.; Naidu, R. Bioavailability and toxicity of cadmium to microorganisms and their activities in soil: A review. *Adv. Environ. Res.* **2003**, *8*, 121–135. [CrossRef]
2. Gong, B.; Nie, W.; Yan, Y.; Gao, Z.; Shi, Q. Unravelling cadmium toxicity and nitric oxide induced tolerance in Cucumis sativus: Insight into regulatory mechanisms using proteomics. *J. Hazard. Mater.* **2017**, *336*, 202–213. [CrossRef] [PubMed]
3. Zhang, Y.; Chen, Y.; Meng, A.; Li, Q.; Cheng, H. Experimental and thermodynamic investigation on transfer of cadmium influenced by sulfur and chlorine during municipal solid waste (MSW) incineration. *J. Hazard. Mater.* **2008**, *153*, 309–319. [CrossRef] [PubMed]
4. Luo, L.; Ma, Y.; Zhang, S.; Wei, D.; Zhu, Y. An inventory of trace element inputs to agricultural soils in China. *J. Environ. Manag.* **2009**, *90*, 2524–2530. [CrossRef] [PubMed]
5. Roberts, T.L. Cadmium and phosphorous fertilizers: The issues and the science. *Procedia Eng.* **2014**, *83*, 52–59. [CrossRef]
6. Zhou, J.; Wan, H.; He, J.; Lyu, D.; Li, H. Integration of cadmium accumulation, subcellular distribution, and physiological responses to understand cadmium tolerance in apple rootstocks. *Front. Plant Sci.* **2017**, *8*, 966. [CrossRef] [PubMed]
7. Clemens, S.; Aarts, M.G.M.; Thomine, S.; Verbruggen, N. Plant science: The key to preventing slow cadmium poisoning. *Trends Plant Sci.* **2013**, *18*, 92–99. [CrossRef]

8. Daud, M.K.; Sun, Y.; Dawood, M.; Hayat, Y.; Variath, M.T.; Wu, Y.X.; Mishkat, U.; Najeeb, U.; Zhu, S. Cadmium-induced functional and ultrastructural alterations in roots of two transgenic cotton cultivars. *J. Hazard. Mater.* **2009**, *161*, 463–473. [CrossRef]
9. Song, Y.; Jin, L.; Wang, X. Cadmium absorption and transportation pathways in plants. *Int. J. Phytoremediat.* **2017**, *19*, 133–141. [CrossRef]
10. Lysenko, E.A.; Klaus, A.A.; Pshybytko, N.L.; Kusnetsov, V.V. Cadmium accumulation in chloroplasts and its impact on chloroplastic processes in barley and maize. *Photosynth. Res.* **2015**, *125*, 291–303. [CrossRef]
11. Hasan, S.A.; Hayat, S.; Ahmad, A. Screening of tomato (*Lycopersicon esculentum*) cultivars against cadmium through shotgun approach. *J. Plant Interact.* **2009**, *3*, 187–201. [CrossRef]
12. Bertoli, A.C.; Cannata, M.G.; Carvalho, R.; Bastos, A.R.R.; Freitas, M.P.; Dos Santos Augusto, A. *Lycopersicon esculentum* submitted to Cd-stressful conditions in nutrition solution: Nutrient contents and translocation. *Ecotoxicol. Environ. Saf.* **2012**, *86*, 176–181. [CrossRef]
13. Potters, G.; Pasternak, T.P.; Guisez, Y.; Palme, K.J.; Jansen, M.A.K. Stress-induced morphogenic responses: Growing out of trouble? *Trends Plant Sci.* **2007**, *12*, 98–105. [CrossRef]
14. Wójcik, M.; D'Haen, V.J.; Tukiendorf, A. Cadmium tolerance in *Thlaspi caerulescens* II. localization of cadmium in *Thlaspi caerulescens*. *Environ. Exp. Bot.* **2005**, *53*, 163–171. [CrossRef]
15. Seth, C.S.; Remans, T.; Keunen, E.; Jozefczak, M.; Gielen, H.; Opdenakker, K.; Weyens, N.; Vangronsveld, J.; Cuypers, A. Phytoextraction of toxic metals: A central role for glutathione. *Plant Cell Environ.* **2012**, *35*, 334–346. [CrossRef] [PubMed]
16. Schützendübel, A.; Schwanz, P.; Teichmann, T.; Gross, K.; Langenfeld-Heyser, R.; Godbold, D.L.; Polle, A. Cadmium-induced changes in antioxidative systems, hydrogen peroxide content, and differentiation in scots pine roots. *Plant Physiol.* **2001**, *127*, 887–898. [CrossRef]
17. Raskin, I. Role of salicylic acid in plants. *Annu. Rev. Plant Physiol. Plant Mol. Biol.* **1992**, *43*, 439–463. [CrossRef]
18. Abreu, M.E.; Munné-Bosch, S. Salicylic acid deficiency in *NahG* transgenic lines and *sid2* mutants increases seed yield in the annual plant *Arabidopsis thaliana*. *J. Exp. Bot.* **2009**, *60*, 1261–1271. [CrossRef]
19. Huang, J.; Gu, M.; Lai, Z.; Fan, B.; Shi, K.; Zhou, Y.H.; Yu, J.Q.; Chen, Z. Functional analysis of the Arabidopsis PAL gene family in plant growth, development, and response to environmental stress. *Plant Physiol.* **2010**, *153*, 1526–1538. [CrossRef]
20. Wildermuth, M.C.; Dewdney, J.; Wu, G.; Ausubel, F.M. Isochorismate synthase is required to synthesize salicylic acid for plant defense. *Nature* **2001**, *414*, 562–565. [CrossRef]
21. Garcion, C.; Lohmann, A.; Lamodiere, E.; Catinot, J.; Buchala, A.; Doermann, P.; Metraux, J.P. Characterization and biological function of the ISOCHORISMATE SYNTHASE2 gene of Arabidopsis. *Plant Physiol.* **2008**, *147*, 1279–1287. [CrossRef] [PubMed]
22. Dempsey, D.A.; Shah, J.; Klessig, D.F. Salicylic acid and disease resistance in plants. *Crit. Rev. Plant Sci.* **1997**, *18*, 547–575. [CrossRef]
23. Kumar, D. Salicylic acid signaling in disease resistance. *Plant Sci.* **2014**, *228*, 127–134. [CrossRef] [PubMed]
24. Pal, M.; Kovacs, V.; Szalai, G.; Soos, V.; Ma, X.; Liu, H.; Mei, H.; Janda, T. Salicylic acid and abiotic stress responses in rice. *J. Agron. Crop Sci.* **2014**, *200*, 1–11. [CrossRef]
25. Chini, A.; Grant, J.J.; Seki, M.; Shinozaki, K.; Loake, G.J. Drought tolerance established by enhanced expression of the *CC–NBS–LRR* gene, *ADR1*, requires salicylic acid, EDS1 and ABI1. *Plant J.* **2004**, *38*, 810–822. [CrossRef]
26. Janda, T.; Szalai, G.; Tari, I.; Páldi, E. Hydroponic treatment with salicylic acid decreases the effects of chilling injury in maize (Zea mays L.) plants. *Planta* **1999**, *208*, 175–180. [CrossRef]
27. Szalai, G.; Janda, T. Effect of salt stress on the salicylic acid synthesis in young maize (Zea mays L.) Plants. *J. Agron. Crop Sci.* **2009**, *195*, 165–171. [CrossRef]
28. Dat, J.F.; Foyer, C.H.; Scott, I.M. Changes in salicylic acid and antioxidants during induced thermotolerance in mustard seedlings. *Plant Physiol.* **1998**, *118*, 1455–1461. [CrossRef]
29. Horváth, E.; Szalai, G.; Janda, T. Induction of abiotic stress tolerance by salicylic acid signaling. *J. Plant Growth Regul.* **2007**, *26*, 290–300. [CrossRef]
30. Liu, C.F.; Guo, J.L.; Cui, Y.L.; Lü, T.F.; Zhang, X.H.; Shi, G.R. Effects of cadmium and salicylic acid on growth, spectral reflectance and photosynthesis of castor bean seedlings. *Plant Soil* **2011**, *344*, 131–141. [CrossRef]

31. Li, Q.; Wang, G.; Wang, Y.; Yang, D.; Guan, C.; Ji, J. Foliar application of salicylic acid alleviate the cadmium toxicity by modulation the reactive oxygen species in potato. *Ecotoxicol. Environ. Saf.* **2019**, *172*, 317–325. [CrossRef] [PubMed]
32. Ahmad, B.; Jaleel, H.; Sadiq, Y.; Khan, M.M.A.; Shabbir, A. Response of exogenous salicylic acid on cadmium induced photosynthetic damage, antioxidant metabolism and essential oil production in peppermint. *Plant Growth Regul.* **2018**, *86*, 273–286. [CrossRef]
33. Ali, E.; Maodzeka, A.; Hussain, N.; Shamsi, I.H.; Jiang, L. The alleviation of cadmium toxicity in oilseed rape (*Brassica napus*) by the application of salicylic acid. *Plant Growth Regul.* **2015**, *75*, 641–655. [CrossRef]
34. Zhang, Y.; Xu, S.; Yang, S.; Chen, Y. Salicylic acid alleviates cadmium-induced inhibition of growth and photosynthesis through upregulating antioxidant defense system in two melon cultivars (*Cucumis melo*, L.). *Protoplasma* **2015**, *252*, 911–924. [CrossRef] [PubMed]
35. Li, X.M.; Ma, L.J.; Bu, N.; Li, Y.Y.; Zhang, L.H. Effects of salicylic acid pre-treatment on cadmium and/or UV-B stress in soybean seedlings. *Biol. Plant.* **2014**, *58*, 195–199. [CrossRef]
36. Raza, S.H.; Shafiq, F. Exploring the role of salicylic acid to attenuate cadmium accumulation in radish (*Raphanus sativus*). *Int. J. Agric. Biol.* **2013**, *15*, 547–552.
37. Ahmad, P.; Nabi, G.; Ashraf, M. Cadmium-induced oxidative damage in mustard (*Brassica juncea* (L.) czern. & coss.) plants can be alleviated by salicylic acid. *S. Afr. J. Bot.* **2011**, *77*, 36–44.
38. Roychoudhury, A.; Ghosh, S.; Paul, S.; Mazumdar, S.; Das, G.; Das, S. Pre-treatment of seeds with salicylic acid attenuates cadmium chloride-induced oxidative damages in the seedlings of mungbean (*Vigna radiata*, L. wilczek). *Acta Physiol. Plant.* **2016**, *38*, 11. [CrossRef]
39. Belkadhi, A.; De Haro, A.; Obregon, S.; Chaïbi, W.; Djebali, W. Positive effects of salicylic acid pretreatment on the composition of flax plastidial membrane lipids under cadmium stress. *Environ. Sci. Pollut. Res.* **2015**, *22*, 1457–1467. [CrossRef]
40. Belkadhi, A.; De Haro, A.; Soengas, P.; Obregón, S.; Cartea, M.E.; Chaïbi, W.; Djebali, W. Salicylic acid increases tolerance to oxidative stress induced by hydrogen peroxide accumulation in leaves of cadmium-exposed flax (*Linum usitatissimum* L.). *J. Plant Interact.* **2014**, *9*, 647–654. [CrossRef]
41. Guo, Q.; Meng, L.; Mao, P.C.; Jia, Y.Q.; Shi, Y.J. Role of exogenous salicylic acid in alleviating cadmium-induced toxicity in Kentucky bluegrass. *Biochem. Syst. Ecol.* **2013**, *50*, 269–276. [CrossRef]
42. Agami, R.A.; Mohamed, G.F. Exogenous treatment with indole-3-acetic acid and salicylic acid alleviates cadmium toxicity in wheat seedlings. *Ecotoxicol. Environ. Saf.* **2013**, *94*, 164–171. [CrossRef] [PubMed]
43. Szalai, G.; Krantev, A.; Yordanova, R.; Popova, L.P.; Janda, T. Influence of salicylic acid on phytochelatin synthesis in Zea mays during Cd stress. *Turk. J. Bot.* **2013**, *37*, 708–714.
44. Belkadhi, A.; De Haro, A.; Soengas, P.; Obregón, S.; Cartea, M.E.; Djebali, W.; Chaïbi, W. Salicylic acid improves root antioxidant defense system and total antioxidant capacities of flax subjected to cadmium. *OMICS J. Integr. Biol.* **2013**, *17*, 398–406. [CrossRef] [PubMed]
45. Belkadhi, A.; Hediji, H.; Abbes, Z.; Djebali, W.; Chaibi, W. Influence of salicylic acid pre-treatment on cadmium tolerance and its relationship with non-protein thiol production in flax root. *Afr. J. Biotechnol.* **2012**, *11*, 9788–9796. [CrossRef]
46. Zhang, F.; Zhang, H.; Xia, Y.; Wang, G.; Xu, L.; Shen, Z. Exogenous application of salicylic acid alleviates cadmium toxicity and reduces hydrogen peroxide accumulation in root apoplasts of Phaseolus aureus and Vicia sativa. *Plant Cell Rep.* **2011**, *30*, 1475–1483. [CrossRef] [PubMed]
47. Belkhadi, A.; Hediji, H.; Abbes, Z.; Nouairi, I.; Barhoumi, Z.; Zarrouk, M.; Chaïbi, W.; Djebali, W. Effects of exogenous salicylic acid pre-treatment on cadmium toxicity and leaf lipid content in Linum usitatissimum, L. *Ecotoxicol. Environ. Saf.* **2010**, *73*, 1004–1011. [CrossRef]
48. Moussa, H.R.; Elgamal, S.M. Effect of salicylic acid pretreatment on cadmium toxicity in wheat. *Biol. Plant.* **2010**, *54*, 315–320. [CrossRef]
49. Dražić, G.; Mihailovic, N. Salicylic acid modulates accumulation of Cd in seedlings of Cd-tolerant and Cd-susceptible soybean genotypes. *Arch. Biol. Sci.* **2009**, *61*, 431–439. [CrossRef]
50. Shi, G.R.; Cai, Q.S.; Liu, Q.Q.; Wu, L. Salicylic acid-mediated alleviation of cadmium toxicity in hemp plants in relation to cadmium uptake, photosynthesis, and antioxidant enzymes. *Acta Physiol. Plant.* **2009**, *31*, 969–977. [CrossRef]
51. Krantev, A.; Yordanova, R.; Janda, T.; Szalai, G.; Popova, L. Treatment with salicylic acid decreases the effect of cadmium on photosynthesis in maize plants. *J. Plant Physiol.* **2008**, *165*, 920–931. [CrossRef] [PubMed]

52. Panda, S.K.; Patra, H.K. Effect of salicylic acid potentiates cadmium-induced oxidative damage in oryza sativa, l. leaves. *Acta Physiol. Plant.* **2007**, *29*, 567–575. [CrossRef]
53. Choudhury, S.; Panda, S.K. Role of salicylic acid in regulating cadmium induced oxidative stress in Oryza sativa L. roots. *Bulg. J. Plant Physiol.* **2004**, *30*, 95–110.
54. Metwally, A.; Finkemeier, I.; Georgi, M.; Dietz, K.J. Salicylic acid alleviates the cadmium toxicity in barley seedlings. *Plant Physiol.* **2003**, *132*, 272–281. [CrossRef] [PubMed]
55. Yotsova, E.K.; Dobrikova, A.G.; Stefanov, M.A.; Kouzmanova, M.; Apostolova, E.L. Improvement of the rice photosynthetic apparatus defence under cadmium stress modulated by salicylic acid supply to roots. *Theor. Exp. Plant Phys.* **2018**, *30*, 57–70. [CrossRef]
56. Gu, C.S.; Yang, Y.H.; Shao, Y.F.; Wu, K.W.; Liu, Z.L. The effects of exogenous salicylic acid on alleviating cadmium toxicity in nymphaea tetragona georgi. *S. Afr. J. Bot.* **2018**, *114*, 267–271. [CrossRef]
57. Lu, Q.; Zhang, T.; Zhang, W.; Su, C.; Yang, Y.; Hu, D.; Xu, Q. Alleviation of cadmium toxicity in Lemna minor by exogenous salicylic acid. *Ecotoxicol. Environ. Saf.* **2018**, *147*, 500–508. [CrossRef] [PubMed]
58. Gondor, O.K.; Pál, M.; Darkó, É.; Janda, T.; Szalai, G. Salicylic acid and sodium salicylate alleviate cadmium toxicity to different extents in maize (Zea mays L.). *PLoS ONE* **2016**, *11*, e0160157. [CrossRef]
59. Singh, I.; Shah, K. Evidences for suppression of cadmium induced oxidative stress in presence of sulphosalicylic acid in rice seedlings. *Plant Growth Regul.* **2015**, *76*, 99–110. [CrossRef]
60. Xu, L.L.; Fan, Z.Y.; Dong, Y.J.; Kong, J.; Bai, X.Y. Effects of exogenous salicylic acid and nitric oxide on physiological characteristics of two peanut cultivars under cadmium stress. *Biol. Plant.* **2015**, *59*, 171–182. [CrossRef]
61. Tamás, L.; Mistrík, I.; Alemayehu, A.; Zelinová, V.; Bočová, B.; Huttová, J. Salicylic acid alleviates cadmium-induced stress responses through the inhibition of Cd-induced auxin-mediated reactive oxygen species production in barley root tips. *J. Plant Physiol.* **2015**, *173*, 1–8. [CrossRef] [PubMed]
62. Wang, Q.; Liang, X.; Dong, Y.; Xu, L.; Zhang, X.; Kong, J.; Liu, S. Effects of exogenous salicylic acid and nitric oxide on physiological characteristics of perennial ryegrass under cadmium stress. *J. Plant Growth Regul.* **2013**, *32*, 721–731. [CrossRef]
63. Saidi, I.; Ayouni, M.; Dhieb, A.; Chtourou, Y.; Chaïbi, W.; Djebali, W. Oxidative damages induced by short-term exposure to cadmium in bean plants: Protective role of salicylic acid. *S. Afr. J. Bot.* **2013**, *85*, 32–38. [CrossRef]
64. Noriega, G.; Caggiano, E.; Lecube, M.L.; Santa-Cruz, D.; Batlle, A.; Tomaro, M.; Balestrasse, K.B. The role of salicylic acid in the prevention of oxidative stress elicited by cadmium in soybean plants. *Biometals* **2012**, *25*, 1155–1165. [CrossRef] [PubMed]
65. Cui, W.; Le, L.; Gao, Z.; Wu, H.; Xie, Y.; Shen, W. Haem oxygenase-1 is involved in salicylic acid-induced alleviation of oxidative stress due to cadmium stress in Medicago sativa. *J. Exp. Bot.* **2012**, *63*, 5521–5534. [CrossRef] [PubMed]
66. Chao, Y.Y.; Chen, C.Y.; Huang, W.D.; Chinghuei, K. Salicylic acid-mediated hydrogen peroxide accumulation and protection against Cd toxicity in rice leaves. *Plant Soil* **2010**, *329*, 327–337. [CrossRef]
67. Guo, B.; Liang, Y.; Zhu, Y. Does salicylic acid regulate antioxidant defense system, cell death, cadmium uptake and partitioning to acquire cadmium tolerance in rice? *J. Plant Physiol.* **2009**, *166*, 20–31. [CrossRef]
68. Guo, B.; Liang, Y.; Zhu, Y.; Zhao, F. Role of salicylic acid in alleviating oxidative damage in rice roots (Oryza sativa) subjected to cadmium stress. *Environ. Pollut.* **2007**, *147*, 743–749. [CrossRef]
69. Bai, X.; Dong, Y.; Kong, J.; Xu, L.; Liu, S. Effects of application of salicylic acid alleviates cadmium toxicity in perennial ryegrass. *Plant Growth Regul.* **2015**, *75*, 695–706. [CrossRef]
70. Pál, M.; Szalai, G.; Horváth, E.; Janda, T.; Páldi, E. Effect of salicylic acid during heavy metal stress. *Acta Biol. Szeged.* **2002**, *46*, 119–120.
71. Wang, Y.Y.; Wang, Y.; Li, G.Z.; Hao, L. Salicylic acid-altering Arabidopsis plant response to cadmium exposure: Underlying mechanisms affecting antioxidation and photosynthesis-related processes. *Ecotoxicol. Environ. Saf.* **2019**, *169*, 645–653. [CrossRef] [PubMed]
72. Guo, B.; Liu, C.; Li, H.; Yi, K.; Ding, N.; Li, N.; Lin, Y.; Fu, Q. Endogenous salicylic acid is required for promoting cadmium tolerance of *Arabidopsis* by modulating glutathione metabolisms. *J. Hazard. Mater.* **2016**, *316*, 77–86. [CrossRef] [PubMed]

73. Guan, C.; Ji, J.; Jia, C.; Guan, W.; Li, X.; Jin, C.; Wang, G. A GSHS-like gene from Lycium chinense maybe regulated by cadmium induced endogenous salicylic acid and overexpression of this gene enhances tolerance to cadmium stress in *Arabidopsis*. *Plant Cell Rep.* **2015**, *34*, 871–884. [CrossRef] [PubMed]
74. Tao, S.; Sun, L.; Ma, C.; Li, L.; Li, G.; Hao, L. Reducing basal salicylic acid enhances Arabidopsis tolerance to lead or cadmium. *Plant Soil* **2013**, *372*, 309–318. [CrossRef]
75. Zawoznik, M.S.; Groppa, M.D.; Tomaro, M.L.; Benavides, M.P. Endogenous salicylic acid potentiates cadmium-induced oxidative stress in *Arabidopsis thaliana*. *Plant Sci.* **2007**, *173*, 190–197. [CrossRef]
76. Guo, B. Role of salicylic acid in mitigating cadmium toxicity in plants. In *Cadmium Toxicity and Tolerance in Plants*; Hasanuzzaman, M., Prasad, M.N.V., Fujita, M., Eds.; Elsevier: Amsterdam, The Netherlands, 2019; Chapter 14, pp. 349–374.
77. Rao, M.V.; Paliyath, G.; Ormrod, D.P. Influence of salicylic acid on H_2O_2 production, oxidative stress, and H_2O_2-metabolizing enzymes salicylic acid-mediated oxidative damage requires H_2O_2. *Plant Physiol.* **1997**, *115*, 137–149. [CrossRef]
78. Sanità di Toppi, L.; Gabbrielli, R. Response to cadmium in higher plants. *Environ. Exp. Bot.* **1999**, *41*, 105–130. [CrossRef]
79. Souza, V.L.; Almeida, A.A.F.D.; Lima, S.G.C.; Cascardo, J.; Silva, D.D.C.; Mangabeira, P.A.O.; Gomes, F.P. Morphophysiological responses and programmed cell death induced by cadmium in *Genipa Americana* L. (Rubiaceae). *Biometals* **2011**, *24*, 59–71. [CrossRef]
80. Qian, H.; Li, J.; Pan, X.; Jiang, H.; Sun, L.; Fu, Z. Photoperiod and temperature influence cadmium's effects on photosynthesis-related gene transcription in *Chlorella vulgaris*. *Ecotoxicol. Environ. Saf.* **2010**, *73*, 1202–1206. [CrossRef]
81. Perfus-Barbeoch, L.; Leonhardt, N.; Vavaddeur, A.; Forestier, C. Heavy metal toxicity: Cadmium permeates through calcium channels and disturbs the plant water status. *Plant J.* **2002**, *32*, 539–548. [CrossRef]
82. Boussama, N.; Ouariti, O.; Suzuki, A.; Ghorbal, M.H. Cd-stress on nitrogen assimilation. *J. Plant Physiol.* **1998**, *155*, 310–317. [CrossRef]
83. Iannone, M.F.; Rosales, E.P.; Groppa, M.D.; Benavides, M.P. Reactive oxygen species formation and cell death in catalase-deficient tobacco leaf disks exposed to cadmium. *Protoplasma* **2010**, *245*, 15–27. [CrossRef] [PubMed]
84. Vicente, R.S.; Plasencia, J. Salicylic acid beyond defence: Its role in plant growth and development. *J. Exp. Bot.* **2011**, *62*, 3321–3338. [CrossRef] [PubMed]
85. Morris, K.; MacKerness, S.A.; Page, T.; John, C.F.; Murphy, A.M.; Carr, J.P.; Buchanan-Wollaston, V. Salicylic acid has a role in regulating gene expression during leaf senescence. *Plant J.* **2000**, *23*, 677–685. [CrossRef] [PubMed]
86. Bowling, S.A.; Clarke, J.D.; Liu, Y.; Klessig, D.F.; Dong, X. The *cpr5* mutant of Arabidopsis expresses both NPR1-dependent and NPR1-independent resistance. *Plant Cell* **1997**, *9*, 1573–1584. [CrossRef] [PubMed]
87. Rate, D.N.; Cuenca, J.V.; Bowman, G.R.; Guttman, D.S.; Greenberg, J.T. The gain-of-function Arabidopsis *acd6* mutant reveals novel regulation and function of the salicylic acid signaling pathway in controlling cell death, defenses and cell growth. *Plant Cell* **1999**, *11*, 1695–1708. [CrossRef]
88. Rate, D.N.; Greenberg, J.T. The *Arabidopsis aberrant growth and death2* mutant shows resistance to *Pseudomonas syringae* and reveals a role for NPR1 in suppressing hypersensitive cell death. *Plant J.* **2001**, *27*, 203–211. [CrossRef]
89. Šašek, V.; Janda, M.; Delage, E.; Puyaubert, J.; Guivarc'h, A.; López Maseda, E.; Dobrev, P.I.; Caius, J.; Bóka, K.; Valentová, O.; et al. Constitutive salicylic acid accumulation in *pi4kIIIβ1β2* Arabidopsis plants stunts rosette but not root growth. *New Phytol.* **2014**, *203*, 805–816. [CrossRef]
90. Janda, M.; Ruelland, E. Magical mystery tour: Salicylic acid signalling. *Environ. Exp. Bot.* **2015**, *114*, 117–128. [CrossRef]
91. Drazic, G.; Mihailovic, N. Modification of cadmium toxicity in soybean seedlings by salicylic acid. *Plant Sci.* **2005**, *168*, 511–517. [CrossRef]
92. Zhu, X.F.; Lei, G.J.; Jiang, T.; Liu, Y.; Li, G.X.; Zheng, S.J. Cell wall polysaccharides are involved in P-deficiency-induced Cd exclusion in *Arabidopsis thaliana*. *Planta* **2012**, *236*, 989–997. [CrossRef] [PubMed]
93. Li, T.; Tao, Q.; Shohag, M.J.I.; Yang, X.; Sparks, D.L.; Liang, Y. Root cell wall polysaccharides are involved in cadmium hyperaccumulation in *Sedum alfredii*. *Plant Soil* **2015**, *389*, 387–399. [CrossRef]

94. Parrotta, L.; Guerriero, G.; Sergeant, K.; Cai, G.; Hausman, J.F. Target or barrier? The cell wall of early- and later-diverging plants vs cadmium toxicity: Differences in the response mechanisms. *Front. Plant Sci.* **2015**, *6*, 133. [CrossRef] [PubMed]
95. Nishimura, M.T.; Stein, M.; Hou, B.H.; Vogel, J.P.; Edwards, H.; Somerville, S.C. Loss of a callose synthase results in salicylic acid-dependent disease resistance. *Science* **2003**, *301*, 969–972. [CrossRef] [PubMed]
96. Napoleão, T.A.; Soares, G.; Vital, C.E.; Bastos, C.; Castro, R.; Loureiro, M.E.; Giordano, A. Methyl jasmonate and salicylic acid are able to modify cell wall but only salicylic acid alters biomass digestibility in the model grass *Brachypodium distachyon*. *Plant Sci.* **2017**, *263*, 46–54. [CrossRef] [PubMed]
97. Guo, B.; Liang, Y.C.; Li, Z.J.; Guo, W. Role of salicylic acid in alleviating cadmium toxicity in rice roots. *J. Plant Nutr.* **2007**, *30*, 427–439. [CrossRef]
98. Shakirova, F.M.; Allagulova, C.R.; Maslennikova, D.R.; Klyuchnikova, E.O.; Avalbaev, A.M.; Bezrukova, M.V. Salicylic acid-induced protection against cadmium toxicity in wheat plants. *Environ. Exp. Bot.* **2016**, *122*, 19–28. [CrossRef]
99. Cohen, C.K.; Fox, T.C.; Garvin, D.F.; Kochian, L.V. The role of iron-deficiency stress responses in stimulating heavy-metal transport in plants. *Plant Physiol.* **1998**, *116*, 1063–1072. [CrossRef]
100. Clemens, S.; Antosiewicz, D.M.; Ward, J.M.; Schachtman, D.P.; Schroeder, J.I. The plant cdna *LCT1* mediates the uptake of calciumand cadmium in yeast. *Proc. Natl. Acad. Sci. USA* **1998**, *95*, 12043–12048. [CrossRef]
101. Fatima, R.N.; Javed, F.; Wahid, A. Salicylic acid modifies growth performance and nutrient status of rice (*Oryza sativa*) under cadmium stress. *Int. J. Agric. Biol.* **2014**, *16*, 1083–1090.
102. Gordon, L.K.; Minibayeva, F.V.; Rakhmatullina, D.F.; Alyabyev, A.J.; Ogorodnikova, T.I.; Loseva, N.L.; Valitova, Y.N. Heat production of wheat roots induced by the disruption of proton gradient by salicylic acid. *Thermochim. Acta* **2004**, *422*, 101–104. [CrossRef]
103. Harper, J.R.; Balke, N.E. Characterization of the inhibition of k absorption in oat roots by salicylic acid. *Plant Physiol.* **1981**, *68*, 1349–1353. [CrossRef] [PubMed]
104. Halliwell, B. Reactive species and antioxidants. Redox biology is a fundamental theme of aerobic life. *Plant Physiol.* **2006**, *141*, 312–322. [CrossRef] [PubMed]
105. Garnier, L.; Simon-Plas, F.; Thuleau, P.; Agnel, J.P.; Blein, J.P.; Ranjeva, R.; Montillet, J.L. Cd affects tobacco cells by a series of three waves of reactive oxygen species that contribute to cytotoxicity. *Plant Cell Environ.* **2006**, *29*, 1956–1969. [CrossRef] [PubMed]
106. Yeh, C.M.; Chien, P.S.; Huang, J. Distinct signaling pathways for induction of MAP kinase activities by cadmium and copper in rice roots. *J. Exp. Bot.* **2007**, *58*, 659–671. [CrossRef] [PubMed]
107. Arasimowicz-Jelonek, M.; Floryszak-Wieczorek, J.; Deckert, J.; Rucińska-Sobkowiak, R.; Gzyl, J.; Pawlak-Sprada, S.; Abramowski, D.; Jelonek, T.; Gwóźdź, E.A. Nitric oxide implication in cadmium-induced programmed cell death in roots and signaling response of yellowem lupine plants. *Plant Physiol. Biochem.* **2012**, *58*, 124–134. [CrossRef] [PubMed]
108. Durner, J.; Klessig, D.F. Salicylic acid is a modulator of tobacco and mammalian catalases. *J. Biol. Chem.* **1996**, *271*, 28492–28501. [CrossRef]
109. Norman, C.; Howell, K.A.; Millar, A.H.; Whelan, J.M.; Day, D.A. Salicylic acid is an uncoupler and inhibitor of mitochondrial electron transport. *Plant Physiol.* **2004**, *134*, 492–501. [CrossRef]
110. Foyer, C.H.; Noctor, G. Redox homeostasis and antioxidant signaling: A metabolic interface between stress perception and physiological responses. *Plant Cell* **2005**, *17*, 1866–1875. [CrossRef]
111. Popova, L.P.; Maslenkova, L.T.; Yordanova, R.Y.; Ivanova, A.P.; Krantev, A.P.; Szalai, G.; Janda, T. Exogenous treatment with salicylic acid attenuates cadmium toxicity in pea seedlings. *Plant Physiol. Biochem.* **2009**, *47*, 224–231. [CrossRef]
112. Szalai, G.; Kellős, T.; Galiba, G.; Kocsy, G. Glutathione as an antioxidant and regulatory molecule in plants subjected to abiotic stresses. *J. Plant Growth Regul.* **2009**, *28*, 66–80. [CrossRef]
113. Chaouch, S.; Queval, G.; Vanderauwera, S.; Mhamdi, A.; Vandorpe, M.; Langlois-Meurinne, M.; Van Breusegem, F.; Saindrenan, P.; Noctor, G. Peroxisomal hydrogen peroxide is coupled to biotic defense responses by ISOCHORISMATE SYNTHASE 1 in a daylength-related manner. *Plant Physiol.* **2010**, *153*, 1692–1705. [CrossRef] [PubMed]
114. Chaouch, S.; Queval, G.; Noctor, G. AtRbohF is a crucial modulator of defence-associated metabolism and a key actor in the interplay between intracellular oxidative stress and pathogenesis responses in Arabidopsis. *Plant J.* **2012**, *69*, 613–627. [CrossRef] [PubMed]

115. Mateo, A.; Funck, D.; Muhlenbock, P.; Kular, B.; Mullineaux, P.M.; Karpinski, S. Controlled levels of salicylic acid are required for optimal photosynthesis and redox homeostasis. *J. Exp. Bot.* **2006**, *57*, 1795–1807. [CrossRef]
116. Yi, H.; Sejir, C.; Amna, M.; Guillaume, Q.; Bernd, Z.; Graham, N. Functional analysis of Arabidopsis mutants points to novel roles for glutathione in coupling H_2O_2 to activation of salicylic acid accumulation and signaling. *Antioxid. Redox Signal.* **2012**, *18*, 2106–2121.
117. Freeman, J.L.; Garcia, D.; Kim, D.; Hopf, A.; Salt, D.E. Constitutively elevated salicylic acid signals glutathione-mediated nickel tolerance in *Thlaspi* nickel hyperaccumulators. *Plant Physiol.* **2005**, *137*, 1082–1091. [CrossRef] [PubMed]
118. Mhamdi, A.; Hager, J.; Chaouch, S.; Queval, G.; Yi, H.; Taconnat, L.; Saindrenan, P.; Gouia, H.; Issakidis-Bourguet, E.; Renou, J.P.; et al. *Arabidopsis* glutathione reductase1 plays a crucial role in leaf responses to intracellular hydrogen peroxide and in ensuring appropriate gene expression through both salicylic acid and jasmonic acid signaling pathways. *Plant Physiol.* **2010**, *153*, 1144–1160. [CrossRef] [PubMed]
119. Michele, R.D.; Vurro, E.; Rigo, C.; Costa, A.; Elviri, L.; Valentin, M.D.; Careri, M.; Zottini, M.; Sanità di Toppi, L.; Lo Schiavo, F. Nitric oxide is involved in cadmium-induced programmed cell death in Arabidopsis suspension cultures. *Plant Physiol.* **2009**, *150*, 217–228. [CrossRef]
120. Kovács, V.; Gondor, O.K.; Szalai, G.; Darkó, É.; Majláth, I.; Janda, T.; Pál, M. Synthesis and role of salicylic acid in wheat varieties with different levels of cadmium tolerance. *J. Hazard. Mater.* **2014**, *280*, 12–19. [CrossRef]
121. Pál, M.; Horváth, E.; Janda, T.; Páldi, E.; Szalai, G. Cadmium stimulates the accumulation of salicylic acid and its putative precursors in maize (Zea mays L.) plants. *Physiol. Plant.* **2005**, *125*, 356–364. [CrossRef]
122. Dean, J.V.; Mohammed, L.A.; Fitzpatrick, T. The formation, vacuolar localization, and tonoplast transport of salicylic acid glucose conjugates in tobacco cell suspension cultures. *Planta* **2005**, *221*, 287–296. [CrossRef] [PubMed]
123. Vanacker, H.; Lu, H.; Rate, D.N.; Greenberg, J.T. A role for salicylic acid and npr1 in regulating cell growth in Arabidopsis. *Plant J.* **2001**, *28*, 209–216. [CrossRef] [PubMed]
124. Fu, Z.Q.; Yan, S.; Saleh, A.; Wang, W.; Ruble, J.; Oka, N.; Mohan, R.; Spoel, S.H.; Tada, Y.; Zheng, N.; et al. NPR3 and NPR4 are receptors for the immune signal salicylic acid in plants. *Nature* **2012**, *486*, 228–232. [CrossRef] [PubMed]
125. Ding, Y.; Sun, T.; Ao, K.; Peng, Y.; Zhang, Y.; Li, X.; Zhang, Y. Opposite roles of salicylic acid receptors NPR1 and NPR 3/ NPR 4 in transcriptional regulation of plant immunity. *Cell* **2018**, *173*, 1454–1467. [CrossRef] [PubMed]
126. Zhang, S.; Klessig, D.F. MAPK cascades in plant defense signaling. *Trends Plant Sci.* **2001**, *6*, 520–527. [CrossRef]
127. Brodersen, P.; Petersen, M.; Nielsen, H.B.; Zhu, S.; Newman, M.A.; Shokat, K.M.; Rietz, S.; Parker, J.; Mundy, J. Arabidopsis MAP kinase 4 regulates salicylic acid- and jasmonic acid/ethylene-dependent responses via EDS1 and PAD4. *Plant J.* **2006**, *47*, 532–546. [CrossRef] [PubMed]
128. Jin, C.W.; Mao, Q.Q.; Luo, B.F.; Lin, X.Y.; Du, S.T. Mutation of *mpk6* enhances cadmium tolerance in *Arabidopsis* plants by alleviating oxidative stress. *Plant Soil* **2013**, *371*, 387–396. [CrossRef]
129. Bovet, L.; Feller, U.; Martinoia, E. Possible involvement of plant ABC transporters in cadmium detoxification: A cDNA sub-microarray approach. *Environ. Int.* **2005**, *31*, 263–267. [CrossRef]
130. Eichhorn, H.; Klinghammer, M.; Becht, P.; Tenhaken, R. Isolation of a novel ABC-transporter gene from soybean induced by salicylic acid. *J. Exp. Bot.* **2006**, *57*, 2193–2201. [CrossRef]
131. Pál, M.; Janda, T.; Szalai, G. Abscisic acid may alter the salicylic acid-related abiotic stress response in maize. *J. Agron. Crop Sci.* **2011**, *197*, 368–377. [CrossRef]

International Journal of
Molecular Sciences

MDPI

Review

Salicylic Acid Binding Proteins (SABPs): The Hidden Forefront of Salicylic Acid Signalling

Igor Pokotylo [1,2], Volodymyr Kravets [1] and Eric Ruelland [2,3,*]

1 V.P. Kukhar Institute of Bioorganic Chemistry and Petrochemistry, National Academy of Sciences of Ukraine, 02094 Kyiv, Ukraine
2 Institut d'Ecologie et des Sciences de l'Environnement de Paris, Université Paris-Est, UPEC, 94010 Créteil, France
3 Institut d'Ecologie et des Sciences de l'Environnement de Paris, CNRS, UMR 7618, 94010 Créteil, France
* Correspondence: eric.ruelland@u-pec.fr or eric.ruelland@upmc.fr

Received: 12 July 2019; Accepted: 4 September 2019; Published: 6 September 2019

Abstract: Salicylic acid (SA) is a phytohormone that plays important roles in many aspects of plant life, notably in plant defenses against pathogens. Key mechanisms of SA signal transduction pathways have now been uncovered. Even though details are still missing, we understand how SA production is regulated and which molecular machinery is implicated in the control of downstream transcriptional responses. The NPR1 pathway has been described to play the main role in SA transduction. However, the mode of SA perception is unclear. NPR1 protein has been shown to bind SA. Nevertheless, NPR1 action requires upstream regulatory events (such as a change in cell redox status). Besides, a number of SA-induced responses are independent from NPR1. This shows that there is more than one way for plants to perceive SA. Indeed, multiple SA-binding proteins of contrasting structures and functions have now been identified. Yet, all of these proteins can be considered as candidate SA receptors and might have a role in multinodal (decentralized) SA input. This phenomenon is unprecedented for other plant hormones and is a point of discussion of this review.

Keywords: Salicylic acid; salicylic acid binding protein; SABP; NPR1; stress response; pathogens

1. Introduction

Salicylic acid (SA, 2-hydroxybenzoic acid) is a phenolic plant hormone. SA has a well- documented role in plant responses to environmental stresses including chilling [1], freezing [2,3], heat [4], heavy metals [5], salt [6], drought [7] and, notably, reactions to biotrophic pathogens [8]. Indeed, SA has a major role in plant innate immunity and systemic acquired resistance (SAR)—a whole-plant resistance triggered by a local infection [9]. SA also plays a role in the regulation of stomatal closure [10] and seed germination [11], among others. Despite these beneficial effects, constitutive over-accumulation of SA stunts plant growth [12,13].

SA is accumulated in plants following stress exposure. In *Arabidopsis thaliana sid2* mutants, SA accumulation induced by *Pseudomonas syringae* pv. *tomato* (Pst) DC3000 expressing *avrRpt2* was abolished [14]. These plants were characterized as isochorismate synthase 1 (ICS1)-deficient [15]. ICS1 is a component of SA biosynthesis pathways (Figure 1). The second ICS-coding gene (*ICS2*) provides a much lower contribution to SA synthesis in Arabidopsis. Both ICS1 and ICS2 proteins are found in chloroplasts [16]. Suppression of the barley *HvICS* gene by RNAi led to an impaired SA accumulation induced by *Fusarium graminearum* infection [17]. Recently, ICS1 was suggested to be post-translationally activated via direct interaction with PHB3—a member of the prohibitin protein family in Arabidopsis [18]. Following the production of isochorismate by ICS, it is believed to be converted to SA by isochorismate pyruvate lyase (IPL). However, the coding gene has not yet been cloned in plants. Recently, an alternative route for the processing of isochorismate was suggested. It could be converted to isochorismate-9-glutamate by avrPphB Susceptible 3 (PBS3, also known as Gretchen Hagen 3.12, GH3.12); isochorismate-9-glutamate would lead to SA either via a passive decay [19] or by Enhanced Pseudomonas Susceptibility 1 (EPS1)—an enzyme with isochorismoyl-glutamate A pyruvoyl-glutamate lyase activity [20] (Figure 1). Interestingly, the activity of GH3.12/PBS3 is inhibited by SA in vitro (see below).

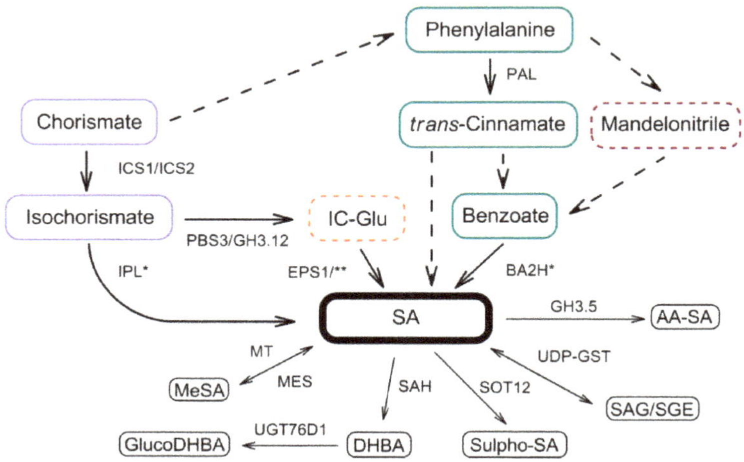

Figure 1. SA biosynthesis pathways in plants. Solid arrows represent single enzymatic steps. Dashed arrows represent multiple consecutive enzymatic steps. AA-SA, amino acid-SA conjugate; BA2H, benzoic acid 2-hydroxylase; DHBA, dihydroxybenzoic acid; EPS1, Enhanced Pseudomonas Susceptibility 1; GH3.5, Gretchen Hagen 3.5; ICS, isochorismate synthase; IC-Glu, isochorismate-9-glutamate; IPL, isochorismate pyruvate lyase; MeSA, methyl-salicylate; MES, methyl-salicylate esterase; MT, metyl transferase; PAL, phenylalanine ammonia-lyase; PBS3, avrPphB susceptible 3; SAG, salicylic acid glucoside; SAH, salicylic acid hydroxylase; SGE, salicylic acid glucose ester; SOT12, sulfotransferase 12; UDP-GST, UDP-glycosyltransferase; UGT76D1, UDP-glycosyltransferase 76D1; *, enzyme not cloned in plants; **, non-enzymatic decay.

Another pathway for SA synthesis involves the phenylalanine ammonia lyase (PAL) converting phenylalanine (Phe) to *trans*-cinnamic acid. This pathway occurs in the cytosol. Four PAL-encoding genes are found in *A. thaliana* [21]. The PAL pathway was shown to be active in poplar infected with *Botryosphaeria dothidea* fungus [22]. Yet, another route for SA biosynthesis involving mandelonitrile, a Phe derivative, was proposed to be functional in peach plants (Figure 1). Mandelonitrile-treated peach plants accumulated more SA and were more resistant to *Plum pox virus* [23]. The synthesis of SA through this pathway occurs through a benzoate intermediate. The exact role of mandelonitrile in SA accumulation is still a matter of discussion.

SA can be converted to a number of derivatives. These molecules have roles as either transportable forms of SA (such as methyl-salicylate, MeSA) or inactive/storage forms of SA (Figure 1). Among the latter are glucosylated SA derivatives—salicylic acid glucoside (SAG) and salicylic acid glucose ester (SGE) [24]. These molecules are stored in the vacuole and can be reversely converted to SA [25]. Hydroxylated SA derivatives—2,3-dihydroxybenzoic acid (DHBA) and 2,5-DHBA—are catabolic SA forms in the cytosol [26]. Yet, glycosylation of these molecules by UGT76D1 (an UDP-glycosyltransferase) was shown to be a part of the immune response to Pst DC3000 in *A. thaliana* and regulated ICS1-dependent SA production [27]. In *A. thaliana*, SA can also be sulfonated by cytosolic SOT12 sulphotransferase. SA sulfonation has a similar positive feedback on SA accumulation in stressed plants [28]. Aspartyl amino acid conjugates of SA are thought to attenuate SA signaling, and they are formed by GH3.5 in *A. thaliana* [29].

Biotic stress-induced SA accumulation is controlled by three protein regulators: Enhanced Disease Susceptibility 1 (EDS1), Phytoalexin Deficient 4 (PAD4) and Senescence Associated Gene 101 (SAG101) [30]. It is known that EDS1 binds at least several TIR-NB-LRR receptors (used by the plant to sense the presence of pathogens) [31] and interacts with either PAD4 or SAG101 [32]. EDS1, PAD4, and SAG101 are all lipase/esterase-like proteins. However, the exact mode of their input to SA synthesis is yet to be established and does not necessarily rely on these enzymatic activities. Recently, an EDS1-PAD4 complex was found to inhibit MYC2 [33], which is a positive regulator of jasmonic acid (JA) pathway signaling that itself antagonizes SA (for review see Ref. [34]). This is thought to be an evolution-selected mechanism to block the action of coronatine—a JA-mimicking compound produced by the bacteria to subdue SA-driven defenses. EDS1 and PAD4 could also have a role independent of SA. For instance, the activation of *FLAVIN-DEPENDENT MONOOXYGENASE1* (*FMO1*) expression in Arabidopsis in response to thaxtomin A (a bacterial toxin) was dependent of EDS1 and PAD4 but independent of ICS1-generated SA [35].

Basal SA accumulation is controlled by multiple inputs. In many SA-overaccumulating mutants reviewed in Ref. [13], the connection between the function of the protein encoded by the altered gene and SA pathway signaling is vague. In rice, the knocking-down of *SEC3A*—coding for an exocyst protein complex component that guides exocytic vesicles to the plasma membrane—led to SA over-accumulation [36]. The *pi4kβ1β2* double mutant, altered in two phosphatidylinositol-4-kinases, also accumulates high SA levels, but the reason for this is still being investigated [12].

After SA is produced, it interacts with Nonexpressor of Pathogenesis-related protein 1 (NPR1)—a key transcriptional regulator of SA signaling. SA affects NPR1 in at least two ways: i. NPR1 directly binds SA [37]; and ii. SA induces redox changes in the cell that conditions NPR1 monomerization [38]. Monomeric NPR1 shuttles to nuclei where it interacts with TGA transcription factors leading to the expression of *PATHOGENESIS-RELATED* (*PR*) genes involved in the set-up of plant immune defenses [13]. The regulation of PR gene transcription by NPR1 has been shown to involve histone acetylation via CBP/p300-family histone acetyltransferases [39].

Intriguingly, the mechanisms of SA perception are not fully understood. For instance, some signaling events are triggered upstream of NPR1 monomerization and/or shuttling into the nucleus. This is the case of the cell redox change, but also of phosphatidylinositol 4-kinase (PI4K) and phospholipase D (PLD) activation [40–42]. Moreover, some responses to SA are observed in NPR1-deficient mutants [43–45]. Therefore, NPR1 cannot be the only SA-binding protein (SABP) in plants. NPR1 paralogs, NPR2 [46], NPR3 and NPR4 [47,48], also bind SA. Moreover, high-throughput biochemical screens have provided a list of almost 100 candidate SABPs [49,50]. Many of these proteins are important enzymes of primary carbon metabolism. This challenges the classical paradigm of hormonal signaling where a ligand is recognized by a single/few receptor(s). Are these SABPs real SA receptors or are they false positive of high through-put omics techniques? If they really bind SA, what are the roles of such binding in SA-signaling pathways? And why do plant cells have so many SABPs. The role(s) of such a multitude of SABPs is still unexplained and offers a new paradigm for hormonal regulation in plants.

2. SA Binding Proteins

The multiplicity of SABPs has already been highlighted as an intriguing phenomenon [51]. For many of the SABPs, the physiological outcome of their interaction with SA is unclear. Moreover, it is not established whether and how the SABPs are integrated into the conventional NPR1 pathway. At least several plant SABPs have their animal orthologs that also bind SA [51]. In animals, SA and its derivatives (e.g., aspirin) act not as hormones but as therapeutic compounds that target proteins and prevent their role as disease components. How animals have developed such sophisticated responses to substances originating from plants is an interesting question by itself. Nevertheless, this suggests that SA can clearly act in a NPR1-free environment [51]. Here, we present an up-to-date vision of these problems, characterise the canonical and newly identified plant SABPs, and discuss how they can be integrated into a unified SA-signaling network.

2.1. SABP1—Catalase

SABP1 was purified from tobacco as a soluble cytosolic protein that binds SA with a K_d of 14 μM [52]. *SABP1* was cloned and the resulting 57 kDa protein was characterized as catalase [53]. Its activity was drastically inhibited by SA but not by inactive SA analogues in vitro [54]. This effect corresponds well to the results obtained using tobacco cell suspensions where SA could inhibit the total catalase activity [55].

In *A. thaliana*, catalase 2 (CAT2) shares 78% sequence identity with tobacco SABP1 (UniProt P49319). SA inhibited CAT2 catalase activity in vitro and total catalase activity in leaves of SA-pre-treated plants. However, the direct binding of SA to CAT2 has not been tested [56]. CAT2 was demonstrated to have a role in SA-mediated resistance to a biotrophic pathogen (Pst DC3000) by inhibiting indole-3-actic acid (IAA, an auxin) and JA accumulation [56]. Inhibition of CAT2 by SA leads to a H_2O_2 increase in plants upon pathogen infection. This promotes sulfenylation (sulfenic acid conjugation) of tryptophan synthetase b subunit1 (TSB1) at Cys308 leading to the inhibition of its activity. This enzyme acts in the IAA biosynthesis pathway and as a result, SA antagonizes IAA accumulation. In parallel, SA diminished the stimulatory effect of CAT2 protein on the in-vitro activity of acyl-CoA oxidases (ACX2/ACX3) implicated in jasmonic acid (JA) biosynthesis. It is not clear if this effect is due to the catalase activity of CAT2. A direct interaction between CAT2 and ACX2/ACX3 has been observed. This interaction was impaired in SA-pre-treated plants [56].

Unlike wild-type (WT) plants, *sid2* mutants accumulated JA in response to Pst DC3000 (reflecting the antagonism of SA against JA accumulation). Such JA accumulation was diminished in *sid2cat2* double mutants, suggesting that CAT2 is indeed a positive regulator of JA production and thus plays a part in the SA-JA antagonism in plants [56]. The *cat1* (78% identity to SABP1) and *cat3* (76% identity to tobacco SABP1) Arabidopsis mutants were not altered in the biotic stress-induction of neither IAA nor JA accumulation [56].

The way SA binds SABP1 or catalase orthologues, from a molecular point of view, has not been deciphered.

2.2. SABP2—MeSA Esterase

SABP2 was similarly purified from tobacco and had a much higher affinity to SA (K_d = 90 nM) when compared to SABP1 [57]. A corresponding ORF encoding a 260-residue α/β fold hydrolase superfamily protein with a calculated molecular mass of 29 kDa was cloned and characterized. Its lipase activity towards *para*-nitrophenyl palmitate (measured by the release of *p*-nitrophenol) was drastically stimulated by SA in vitro. The silencing of SABP2 using the RNAi technique resulted in the lowering of SA-induced *PR1* expression and resistance level against *Tobacco mosaic virus* (TMV) [58]. Later, it was established that SABP2 activity was that of a methyl salicylate (MeSA) esterase, converting MeSA into SA, an end-product inhibitor of such activity. The co-crystallization of SABP2 with SA revealed that it is positioned carboxylate group-first in the active site [59]. The inhibition of methyl esterase (MES) activity of SABP2 by SA could be a mechanism to fine-tune active SA concentration in the cell.

SABP2 can also convert acibenzolar-S-methyl (a functional analogue of SA) into acibenzolar. Silencing of SABP2 in tobacco results in the loss of an acibenzolar-S-methyl effect on the induction of *PR1* expression and the onset of SAR [60].

MeSA (produced by salicylate carboxyl methyltransferases) is inactive but is more hydrophobic than SA and it easily penetrates cell membranes. Therefore, MeSA is, in line with other molecules such as pipecolic acid [61], considered to be a mobile signal of the SAR. Produced in infected cells, MeSA reaches distant leaves by phloem transport. In these so-called systemic tissues, it is converted into active SA via MES activity. The SA thus produced triggers-preventive defense responses in these distal leaves. SABP2 was established as a key enzyme of SAR in tobacco [62,63]. In potato, an orthologue of SABP2, StMES1, has been cloned. The enzymatic activity of StMES1 was inhibited by SA in vitro and its role was, in a similar way, linked to SAR development [64].

Two methyl esterase-encoding genes were identified in poplar [65]. In *A. thaliana*, 18 SABP2 orthologues were identified and at least five proteins (AtMES1,-2,-4,-7,and -9) were shown to possess an esterase activity acting on MeSA that was inhibited by SA in vitro [66]. In these two species, MES is also an important component of SAR [22,66].

2.3. SABP3—β Carbonic Anhydrase

SABP3 was identified in tobacco chloroplasts as β carbonic anhydrase (βCA). SABP3 binds SA with moderate affinity (K_d = 3.7 μM) [67]. CAs are ubiquitous and evolutionary-conserved enzymes that catalyse the interconversion of CO_2 and bicarbonate (HCO_3^-). CAs have roles in photosynthesis, respiration, stomata movements, and lipid biosynthesis among others [68]. From an immunity point of view, SABP3 was required for the hypersensitive response (HR) in tobacco leaves infiltrated with *A. tumefaciens* expressing *Pto:avrPto* (R-avr gene pair). Two recombinant tobacco proteins—SABP3/βCA1 and βCA2—were shown to bind SA but not 4-hydroxybenzoic acid (4-HBA, an inactive isomer). Based on the fact that SABP3 could complement the phenotype of an oxidative stress-sensitive strain of *Saccharomyces cerevisiae* [67], it was suggested that SABP3 might have antioxidative properties.

In *A. thaliana*, the affinity of AtSABP3/βCA1 (a SABP3 orthologue) to SA and its CA activity is diminished in the presence of S-nitrosoglutathione (GSNO), a NO donor. These effects were not observed in a C280S AtSABP3-mutated protein, suggesting that they are due to the nitrosylation of Cys280 [69]. AtSABP3 knockout mutants have their resistance compromised towards Pst DC3000 (*avrB*). The resistance phenotype was restored when complemented by AtSABP3 but not by C280S AtSABP3 [69]. In this manner, AtSABP3 nitrosylation is clearly a requirement for its role in immunity. In contrast, the role of the binding of SA to AtSABP3 requires further research.

Indeed, there are contrasting data about SA influence on CA activity in plants. SA did not affect the CA activity of purified SABP3 from tobacco chloroplasts at physiological concentrations. The inhibition only occurred at concentrations as high as 3 mM, while binding occurred at lower concentrations [67]. A significant increase in CA activity was reported in leaves of SA-treated peppermint (*Mentha piperita*) [70]. In contrast, in *A. thaliana*, CA activity was reported to decrease following treatment with SA or with benzothiadiazole (a functional SA analogue), or after inoculation with *P. syringae* [71].

SABP3 orthologs from Arabidopsis and *Chenopodium quinoa* were found to physically interact with HCPro—a viral protein of *Turnip mosaic virus* (TuMV). This protein has RNA silencing suppressor (RSS) activity and counteracts host's anti-viral RNA interference. The transient expression of HCPro antagonized AtSABP3 transcripts and protein accumulation in Arabidopsis [72]. AtSABP3 was thus established as a component of anti-viral defense. Whether this role of AtSABP3 is dependent on SA binding is still unknown. Unexpectedly, both AtSABP3-knockout and AtSABP3-overexpressing lines of Arabidopsis were compromised for their resistance to TuMV [72].

AtSABP3 bearing no signal peptide was shown to interact with NPR1 in a yeast two-hybrid assay. Intriguingly, this result was observed only when growth plates were supplemented with SA (but not with inactive isomer 4-HBA), suggesting that this interaction was SA dependent [71]. An AtSABP3-NPR1 interaction was also demonstrated in tobacco leaves in planta. Bimolecular fluorescence complementation revealed that this interaction occurred in the nucleus and the perinuclear region. The transient expression of GFP-NPR1 and MBP-AtSABP3 constructs led to co-purification of GFP-NPR1 together with MBP-AtSABP3 on amylose resin [71]. In the same study, interactions with NPR1 and NRB4 (a protein that is perhaps involved in SA perception), were similarly reported for at least several other cloned fragments/alternative splice variants of proteins representing the βCA family in Arabidopsis. These data suggest that βCA family members could be a part of SA signalling. However, an Arabidopsis *βca1* mutant deficient in AtSABP3 (or quintuple *βca1,2,3,4,6* mutant for that matter) was only partially insensitive to exogenous SA as suggested by pathogen resistance and

PR1 expression responses. The homozygous mutation in *βCA5* could not be tested due to plants sterility [71].

Note that amino acid residues required for SA binding to SABP3 and orthologues have not yet been identified.

2.4. NPR1/2/3/4—Signalling Proteins

NPR1 binds SA with high affinity in vitro (K_d = 140 nM) [37]. SA binding to NPR1 is Cu^{2+}-dependent, implicates Cys521 and Cys529, and results in conformational changes of NPR1—a mechanism that could stand behind its role as a transcription cofactor with TGA [37].

NPR1 exists in the cytosol as an oligomer due to disulphide bridges between Cys82 and Cys216 of different subunits. In order to shuttle to the nucleus and act on regulating gene expression, NPR1 has to be monomerized [38]. This monomerization requires the reduction of the disulphide bonds. SA binding has been shown to facilitate de-oligomerization of recombinant NPR1. Yet, SA alone is not sufficient to trigger NPR1 monomerization [37]. The upstream SA-driven redox events that allow the reduction of the disulphide bonds in NPR1 are not fully understood. Two cytosolic thioredoxins, TRX-h3 and TRX-h5, have been shown to interact with NPR1 since they were pulled-down by the immobilised His-tagged N-terminal part of NPR1. Furthermore, co-immunoprecipitation of TRX-h5 with a NPR1-TAP fusion protein was stimulated by SA [73]. Thioredoxins are redox regulators and can reduce disulphide bridges [74,75]. Co-incubation of cell lysates with recombinant TRX-h5 led to an increase in NPR1-GFP monomers, thus suggesting that TRX-h5 plays a part in NPR1 monomerization. Both *trx-h3* and *trx-h5* mutants were compromised in *PR1* induction by exogenous SA [73]. In contrast, NPR1 monomerization is negatively regulated by S-nitrosylation at Cys156 [73].

In Arabidopsis, NPR3, NPR4 (both having 39% identity to NPR1) [48] and NPR2 (62% identity to NPR1) [46] are paralogues of NPR1. They have all been shown to bind SA. The affinities of NPR3 (K_d = 176nM) and of NPR4 (K_d = 23nM) to SA are known [76]. NPR3 [77], NPR4 [78] and NPR2 [46], similarly to NPR1, act in the nucleus. Interestingly, NPR3 and NPR4 were demonstrated to bind NPR1 in a SA-dependent manner in vitro. Based on the fact that the ability of cullin 3 (CUL3, a component of protein E3 ligase complex) to pull-down NPR1-GFP was reduced in a *npr3 npr4* genetic background, it was suggested that NPR3 and NPR4 act as adaptors for SA-dependent proteasome-mediated degradation of NPR1 [48]. In a similar manner, NPR3 and NPR4 were suggested to mediate the proteasome-dependent degradation of EDS1 [79].

NPR3 and NPR4 have been shown to have roles as transcriptional co-repressors that function in parallel to NPR1. They act as negative regulators of immunity. *SARD1* and *WRKY70* genes are, for instance, under the negative transcriptional control of NPR3/NPR4 [76]. Co-transformation of Arabidopsis protoplasts expressing a luciferase reporter gene under the control of either SARD1 or WRKY70 promoters with plasmids overexpressing either NPR3 or NPR4 resulted in the inhibition of luciferase expression. This effect was diminished when TGA-binding motifs were mutated in either SARD1 or WRKY70 promoter regions. SARD1 and WRKY70 expression were also partly de-repressed in TGA2/TGA5/TGA6-deficient plants. This suggests that TGA transcription factors are implicated in the negative control of their expression by NPR3 or NPR4. More importantly, SA could antagonise the observed transcriptional repression activity of NPR4 [76].

In potato, the StNPR3L (NPR3-like protein) was shown to interact with the transcription factor StbZIP61 and inhibit its transcriptional activation activity in an SA-dependent manner [80]. This could have a role in the regulation of SA accumulation in infected plants since the expression of *StICS1*, a SA biosynthesis gene, positively correlated to *StbZIP61* expression as suggested by RNAi and mutant studies [80].

NPR2 physically interacts with NPR1 in vitro, and in planta the overexpression of NPR2 could partly complement the *NPR1*-deficient phenotype [46]. The role of SA binding in this interaction is unknown.

2.5. Glutathione S-transferase

Several *A. thaliana* glutathione *S*-transferase (GST) isoenzymes (GSTF2, GSTF8, GSTF10, GSTF11) have been shown to bind SA [81]. The enzymatic activity of GSTF10, GSTF11 and that of GSTF8 (to a lesser extent) were inhibited by SA in vitro. GSTs comprise a large group of enzymes that catalyse at least several reactions in connection to glutathione conjugation. Promoters of many GST-encoding genes contain disease-related W-boxes and WT-box cis-regulatory elements. GSTs are a part of plant immunity with roles in glucosinolate (antimicrobial compound) metabolism and detoxification of mycotoxins among others [82].

2.6. Thioredoxins

The chloroplastic thioredoxin-m1 (TRXm1) was shown to bind SA in a high-throughput screen and later confirmed by surface plasmon resonance (SPR) analysis [50]. Reduced thioredoxins can reduce disulphides of target proteins [74,75]. The reduction of TRX arises either from NADPH through NADPH thioredoxin reductases or from ferredoxin via ferredoxin thioredoxin reductases. The effect of SA binding on TRXm1 activity has not been yet established. As mentioned earlier, the activity of other thioredoxins, the cytosol located TRX-h3 and TRX-h5, is required for NPR1 monomerization [73]. It could be interesting to investigate whether these cytosolic TRXs are also SA-binding proteins.

2.7. GH3—Acyl Acid Amido Synthetase

GH3.12/PBS3 from *A. thaliana* conjugates specific amino acids to acyl substrates (e.g., 4-substituted benzoates) in an Mg^{2+}- and ATP-dependent manner. This enzyme binds SA in a ternary complex with AMP as shown by its crystallographic structure [83]. Due to the observed position in the active pocket, SA was suggested not to be a substrate but an inhibitor of AtGH3.12/PBS3. This is consistent with the fact that SA indeed inhibits AtGH3.12/PBS3 activity in vitro [84]. Conjugation of amino acids to plant hormones, such as jasmonic acid or auxins, is a common strategy aimed at controlling their active level in plant cells [85]. PBS3 was also shown to be an important enzyme of SA biosynthesis [19,20]. The role of AtGH3.12/PBS3 in SA signalling is, however, unclear. The *P. syringae*-induced free SA accumulation in *pbs3* mutants was actually higher than that of WT plants, but it was accompanied by the reduced pathogen resistance and retarded *PR1* expression in such plants. These effects were accompanied by the drastic diminution of SA-*O*-β-glucoside (SAG) accumulation [86]. The connection between SAG production and AtGH3.12/PBS3 has not been established yet.

Recently, PBS3 was shown to directly interact with EDS1—one of the three key protein regulators of SA pathway signalling (see above). In tobacco leaves, the interaction occurred both in the nucleus and cytoplasm as revealed by bimolecular fluorescence complementation assays. No interaction was observed for PBS3 and PAD4 or SAG101. Using an inhibitor approach, authors concluded that PBS3 could control EDS1 abundance in a post-translational manner by preventing its proteasome degradation—most likely by the 26S proteasome [79]. The role of SA binding to PBS3 was not investigated in this context.

It was reported that another GH3 family member, AtGH3.5, could accept SA as a substrate. In AtGH3.5-overexpressing Arabidopsis lines, a significant increase in the content of SA-aspartyl was registered [29]. The very same enzyme also produces inactive aspartyl conjugates of indole-3-acetic acid (IAA)—an auxin hormone. In this manner, AtGH3.5 could play a part in SA-IAA crosstalk.

2.8. GAPDH

Several isoforms (subunits) of *A. thaliana* GAPDH (glyceraldehyde 3-phosphate dehydrogenase)—GAPA-1, GAPA-2, GAPC-1, and GAPC-2—bind SA as demonstrated using SPR [87]. Among the detected SA-binding GAPDH isoforms, some are cytosolic (GAPC-1, GAPC-2) while others are plastidial (GAPA-1, GAPA-2) enzymes. They play essential roles in glycolysis (cytosolic) and Calvin cycle (plastidial), respectively. However, the effect of SA binding on GAPDH function is unknown.

Some GAPDH isoenzymes have been shown to be multifunctional. AtGAPC-1 is required for *Tomato bushy stunt virus* (TBSV) asymmetric replication via its direct association with a negative RNA strand of the virus. The binding of SA to AtGAPC-1 inhibits the association of the enzyme to the virus RNA. This was suggested to be a part of plant anti-viral defenses [87].

Interestingly, human GAPDH also binds SA and is similarly implicated in the regulation of the replication of some viruses [88]. SA binding suppresses HsGAPDH translocation to the nucleus. Whether a similar mechanism affecting GAPDH localization is employed in plants is yet to be established.

2.9. Alpha-ketoglutarate Dehydrogenase—Krebs Cycle Enzyme

The E2 subunit of the α-ketoglutarate dehydrogenase (αKGDE2) enzyme complex acts in the tricarboxylic acid (Krebs) cycle in mitochondria. This protein was shown to bind SA in two independent assays (photoaffinity labelling and SPR) both in Arabidopsis [81] and tomato [89]. The αKGDE2 activity was reduced almost by half in isolated mitochondria sampled from tomato leaves pre-treated with SA for 24 h. This could be due to either a transcriptional or translational regulation by SA and not to a direct effect of the binding of SA on the protein. The silencing of αKGDE2 resulted in the increase of tomato resistance to TMV. Interestingly, the treatment with SA could similarly induce resistance to TMV in WT plants but it could not enhance the resistance phenotype of αKGDE2-silenced plants [89]. Based on these facts, a suppression of αKGDE2 by SA was suggested to be a part of plant antivirus defenses.

2.10. Thimet Oligopeptidases + TPPII Exopeptidase—Proteolysis

In *A. thaliana*, there are three thimet oligopeptidases (TOP, zinc-dependent metalloendopeptidases) [90]. Two of them, TOP1 and TOP2, bind SA [49]. SA inhibited peptidase activities of TOP1 and, to a lesser extent, of TOP2, in vitro. For TOP1, kinetics indicated a non-competitive mechanism. SA treatment also inhibited the bulk peptidase activity in plant extracts as measured by the release of a fluorescent peptide marker. TOP1 contains a signal peptide and TOP1–GFP was found to be localized in chloroplasts. The inhibitory effect of SA on a truncated form of TOP1, lacking 110 N–terminal residues spanning the signal peptide, was much weaker. This suggests that SA could selectively affect TOP1 activity based on its localization. Based on mutant studies, both TOP1 and TOP2 were required for plant response to either Pst *avrRpt2* or Pst *avrRps4*. However, when tested with Pst *avrRpm1*, Pst *avrPphB* or *Pseudomonas syringae* pv. *maculicola* (Psm), no differences to WT plants were observed. Moreover, the assessed level of programmed cell death was actually lower in *top1-3* mutants inoculated with Pst avrRpt2 compared to WT [49].

TOP1 and TOP2 were found to produce homo- and heterodimers. The formation of TOP2-TOP2 and TOP1-TOP2 dimers were diminished by the addition of SA in isolated *A. thaliana* protoplasts. Authors suggested that the effect of SA could be due to redox changes since the effect of dithiothreitol, a strong reductant, led to a strong shift towards the presence of TOP1 and TOP2 monomeric forms in vitro [91]. The functional role of TOP dimers is unclear.

Another enzyme implicated in proteolysis that binds SA is tripeptidyl peptidase II [50].

2.11. MORC Proteins—Epigenetic Regulation

Microrchidia (MORC) proteins comprise a group of peculiar DNA-binding enzymes with ATPase, endonuclease and topoisomerase activities. These proteins can be potentially involved in epigenetic gene silencing [92]. In tomato, SlMORC1 binds SA as demonstrated by SPR analysis [93]. This interaction resulted in altered activities of SlMORC1 in vitro: SA suppressed ATPase and decatenation activities but not the DNA relaxation activity of SlMORC1.

2.12. HMGB3—DAMP Protein

High Mobility Group Box 3 (HMGB3) was shown to bind SA using SPR analysis [94]. The binding affinity of this protein to the immobilized 3-aminoethyl SA was very high (K_d = 1.5 nM). This protein was found to be a DAMP (damage-associated molecular pattern molecules) acting via BAK1 and BKK1 receptor kinases. Exogenous application of purified recombinant HMGB3 induced plant immune responses and was enough to improve Arabidopsis resistance to *B. cinerea*. HMGB3, when applied together with 1 μM SA, lost its effectiveness as a DAMP.

3. Response of SABPs to Treatments Linked to SA/Biotic Stress

To sum up the above sections and to find common regulatory patterns, if any, among SABPs, we mined transcriptomics data for genes encoding the SABPs described above (Figure 2). In panel A, we show the effect of SA on the protein, when it is known. In our list are proteins from *A. thaliana*. However, the effects of SA on CAT2, MES9 and βCA1 were extrapolated from those observed for tobacco orthologs (SABP1, SABP2 and SABP3, respectively). SA treatment led to an inhibition of enzymatic activity of a number of SABPs in vitro. The activity of NtSABP3 was not affected by SA while the effect of SA on transcriptomic activities of NPR proteins was not considered. No example of SA activating a SABP enzymatic activity in vitro is currently available to us.

In panel B (Figure 2), we show two sets of transcriptomics data. A first set was used to draw the dendrogram showing the hierarchical clustering of SABPs based on their expression across 111 conditions. In these experiments, *A. thaliana* plants were challenged with bacterial (*Pseudomonas* spp.), fungal (*Sclerotinia sclerotiorum*, *Golovinomyces orontii*, *Hyaloperonospora arabidopsidis*) and viral (TuMV) pathogens. All of these conditions should directly implicate SA signaling responses. We could separate three clusters. The alpha cluster represents the genes whose expression was mostly inhibited by the treatments; the gamma clusters represent genes whose expression was induced by the treatments while the beta cluster represents genes exhibiting an intermediate situation.

In the second set of transcriptomics data (heat map), we illustrate how the expression of SABPs change when different aspects of SA signalling are stimulated. Here we included the SA treatment per se, but also treatments with methyl jasmonate (MeJa, a SA-antagonist hormone), model pathogens, and a flg22 elicitor—a fragment of flagellin, a protein from bacterial flagellum that triggers immunity responses. *NPR3* and *NPR4* are two genes early stimulated by SA (3 h). After a longer time (24 h), SA could both stimulate (*NRP* genes, *GH3.12*) and inhibit (*GSTF11*, *CA1*, *GAPA-2*) the expression of SABPs. No SABPs were reactive to MeJa. We could not observe significant differences in expression patterns of SABPs following treatments with Pst and Psm bacteria—the latter bearing the *avrRMP1* avirulence gene that activates effector-triggered immunity (ETI). The transcriptional responses to *S. sclerotiorum* fungus and TuMV were a tad divergent. In virus-treated plants, a negative effect on *GAPA-2*, *TRX-m1* and *CA1* expression was relieved. The same can be said for plants treated with flg22. It should be noted that across all modelled infections, the stimulation of *NPR* genes and *GH3.12* was quite consistent.

Therefore, similarly to what was observed in the clustering analysis, binding to SA concerns proteins that are both positively and negatively expressed in response to elicitation related to biotic stresses. All the SABP-coding genes behaved differently, but at least some SABPs are well synchronised at the transcriptomic level. For instance, NPR proteins, especially NPR2 and NPR3, appear to cluster together (gamma2), thus suggesting that they are involved in the same signaling cascade. Indeed, NPR2, NPR3 and NPR4 have been shown to interact physically with NPR1 [46,48]. The same transcriptomic connections are true for βCA1, GAPA-1 and TRXm1 in the alpha cluster, and KGDE2 with GAPC1 in the gamma1 cluster. How these connections at the transcriptomic level translate into interplay between protein functions is yet to be established.

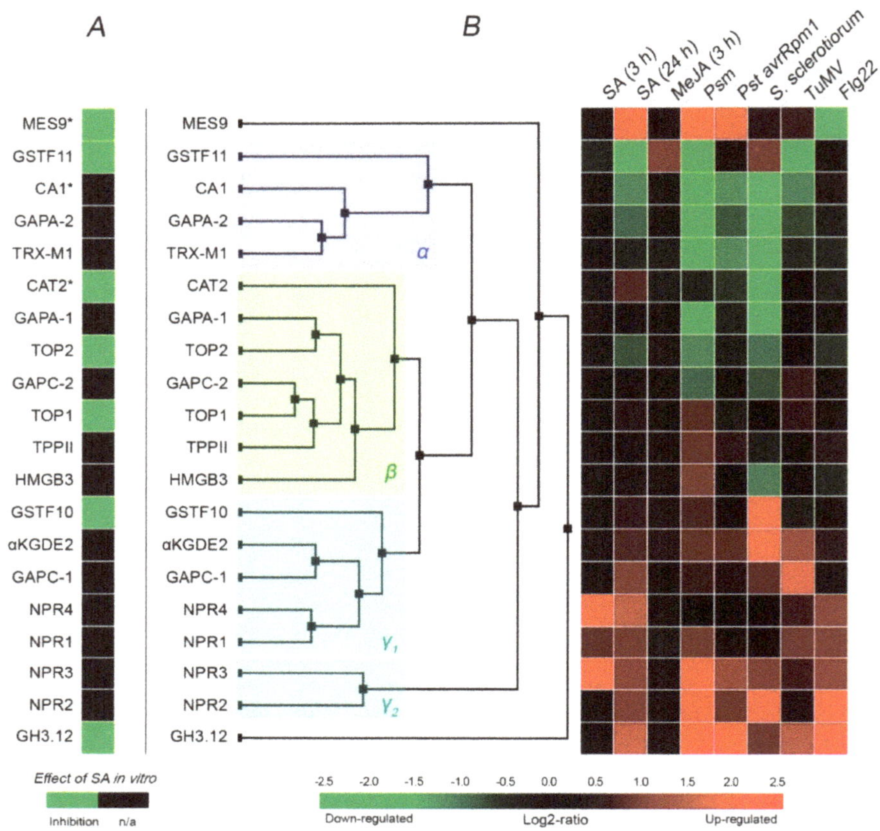

Figure 2. Effect of SA on the in-vitro catalytic activity of SABPs from *A. thaliana* (**A**) and transcriptional responses of corresponding coding genes (**B**). *, effect observed in tobacco orthologs. Transcriptomic data was mined using Genevestigator [95]. Experiment IDs: SA 3 h, AT-00113; SA 24 h, AT-00320; MeJA 3 h, AT-00110; Psm, AT-00406; Pst avrRpm1, AT-00106; *S. sclerotiorum*, AT-00681; TuMV, AT-00324; flg22, AT-00392. Note that the hierarchical clustering was performed on more experiments than the one used for the heat map (see the main text).

It is also interesting to note that for GAPDH family members, some appear to be strongly inhibited in responses to immunity-related stresses (such as the plastidial GAPA2) while others are induced (such as cytosolic GAPC1). These enzymes are likely not to have the same role in SA-signaling pathways. Besides, there is no strong correlation between the effect of SA on the protein (panel A) and the way their genes react to immunity-related inputs (panel B): Proteins that are inhibited by SA are found in alpha and gamma clusters.

Interestingly, a human GAPDH gene is commonly used as a "housekeeping" reference gene in quantitative RT-PCR analyses [96]. The data presented in this paper, however, show that the same cannot be translated to GAPDH isogenes in Arabidopsis since a strong transcriptomic reaction to either viral (GAPC-1) or bacterial/fungal infections (GAPA-2) was observed.

4. Molecular Mechanisms of SA-Protein Interactions

Some plant SABPs have been crystallized with SA. Since no conserved SA binding motif is known, the analysis of crystallography data could help to better understand how binding occurs. Here we have focused on two plant proteins, AtGH3.12 and NtSABP2 (Figure 3).

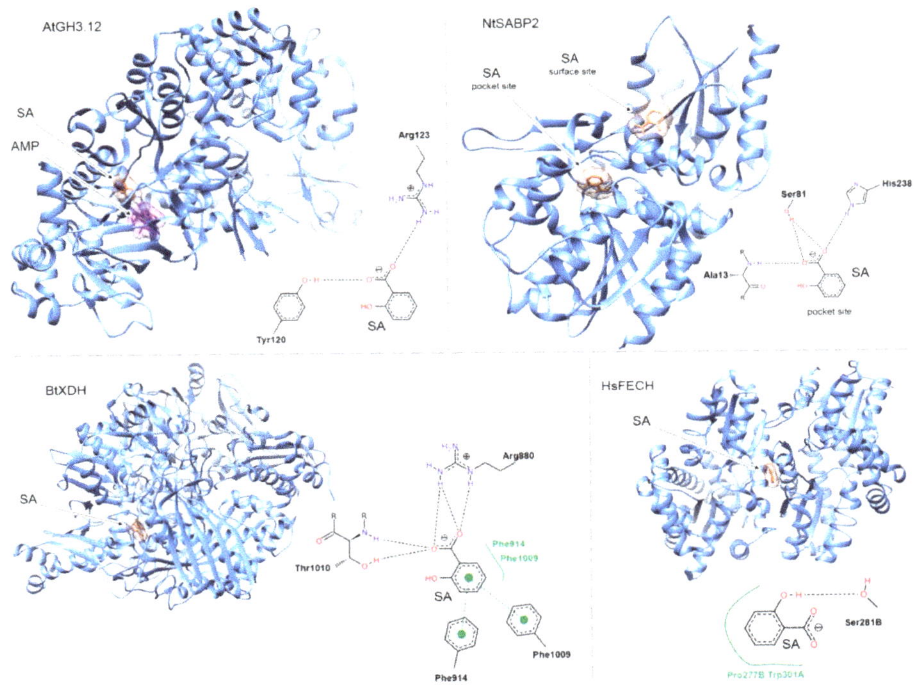

Figure 3. Molecular interactions between SA and selected SABPs. Black dashed links represent hydrogen bonds. Green dashed lines represent pi-pi stacking. Green solid lines represent hydrophobic interactions. Molecular graphic images were produced using UCSF Chimera [97].

In AtGH3.12, SA binding occurs in the active site. The carboxyl group of SA forms hydrogen bonds with side chains of Arg123 and Tyr120. Using the UCSF Chimera software [97], we could identify that SA also forms contacts (likely nonpolar interactions) with Gly326 and the side chain of Ile217 (Figure S1). While SA was co-crystallized with AMP in the active site of AtGH3.12, there is no direct interaction between SA and this cofactor (Figure S1).

For NtSABP2, two SA-binding sites have been found; an inner pocket, in the active site and a surface pocket that has been suggested to be a crystallographic artefact [59]. In the inner pocket, the carboxyl group of SA forms hydrogen bonds with Ala13, Ser81 and His238 residues (Figure 3). An interaction between the SA carboxyl group carbon chain and the carbon chain of Ser81 was predicted using UCSF Chimera (Figure S1). As for the surface-binding pocket, the carboxyl group of SA forms a hydrogen bond with Lys159, while interactions were predicted with Lys159, Leu132 and His158. As for the latter, the aromatic ring of SA is in a parallel plane to the imidazole ring of His158 (Figure S1).

Interestingly, many animal and human proteins have been co-crystallized with SA (or, alternatively, with acetyl SA—an aspirin). Similar mechanisms are apparently involved in the binding. Bovine milk xanthine dehydrogenase binds SA via hydrogen bonds formed between Thr1010, Arg880 and the carboxyl group of SA. Moreover, clear π–π stacking is formed between the aromatic ring of SA and aromatic rings of Phe914 and Phe1009 (Figure 3). In human ferrochelatase (FECH), however, it is a side hydroxyl group, and not the carboxylic group, that forms a hydrogen bond to Ser281 (Figure 3). In this protein (a homodimer), SA binds directly at the dimer interface, implicating hydrophobic amino acid residues from both monomers (Figure S1). It is worth mentioning that SA inhibits FECH activity in vitro. This could be due to induced conformational changes since the gel filtration elution profile was altered in the presence of SA [98].

Interestingly, for NPR1, SA binding was suggested to involve Cys521 and Cys529 based on protein mutation studies and requires the presence of Cu^{2+} ions. However, NPR1 has not yet been crystallized with SA [37].

In conclusion, SA binds to proteins both in plants and animals using similar mechanisms. As a common rule, binding occurs by hydrogen bond formation between the carboxyl group of SA and side chains of various amino acid residues (Arg, Tyr, Ala, Ser, His). Such interaction could be strengthened by nonpolar/π-π contacts made by the aromatic ring of SA. The side hydroxyl group of SA can also be involved. The change in its relative position is enough to alter the binding affinity of SA/4-HBA to proteins [81,89].

Many of the SABPs were identified by SPR analysis, a method where an immobilized 3-animoethyl SA is used as a ligand. This implies that interactions should occur on the surface of the proteins, at least in the SABPs identified with this technique.

5. Discussion and Conclusions

SA interacts with multiple plant proteins. These include the canonical SA receptor, NPR1, and its paralogs, but also many other proteins with diverse roles in cell regulation. The role of these interactions is still not known. NPR1 is, without a doubt, a bottleneck of SA signaling, since in NPR1-deficient plants, SA-induced responses, e.g., the expression of PR genes [99], are drastically altered. The goal of this review was to give a physiological reasoning for SA binding to other plant proteins as a mode of multinodal input (Figure 4). In a canonical situation (panel A), a hormone binds a receptor (one protein or proteins of the same family) and activates a signaling cascade, leading to cell responses. This is true for many plant hormones (e.g., auxins and jasmonates). Based on available data, the SA-signaling pathway appears to be different (panel B). A number of structurally unrelated receptors exist in parallel. Each receptor is a node that will lead either to the same cascade, acting synergistically, or activate separate cascades.

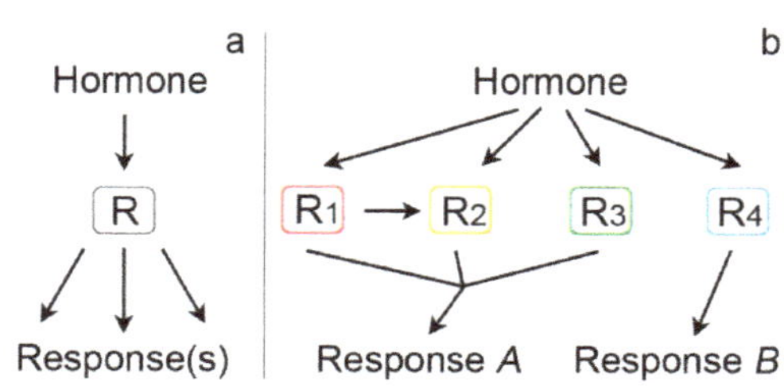

Figure 4. Model of conventional hormonal input via a single receptor (R) leading to downstream responses (**a**); model of SA multinodal input via multiple SABPs acting as independent receptors (**b**). Some receptors in model B may be in a functional connection (e.g., a putative SA-binding TRX acting on NPR1) and act in a single reception pathway.

This model is based on the fact that many SABPs (acting as potential SA receptors) are important enzymes where, at least for some of them, SA binding results in a modification of their activity in vitro (Figure 2). In most, if not all examples available to us where SA inhibits enzymes, SA is likely to obstruct substrate–enzyme interactions or take the place of a necessary cofactor.

SABPs are often connected to plant immunity. For instance, the silencing of αKGDE2 (a SABP) led to an increase in plant-virus resistance [89]. However, since many SABPs are crucial enzymes of basal metabolism, knocking them out could result in unspecific growth aberrations. Instead, point mutations that preserve enzymatic function but disrupt SA interactions should be introduced in planta to test the role of SA binding. Finding SA interaction sites in proteins with no crystallization data available is an intriguing task. This approach would also help to validate SA binding to SABPs in planta.

SABPs could play one of the two probable roles: 1.) Act in support to the NPR1 pathway or 2.) Act in parallel to NPR1 (Figure 5). Regarding the first option, the mechanisms of NPR1 monomerization are indeed not fully understood, although we know that they require upstream redox changes to occur. The proteins involved in this process are still a matter of further investigation. Thioredoxins are perfect candidates for disulphide bond reduction in NPR1 oligomers. TRXm1 has been found to be an SABP, however, this protein is in chloroplasts, while cytosolic TRXh5, implicated in NPR1 monomerization, is not a SABP. AtCAT2 is a homolog of tobacco SABP1—both enzymes could also be a part of cell redox regulation (Figure 5). At some point, NPR1 has to be degraded to attenuate the response, and this involves NPR3 and NPR4 [48], but, in addition, TOP1, TOP2 and TPPII—all proteolytic SABPs—could arguably play a part. Another question is what receptor allows the SA-driven activation of PI4K and PLD, which are both linked to NPR1 [40–42].

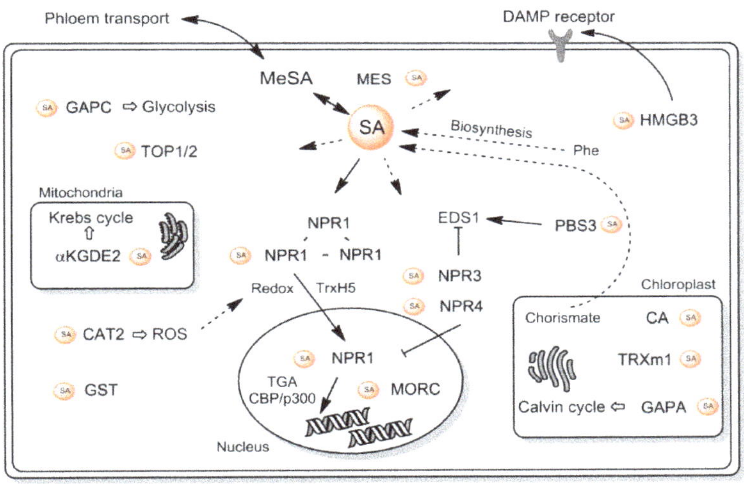

Figure 5. Schematic representation of putative functional roles of SABPs in cell metabolism. Phe, phenylalanine; DAMP, damage-associated molecular pattern; ROS, reactive oxygen species. See the main text for protein abbreviations. Solid arrows represent single enzymatic steps. Dashed arrows represent multiple consecutive enzymatic steps. Hollow arrows represent functional connections to cell activities.

Alternatively, SABPs could act independently of NPR1. Indeed, at least some of the plant reactions to SA occur in an *npr1* genetic background [43–45], but our knowledge of the molecules involved in the NPR1-independent pathway is next to none.

At least two enzymes directly involved in energy metabolism have been found to bind SA—GAPDH (glycolysis) and alpha-ketoglutarate dehydrogenase (Krebs cycle). This could be a way for SA to control the stress-to-growth transition of cell metabolism. Alternatively, since cytosolic GAPDH has been shown to be a dual activity enzyme [87], SA could be a trigger for the transition.

SA has an activity in pollen tubes and in isolated organelles. In these systems, the NPR1 pathway is absent. In Arabidopsis, exogenous SA and MeSA had an opposite effect on pollen tube elongation. The inhibitory effect of SA could be due to changes in clathrin-mediated endocytosis. SA inhibited the internalization of FM4-64 dye while this effect was abolished in *chc2-2* (clathrin heavy chain) mutants [45]. MeSA methylesterase- and SA methyltransferase-GFP constructs were both located in growing pollen tips, suggesting that SA is employed in the control of polarized growth.

SA stimulates the activity of mitochondrial succinate dehydrogenase (SDH, respiratory Complex II) and H_2O_2 accumulation in isolated mitochondria [100]. The authors concluded that SA could act at the ubiquinone binding site of respiratory Complex II. Since SDH-deficient plants had diminished SA responses and SA-induced ROS production—this can be the bona fide mechanism of SA perception. At the moment, no SABPs have been formally identified to mediate the above-mentioned effects of SA.

Affinity to SA can differ by up to 1000-fold in SABPs (e.g., SABP1 K_d = 14 μM; NPR4 K_d = 23 nM). So, at any given moment, based on current cell SA concentration, SA will interact with a limited set of SABPs. In this manner, a regulatory input of SA will differ depending on SA concentration. In Arabidopsis, the basal SA level is around 1 μM, thus some SABPs will bind SA even at basal concentrations, while others will interact only when SA levels rise following stress exposure. The abundance of SABPs (e.g., by transcriptional regulation) is also subjected to regulation (Figure 2).

The uncertainty of a role for SABPs in SA signaling stipulates the need to study SA–protein interaction in planta. In such experiments, the use of isotope-labeled SA is preferred while alternatively photoaffinity labelling [81] could be adopted for protoplast experiments.

Intriguingly, components of the SA-signaling network are still being revealed as we speak. In recent publications, a role for GH3.12/PBS3 was highlighted. PBS3, whose enzymatic activity is regulated by SA [84], was shown to be both an important enzyme of SA biosynthesis [19,20] and signaling (Figure 5), controlling EDS1 [79].

Supplementary Materials: Supplementary materials can be found at http://www.mdpi.com/1422-0067/20/18/4377/s1.

Author Contributions: I.P. wrote the manuscript and generated the figures. V.K. corrected and added insights to the manuscript. E.R. supervised and reviewed the manuscript.

Funding: This research was funded by NAS of Ukraine, grant number 2.1.10.32-15 and PHC DNIPRO program. I.P. benefitted from a Marie Curie/Université Paris-Est Créteil post doc fellowship. We would like to thank Michael Hodges, CNRS UMR9213 Institute of Plant Sciences Paris-Saclay, for proofreading of the manuscript.

Conflicts of Interest: The authors declare no conflict of interest.

References

1. Dong, C.-J.; Li, L.; Shang, Q.-M.; Liu, X.-Y.; Zhang, Z.-G. Endogenous salicylic acid accumulation is required for chilling tolerance in cucumber (*Cucumis sativus* L.) seedlings. *Planta* **2014**, *240*, 687–700. [CrossRef] [PubMed]
2. Shin, H.; Min, K.; Arora, R. Exogenous salicylic acid improves freezing tolerance of spinach (*Spinacia oleracea* L.) leaves. *Cryobiology* **2018**, *81*, 192–200. [CrossRef] [PubMed]
3. Wang, W.; Wang, X.; Huang, M.; Cai, J.; Zhou, Q.; Dai, T.; Cao, W.; Jiang, D. Hydrogen Peroxide and Abscisic Acid Mediate Salicylic Acid-Induced Freezing Tolerance in Wheat. *Front. Plant. Sci.* **2018**, *9*, 1137. [CrossRef] [PubMed]
4. Feng, B.; Zhang, C.; Chen, T.; Zhang, X.; Tao, L.; Fu, G. Salicylic acid reverses pollen abortion of rice caused by heat stress. *BMC Plant Biol.* **2018**, *18*, 245. [CrossRef] [PubMed]
5. Moravcová, Š.; Tůma, J.; Dučaiová, Z.K.; Waligórski, P.; Kula, M.; Saja, D.; Słomka, A.; Bąba, W.; Libik-Konieczny, M. Influence of salicylic acid pretreatment on seeds germination and some defence mechanisms of Zea mays plants under copper stress. *Plant Physiol. Biochem.* **2018**, *122*, 19–30. [CrossRef] [PubMed]
6. Babourina, O.; Rengel, Z.; Jayakannan, M.; Bose, J.; Shabala, S.; Poschenrieder, C.; Massart, A. The NPR1-dependent salicylic acid signalling pathway is pivotal for enhanced salt and oxidative stress tolerance in *Arabidopsis*. *J. Exp. Bot.* **2015**, *66*, 1865–1875. [CrossRef]
7. La, V.H.; Lee, B.-R.; Islam, M.T.; Park, S.-H.; Jung, H.-I.; Bae, D.-W.; Kim, T.-H. Characterization of salicylic acid-mediated modulation of the drought stress responses: Reactive oxygen species, proline, and redox state in *Brassica napus*. *Environ. Exp. Bot.* **2019**, *157*, 1–10. [CrossRef]

8. Yang, L.; Li, B.; Zheng, X.-Y.; Li, J.; Yang, M.; Dong, X.; He, G.; An, C.; Deng, X.W. Salicylic acid biosynthesis is enhanced and contributes to increased biotrophic pathogen resistance in *Arabidopsis* hybrids. *Nat. Commun.* **2015**, *6*, 7309. [CrossRef]
9. Bernsdorff, F.; Döring, A.-C.; Gruner, K.; Schuck, S.; Bräutigam, A.; Zeier, J. Pipecolic Acid Orchestrates Plant Systemic Acquired Resistance and Defense Priming via Salicylic Acid-Dependent and -Independent Pathways. *Plant Cell* **2016**, *28*, 102–129. [CrossRef]
10. Prodhan, M.Y.; Munemasa, S.; Nahar, M.N.-E.-N.; Nakamura, Y.; Murata, Y. Guard Cell Salicylic Acid Signaling Is Integrated into Abscisic Acid Signaling via the Ca^{2+}/CPK-Dependent Pathway. *Plant Physiol.* **2018**, *178*, 441–450. [CrossRef]
11. Lee, S.; Kim, S.-G.; Park, C.-M. Salicylic acid promotes seed germination under high salinity by modulating antioxidant activity in *Arabidopsis*. *New Phytol.* **2010**, *188*, 626–637. [CrossRef] [PubMed]
12. Šašek, V.; Janda, M.; Delage, E.; Puyaubert, J.; Guivarc'h, A.; López Maseda, E.; Dobrev, P.I.; Caius, J.; Bóka, K.; Valentová, O.; et al. Constitutive salicylic acid accumulation in pi4kIIIβ1β2 *Arabidopsis* plants stunts rosette but not root growth. *New Phytol.* **2014**, *203*, 805–816. [CrossRef] [PubMed]
13. Janda, M.; Ruelland, E. Magical mystery tour: Salicylic acid signalling. *Environ. Exp. Bot.* **2015**, *114*, 117–128. [CrossRef]
14. Nawrath, C.; Métraux, J.-P. Salicylic Acid Induction-Deficient Mutants of *Arabidopsis* Express PR-2 and PR-5 and Accumulate High Levels of Camalexin after Pathogen Inoculation. *Plant Cell* **1999**, *11*, 1393–1404. [CrossRef] [PubMed]
15. Wildermuth, M.C.; Dewdney, J.; Wu, G.; Ausubel, F.M. Isochorismate synthase is required to synthesize salicylic acid for plant defence. *Nature* **2001**, *414*, 562. [CrossRef] [PubMed]
16. Garcion, C.; Lohmann, A.; Lamodière, E.; Catinot, J.; Buchala, A.; Doermann, P.; Métraux, J.-P. Characterization and biological function of the ISOCHORISMATE SYNTHASE2 gene of *Arabidopsis*. *Plant Physiol.* **2008**, *147*, 1279–1287. [CrossRef]
17. Hao, Q.; Wang, W.; Han, X.; Wu, J.; Lyu, B.; Chen, F.; Caplan, A.; Li, C.; Wu, J.; Wang, W.; et al. Isochorismate-based salicylic acid biosynthesis confers basal resistance to *Fusarium graminearum* in barley. *Mol. Plant Pathol.* **2018**, *19*, 1995–2010. [CrossRef]
18. Seguel, A.; Jelenska, J.; Herrera-Vásquez, A.; Marr, S.K.; Joyce, M.B.; Gagesch, K.R.; Shakoor, N.; Jiang, S.-C.; Fonseca, A.; Wildermuth, M.C.; et al. PROHIBITIN3 Forms Complexes with ISOCHORISMATE SYNTHASE1 to Regulate Stress-Induced Salicylic Acid Biosynthesis in *Arabidopsis*. *Plant Physiol.* **2018**, *176*, 2515–2531. [CrossRef]
19. Rekhter, D.; Lüdke, D.; Ding, Y.; Feussner, K.; Zienkiewicz, K.; Lipka, V.; Wiermer, M.; Zhang, Y.; Feussner, I. Isochorismate-derived biosynthesis of the plant stress hormone salicylic acid. *Science* **2019**, *365*, 498–502. [CrossRef]
20. Torrens-Spence, M.P.; Bobokalonova, A.; Carballo, V.; Glinkerman, C.M.; Pluskal, T.; Shen, A.; Weng, J.-K. PBS3 and EPS1 complete salicylic acid biosynthesis from isochorismate in *Arabidopsis*. *bioRxiv* **2019**, 601948. [CrossRef]
21. Huang, J.; Gu, M.; Lai, Z.; Fan, B.; Shi, K.; Zhou, Y.-H.; Yu, J.-Q.; Chen, Z. Functional analysis of the *Arabidopsis* PAL gene family in plant growth, development, and response to environmental stress. *Plant Physiol.* **2010**, *153*, 1526–1538. [CrossRef] [PubMed]
22. Li, Y.-X.; Zhang, W.; Dong, H.-X.; Liu, Z.-Y.; Ma, J.; Zhang, X.-Y. Salicylic acid in *Populus tomentosa* is a remote signalling molecule induced by *Botryosphaeria dothidea* infection. *Sci. Rep.* **2018**, *8*, 14059. [CrossRef] [PubMed]
23. Diaz-Vivancos, P.; Bernal-Vicente, A.; Cantabella, D.; Petri, C.; Hernández, J.A. Metabolomics and Biochemical Approaches Link Salicylic Acid Biosynthesis to Cyanogenesis in Peach Plants. *Plant. Cell Physiol.* **2017**, *58*, 2057–2066. [CrossRef] [PubMed]
24. Thompson, A.M.G.; Iancu, C.V.; Neet, K.E.; Dean, J.V.; Choe, J.-Y. Differences in salicylic acid glucose conjugations by UGT74F1 and UGT74F2 from *Arabidopsis thaliana*. *Sci. Rep.* **2017**, *7*, 46629. [CrossRef] [PubMed]
25. Dean, J.V.; Mohammed, L.A.; Fitzpatrick, T. The formation, vacuolar localization, and tonoplast transport of salicylic acid glucose conjugates in tobacco cell suspension cultures. *Planta* **2005**, *221*, 287–296. [CrossRef]

26. Zhang, K.; Halitschke, R.; Yin, C.; Liu, C.-J.; Gan, S.-S. Salicylic acid 3-hydroxylase regulates *Arabidopsis* leaf longevity by mediating salicylic acid catabolism. *Proc. Natl. Acad. Sci. USA* **2013**, *110*, 14807–14812. [CrossRef]
27. Huang, X.-X.; Zhu, G.-Q.; Liu, Q.; Chen, L.; Li, Y.-J.; Hou, B.-K. Modulation of Plant Salicylic Acid-Associated Immune Responses via Glycosylation of Dihydroxybenzoic Acids. *Plant Physiol.* **2018**, *176*, 3103–3119. [CrossRef]
28. Baek, D.; Pathange, P.; Chung, J.-S.; Jiang, J.; Gao, L.; Oikawa, A.; Hirai, M.Y.; Saito, K.; Pare, P.W.; Shi, H. A stress-inducible sulphotransferase sulphonates salicylic acid and confers pathogen resistance in *Arabidopsis*. *Plant Cell Environ.* **2010**, *33*, 1383–1392. [CrossRef]
29. Westfall, C.S.; Sherp, A.M.; Zubieta, C.; Alvarez, S.; Schraft, E.; Marcellin, R.; Ramirez, L.; Jez, J.M. *Arabidopsis thaliana* GH3.5 acyl acid amido synthetase mediates metabolic crosstalk in auxin and salicylic acid homeostasis. *Proc. Natl. Acad. Sci. USA* **2016**, *113*, 13917–13922. [CrossRef]
30. Makandar, R.; Nalam, V.J.; Chowdhury, Z.; Sarowar, S.; Klossner, G.; Lee, H.; Burdan, D.; Trick, H.N.; Gobbato, E.; Parker, J.E.; et al. The Combined Action of ENHANCED DISEASE SUSCEPTIBILITY1, PHYTOALEXIN DEFICIENT4, and SENESCENCE-ASSOCIATED101 Promotes Salicylic Acid-Mediated Defenses to Limit *Fusarium graminearum* Infection in *Arabidopsis thaliana*. *Mol. Plant Microbe Interact.* **2015**, *28*, 943–953. [CrossRef]
31. Bhattacharjee, S.; Halane, M.K.; Kim, S.H.; Gassmann, W. Pathogen Effectors Target *Arabidopsis* EDS1 and Alter Its Interactions with Immune Regulators. *Science* **2011**, *334*, 1405–1408. [CrossRef] [PubMed]
32. Zhu, S.; Jeong, R.-D.; Venugopal, S.C.; Lapchyk, L.; Navarre, D.; Kachroo, A.; Kachroo, P. SAG101 Forms a Ternary Complex with EDS1 and PAD4 and Is Required for Resistance Signaling against Turnip Crinkle Virus. *PLoS Pathog.* **2011**, *7*, e1002318. [CrossRef] [PubMed]
33. Cui, H.; Qiu, J.; Zhou, Y.; Bhandari, D.D.; Zhao, C.; Bautor, J.; Parker, J.E. Antagonism of Transcription Factor MYC2 by EDS1/PAD4 Complexes Bolsters Salicylic Acid Defense in *Arabidopsis* Effector-Triggered Immunity. *Mol. Plant* **2018**, *11*, 1053–1066. [CrossRef] [PubMed]
34. Li, N.; Han, X.; Feng, D.; Yuan, D.; Huang, L.-J. Signaling Crosstalk between Salicylic Acid and Ethylene/Jasmonate in Plant Defense: Do We Understand What They Are Whispering? *Int. J. Mol. Sci.* **2019**, *20*, 671. [CrossRef] [PubMed]
35. Joglekar, S.; Suliman, M.; Bartsch, M.; Halder, V.; Maintz, J.; Bautor, J.; Zeier, J.; Parker, J.E.; Kombrink, E. Chemical Activation of EDS1/PAD4 Signaling Leading to Pathogen Resistance in *Arabidopsis*. *Plant Cell Physiol.* **2018**, *59*, 1592–1607. [CrossRef] [PubMed]
36. Ma, J.; Chen, J.; Wang, M.; Ren, Y.; Wang, S.; Lei, C.; Cheng, Z.; Sodmergen. Disruption of OsSEC3A increases the content of salicylic acid and induces plant defense responses in rice. *J. Exp. Bot.* **2018**, *69*, 1051–1064. [CrossRef]
37. Wu, Y.; Zhang, D.; Chu, J.Y.; Boyle, P.; Wang, Y.; Brindle, I.D.; De Luca, V.; Després, C. The *Arabidopsis* NPR1 Protein Is a Receptor for the Plant Defense Hormone Salicylic Acid. *Cell Rep.* **2012**, *1*, 639–647. [CrossRef] [PubMed]
38. Mou, Z.; Fan, W.; Dong, X. Inducers of Plant Systemic Acquired Resistance Regulate NPR1 Function through Redox Changes. *Cell* **2003**, *113*, 935–944. [CrossRef]
39. Kwon, D.-J.; Jin, H.; Kang, M.-J.; Yun, S.-H.; Choi, S.-M.; Noh, Y.-S.; Noh, B. Salicylic acid-induced transcriptional reprogramming by the HAC–NPR1–TGA histone acetyltransferase complex in *Arabidopsis*. *Nucleic Acids Res.* **2018**, *46*, 11712–11725. [CrossRef]
40. Krinke, O.; Flemr, M.; Vergnolle, C.; Collin, S.; Renou, J.-P.; Taconnat, L.; Yu, A.; Burketová, L.; Valentová, O.; Zachowski, A.; et al. Phospholipase D Activation Is an Early Component of the Salicylic Acid Signaling Pathway in *Arabidopsis* Cell Suspensions. *Plant Physiol.* **2009**, *150*, 424–436. [CrossRef]
41. Krinke, O.; Ruelland, E.; Valentová, O.; Vergnolle, C.; Renou, J.-P.; Taconnat, L.; Flemr, M.; Burketová, L.; Zachowski, A. Phosphatidylinositol 4-Kinase Activation Is an Early Response to Salicylic Acid in *Arabidopsis* Suspension Cells. *Plant Physiol.* **2007**, *144*, 1347–1359. [CrossRef] [PubMed]
42. Janda, M.; Šašek, V.; Chmelařová, H.; Andrejch, J.; Nováková, M.; Hajšlová, J.; Burketová, L.; Valentová, O. Phospholipase D affects translocation of NPR1 to the nucleus in *Arabidopsis thaliana*. *Front. Plant Sci.* **2015**, *6*, 59. [CrossRef] [PubMed]

43. Canet, J.V.; Dobon, A.; Roig, A.; Tornero, P. Structure-function analysis of npr1 alleles in *Arabidopsis* reveals a role for its paralogs in the perception of salicylic acid. *Plant Cell Environ.* **2010**, *33*, 1911–1922. [CrossRef] [PubMed]
44. Blanco, F.; Garretón, V.; Frey, N.; Dominguez, C.; Pérez-Acle, T.; Van der Straeten, D.; Jordana, X.; Holuigue, L. Identification of NPR1-Dependent and Independent Genes Early Induced by Salicylic Acid Treatment in *Arabidopsis*. *Plant Mol. Biol.* **2005**, *59*, 927–944. [CrossRef] [PubMed]
45. Rong, D.; Luo, N.; Mollet, J.C.; Liu, X.; Yang, Z. Salicylic Acid Regulates Pollen Tip Growth through an NPR3/NPR4-Independent Pathway. *Mol. Plant.* **2016**, *9*, 1478–1491. [CrossRef] [PubMed]
46. Castelló, M.J.; Medina-Puche, L.; Lamilla, J.; Tornero, P. NPR1 paralogs of *Arabidopsis* and their role in salicylic acid perception. *PLoS ONE* **2018**, *13*, e0209835. [CrossRef] [PubMed]
47. Moreau, M.; Tian, M.; Klessig, D.F. Salicylic acid binds NPR3 and NPR4 to regulate NPR1-dependent defense responses. *Cell Res.* **2012**, *22*, 1631. [CrossRef] [PubMed]
48. Fu, Z.Q.; Yan, S.; Saleh, A.; Wang, W.; Ruble, J.; Oka, N.; Mohan, R.; Spoel, S.H.; Tada, Y.; Zheng, N.; et al. NPR3 and NPR4 are receptors for the immune signal salicylic acid in plants. *Nature* **2012**, *486*, 228–232. [CrossRef] [PubMed]
49. Moreau, M.; Westlake, T.; Zampogna, G.; Popescu, G.; Tian, M.; Noutsos, C.; Popescu, S. The *Arabidopsis* oligopeptidases TOP1 and TOP2 are salicylic acid targets that modulate SA-mediated signaling and the immune response. *Plant J.* **2013**, *76*, 603–614. [CrossRef] [PubMed]
50. Manohar, M.; Tian, M.; Moreau, M.; Park, S.-W.; Choi, H.W.; Fei, Z.; Friso, G.; Asif, M.; Manosalva, P.; von Dahl, C.C.; et al. Identification of multiple salicylic acid-binding proteins using two high throughput screens. *Front. Plant Sci.* **2015**, *5*. [CrossRef] [PubMed]
51. Klessig, D.F.; Tian, M.; Choi, H.W. Multiple Targets of Salicylic Acid and Its Derivatives in Plants and Animals. *Front. Immunol.* **2016**, *7*. [CrossRef] [PubMed]
52. Chen, Z.; Klessig, D.F. Identification of a soluble salicylic acid-binding protein that may function in signal transduction in the plant disease-resistance response. *Proc. Natl. Acad. Sci. USA* **1991**, *88*, 8179–8183. [CrossRef] [PubMed]
53. Chen, Z.; Ricigliano, J.W.; Klessig, D.F. Purification and characterization of a soluble salicylic acid-binding protein from tobacco. *Proc. Natl. Acad. Sci. USA* **1993**, *90*, 9533–9537. [CrossRef] [PubMed]
54. Chen, Z.; Silva, H.; Klessig, D. Active oxygen species in the induction of plant systemic acquired resistance by salicylic acid. *Science* **1993**, *262*, 1883–1886. [CrossRef] [PubMed]
55. Conrath, U.; Chen, Z.; Ricigliano, J.R.; Klessig, D.F. Two inducers of plant defense responses, 2,6-dichloroisonicotinec acid and salicylic acid, inhibit catalase activity in tobacco. *Proc. Natl. Acad. Sci. USA* **1995**, *92*, 7143–7147. [CrossRef] [PubMed]
56. Yuan, H.-M.; Liu, W.-C.; Lu, Y.-T. CATALASE2 Coordinates SA-Mediated Repression of Both Auxin Accumulation and JA Biosynthesis in Plant Defenses. *Cell Host Microbe* **2017**, *21*, 143–155. [CrossRef] [PubMed]
57. Du, H.; Klessig, D.F. Identification of a Soluble, High-Affinity Salicylic Acid-Binding Protein in Tobacco. *Plant Physiol.* **1997**, *113*, 1319–1327. [CrossRef] [PubMed]
58. Kumar, D.; Klessig, D.F. High-affinity salicylic acid-binding protein 2 is required for plant innate immunity and has salicylic acid-stimulated lipase activity. *Proc. Natl. Acad. Sci. USA* **2003**, *100*, 16101–16106. [CrossRef]
59. Forouhar, F.; Yang, Y.; Kumar, D.; Chen, Y.; Fridman, E.; Park, S.W.; Chiang, Y.; Acton, T.B.; Montelione, G.T.; Pichersky, E.; et al. Structural and biochemical studies identify tobacco SABP2 as a methyl salicylate esterase and implicate it in plant innate immunity. *Proc. Natl. Acad. Sci. USA* **2005**, *102*, 1773–1778. [CrossRef]
60. Tripathi, D.; Jiang, Y.-L.; Kumar, D. SABP2, a methyl salicylate esterase is required for the systemic acquired resistance induced by acibenzolar-S-methyl in plants. *FEBS Lett.* **2010**, *584*, 3458–3463. [CrossRef]
61. Hartmann, M.; Zeier, T.; Bernsdorff, F.; Reichel-Deland, V.; Kim, D.; Hohmann, M.; Scholten, N.; Schuck, S.; Bräutigam, A.; Hölzel, T.; et al. Flavin Monooxygenase-Generated N-Hydroxypipecolic Acid Is a Critical Element of Plant Systemic Immunity. *Cell* **2018**, *173*, 456–469.e416. [CrossRef] [PubMed]
62. Kumar, D.; Gustafsson, C.; Klessig, D.F. Validation of RNAi silencing specificity using synthetic genes: salicylic acid-binding protein 2 is required for innate immunity in plants. *Plant J.* **2006**, *45*, 863–868. [CrossRef] [PubMed]

63. Park, S.-W.; Liu, P.-P.; Forouhar, F.; Vlot, A.C.; Tong, L.; Tietjen, K.; Klessig, D.F. Use of a synthetic salicylic acid analog to investigate the roles of methyl salicylate and its esterases in plant disease resistance. *J. Biol. Chem.* **2009**, *284*, 7307–7317. [CrossRef] [PubMed]
64. Manosalva, P.M.; Park, S.-W.; Forouhar, F.; Tong, L.; Fry, W.E.; Klessig, D.F. Methyl Esterase 1 (StMES1) Is Required for Systemic Acquired Resistance in Potato. *Mol. Plant Microbe Interact.* **2010**, *23*, 1151–1163. [CrossRef] [PubMed]
65. Zhao, N.; Guan, J.; Forouhar, F.; Tschaplinski, T.J.; Cheng, Z.-M.; Tong, L.; Chen, F. Two poplar methyl salicylate esterases display comparable biochemical properties but divergent expression patterns. *Phytochemistry* **2009**, *70*, 32–39. [CrossRef] [PubMed]
66. Vlot, A.C.; Liu, P.-P.; Cameron, R.K.; Park, S.-W.; Yang, Y.; Kumar, D.; Zhou, F.; Padukkavidana, T.; Gustafsson, C.; Pichersky, E.; et al. Identification of likely orthologs of tobacco salicylic acid-binding protein 2 and their role in systemic acquired resistance in *Arabidopsis thaliana*. *Plant J.* **2008**, *56*, 445–456. [CrossRef] [PubMed]
67. Slaymaker, D.H.; Navarre, D.A.; Clark, D.; del Pozo, O.; Martin, G.B.; Klessig, D.F. The tobacco salicylic acid-binding protein 3 (SABP3) is the chloroplast carbonic anhydrase, which exhibits antioxidant activity and plays a role in the hypersensitive defense response. *Proc. Natl. Acad. Sci. USA* **2002**, *99*, 11640–11645. [CrossRef]
68. DiMario, R.J.; Clayton, H.; Mukherjee, A.; Ludwig, M.; Moroney, J.V. Plant Carbonic Anhydrases: Structures, Locations, Evolution, and Physiological Roles. *Mol. Plant* **2017**, *10*, 30–46. [CrossRef]
69. Wang, Y.-Q.; Feechan, A.; Yun, B.-W.; Shafiei, R.; Hofmann, A.; Taylor, P.; Xue, P.; Yang, F.-Q.; Xie, Z.-S.; Pallas, J.A.; et al. S-Nitrosylation of AtSABP3 Antagonizes the Expression of Plant Immunity. *J. Biol. Chem.* **2009**, *284*, 2131–2137. [CrossRef]
70. Ahmad, B.; Jaleel, H.; Sadiq, Y.; Khan, M.M.A.; Shabbir, A. Response of exogenous salicylic acid on cadmium induced photosynthetic damage, antioxidant metabolism and essential oil production in peppermint. *Plant Growth Regul.* **2018**, *86*, 273–286. [CrossRef]
71. Medina-Puche, L.; Castelló, M.J.; Canet, J.V.; Lamilla, J.; Colombo, M.L.; Tornero, P. β-carbonic anhydrases play a role in salicylic acid perception in *Arabidopsis*. *PLoS ONE* **2017**, *12*, e0181820. [CrossRef] [PubMed]
72. Poque, S.; Wu, H.-W.; Huang, C.-H.; Cheng, H.-W.; Hu, W.-C.; Yang, J.-Y.; Wang, D.; Yeh, S.-D. Potyviral gene-silencing suppressor HCPro interacts with salicylic acid (SA)-binding protein 3 to weaken SA-mediated defense responses. *Mol. Plant Microbe Interact.* **2017**, *31*, 86–100. [CrossRef] [PubMed]
73. Tada, Y.; Spoel, S.H.; Pajerowska-Mukhtar, K.; Mou, Z.; Song, J.; Wang, C.; Zuo, J.; Dong, X. Plant Immunity Requires Conformational Charges of NPR1 via S-Nitrosylation and Thioredoxins. *Science* **2008**, *321*, 952–956. [CrossRef] [PubMed]
74. Geigenberger, P.; Thormählen, I.; Daloso, D.M.; Fernie, A.R. The Unprecedented Versatility of the Plant Thioredoxin System. *Trends Plant Sci.* **2017**, *22*, 249–262. [CrossRef]
75. Ruelland, E.; Miginiac-Maslow, M. Regulation of chloroplast enzyme activities by thioredoxins: Activation or relief from inhibition? *Trends Plant Sci.* **1999**, *4*, 136–141. [CrossRef]
76. Ding, Y.; Sun, T.; Ao, K.; Peng, Y.; Zhang, Y.; Li, X.; Zhang, Y. Opposite Roles of Salicylic Acid Receptors NPR1 and NPR3/NPR4 in Transcriptional Regulation of Plant Immunity. *Cell* **2018**, *173*, 1454.e1415–1467.e1415. [CrossRef] [PubMed]
77. Zhang, Y.; Cheng, Y.T.; Qu, N.; Zhao, Q.; Bi, D.; Li, X. Negative regulation of defense responses in *Arabidopsis* by two NPR1 paralogs. *Plant. J.* **2006**, *48*, 647–656. [CrossRef]
78. Liu, G.; Holub, E.B.; Alonso, J.M.; Ecker, J.R.; Fobert, P.R. An *Arabidopsis* NPR1-like gene, NPR4, is required for disease resistance. *Plant. J.* **2005**, *41*, 304–318. [CrossRef]
79. Chang, M.; Zhao, J.; Chen, H.; Li, G.; Chen, J.; Li, M.; Palmer, I.A.; Song, J.; Alfano, J.R.; Liu, F.; et al. PBS3 Protects EDS1 from Proteasome-Mediated Degradation in Plant Immunity. *Mol. Plant* **2019**, *12*, 678–688. [CrossRef]
80. Zhou, X.-T.; Jia, L.-J.; Wang, H.-Y.; Zhao, P.; Wang, W.-Y.; Liu, N.; Song, S.-W.; Wu, Y.; Su, L.; Zhang, J.; et al. The potato transcription factor StbZIP61 regulates dynamic biosynthesis of salicylic acid in defense against *Phytophthora infestans* infection. *Plant. J.* **2018**, *95*, 1055–1068. [CrossRef]
81. Tian, M.; von Dahl, C.C.; Liu, P.-P.; Friso, G.; van Wijk, K.J.; Klessig, D.F. The combined use of photoaffinity labeling and surface plasmon resonance-based technology identifies multiple salicylic acid-binding proteins. *Plant J.* **2012**, *72*, 1027–1038. [CrossRef] [PubMed]

82. Gullner, G.; Komives, T.; Király, L.; Schröder, P. Glutathione S-Transferase Enzymes in Plant-Pathogen Interactions. *Front. Plant Sci.* **2018**, *9*. [CrossRef] [PubMed]
83. Westfall, C.S.; Zubieta, C.; Herrmann, J.; Kapp, U.; Nanao, M.H.; Jez, J.M. Structural Basis for Prereceptor Modulation of Plant Hormones by GH3 Proteins. *Science* **2012**, *336*, 1708–1711. [CrossRef]
84. Okrent, R.A.; Brooks, M.D.; Wildermuth, M.C. *Arabidopsis* GH3.12 (PBS3) conjugates amino acids to 4-substituted benzoates and is inhibited by salicylate. *J. Biol. Chem.* **2009**, *284*, 9742–9754. [CrossRef] [PubMed]
85. Schuman, M.C.; Meldau, S.; Gaquerel, E.; Diezel, C.; McGale, E.; Greenfield, S.; Baldwin, I.T. The Active Jasmonate JA-Ile Regulates a Specific Subset of Plant Jasmonate-Mediated Resistance to Herbivores in Nature. *Front. Plant Sci.* **2018**, *9*, 787. [CrossRef] [PubMed]
86. Nobuta, K.; Okrent, R.A.; Stoutemyer, M.; Rodibaugh, N.; Kempema, L.; Wildermuth, M.C.; Innes, R.W. The GH3 acyl adenylase family member PBS3 regulates salicylic acid-dependent defense responses in Arabidopsis. *Plant Physiol.* **2007**, *144*, 1144–1156. [CrossRef] [PubMed]
87. Tian, M.; Sasvari, Z.; Gonzalez, P.A.; Friso, G.; Rowland, E.; Liu, X.-M.; van Wijk, K.J.; Nagy, P.D.; Klessig, D.F. Salicylic Acid Inhibits the Replication of Tomato bushy stunt virus by Directly Targeting a Host Component in the Replication Complex. *Mol. Plant Microbe Interact.* **2015**, *28*, 379–386. [CrossRef]
88. Choi, H.W.; Tian, M.; Manohar, M.; Harraz, M.M.; Park, S.-W.; Schroeder, F.C.; Snyder, S.H.; Klessig, D.F. Human GAPDH Is a Target of Aspirin's Primary Metabolite Salicylic Acid and Its Derivatives. *PLoS ONE* **2015**, *10*, e0143447. [CrossRef]
89. Liao, Y.; Tian, M.; Zhang, H.; Li, X.; Wang, Y.; Xia, X.; Zhou, J.; Zhou, Y.; Yu, J.; Shi, K.; et al. Salicylic acid binding of mitochondrial alpha-ketoglutarate dehydrogenase E2 affects mitochondrial oxidative phosphorylation and electron transport chain components and plays a role in basal defense against tobacco mosaic virus in tomato. *New Phytol.* **2015**, *205*, 1296–1307. [CrossRef]
90. Wang, R.; Rajagopalan, K.; Sadre-Bazzaz, K.; Moreau, M.; Klessig, D.F.; Tong, L. Structure of the *Arabidopsis thaliana* TOP2 oligopeptidase. *Acta Crystallogr. Sect. F Struct. Biol. Commun.* **2014**, *70*, 555–559. [CrossRef]
91. Westlake, T.J.; Ricci, W.A.; Popescu, G.V.; Popescu, S.C. Dimerization and thiol sensitivity of the salicylic acid binding thimet oligopeptidases TOP1 and TOP2 define their functions in redox-sensitive cellular pathways. *Front. Plant Sci.* **2015**, *6*. [CrossRef] [PubMed]
92. Moissiard, G.; Bischof, S.; Husmann, D.; Pastor, W.A.; Hale, C.J.; Yen, L.; Stroud, H.; Papikian, A.; Vashisht, A.A.; Wohlschlegel, J.A.; et al. Transcriptional gene silencing by *Arabidopsis* microrchidia homologues involves the formation of heteromers. *Proc. Natl. Acad. Sci. USA* **2014**, *111*, 7474–7479. [CrossRef] [PubMed]
93. Manohar, M.; Choi, H.W.; Manosalva, P.; Austin, C.A.; Peters, J.E.; Klessig, D.F. Plant and Human MORC Proteins Have DNA-Modifying Activities Similar to Type II Topoisomerases, but Require One or More Additional Factors for Full Activity. *Mol. Plant Microbe Interact.* **2017**, *30*, 87–100. [CrossRef] [PubMed]
94. Choi, H.W.; Manohar, M.; Manosalva, P.; Tian, M.; Moreau, M.; Klessig, D.F. Activation of Plant Innate Immunity by Extracellular High Mobility Group Box 3 and Its Inhibition by Salicylic Acid. *PLoS Pathog.* **2016**, *12*, e1005518. [CrossRef] [PubMed]
95. Hruz, T.; Laule, O.; Szabo, G.; Wessendorp, F.; Bleuler, S.; Oertle, L.; Widmayer, P.; Gruissem, W.; Zimmermann, P. Genevestigator V3: A Reference Expression Database for the Meta-Analysis of Transcriptomes. *Adv. Bioinform.* **2008**, *2008*, 5. [CrossRef] [PubMed]
96. Barber, R.D.; Harmer, D.W.; Coleman, R.A.; Clark, B.J. GAPDH as a housekeeping gene: analysis of GAPDH mRNA expression in a panel of 72 human tissues. *Physiol. Genom.* **2005**, *21*, 389–395. [CrossRef] [PubMed]
97. Pettersen, E.F.; Goddard, T.D.; Huang, C.C.; Couch, G.S.; Greenblatt, D.M.; Meng, E.C.; Ferrin, T.E. UCSF Chimera—A visualization system for exploratory research and analysis. *J. Comput. Chem.* **2004**, *25*, 1605–1612. [CrossRef]
98. Gupta, V.; Liu, S.; Ando, H.; Ishii, R.; Tateno, S.; Kaneko, Y.; Yugami, M.; Sakamoto, S.; Yamaguchi, Y.; Nureki, O.; et al. Salicylic Acid Induces Mitochondrial Injury by Inhibiting Ferrochelatase Heme Biosynthesis Activity. *Mol. Pharmacol.* **2013**, *84*, 824–833. [CrossRef]

99. Delaney, T.P.; Friedrich, L.; Ryals, J.A. *Arabidopsis* signal transduction mutant defective in chemically and biologically induced disease resistance. *Proc. Natl. Acad. Sci. USA* **1995**, *92*, 6602–6606. [CrossRef]
100. Belt, K.; Huang, S.; Thatcher, L.F.; Casarotto, H.; Singh, K.B.; Van Aken, O.; Millar, A.H. Salicylic Acid-Dependent Plant Stress Signaling via Mitochondrial Succinate Dehydrogenase. *Plant Physiol.* **2017**, *173*, 2029–2040. [CrossRef]

International Journal of
Molecular Sciences

MDPI

Review

The Multifaceted Roles of Plant Hormone Salicylic Acid in Endoplasmic Reticulum Stress and Unfolded Protein Response

Péter Poór *, Zalán Czékus, Irma Tari and Attila Ördög

Department of Plant Biology, University of Szeged, Közép fasor 52, H-6726 Szeged, Hungary
* Correspondence: poorpeti@bio.u-szeged.hu; Tel.: +36-62-544-307

Received: 18 October 2019; Accepted: 19 November 2019; Published: 21 November 2019

Abstract: Different abiotic and biotic stresses lead to the accumulation of unfolded and misfolded proteins in the endoplasmic reticulum (ER), resulting in ER stress. In response to ER stress, cells activate various cytoprotective responses, enhancing chaperon synthesis, protein folding capacity, and degradation of misfolded proteins. These responses of plants are called the unfolded protein response (UPR). ER stress signaling and UPR can be regulated by salicylic acid (SA), but the mode of its action is not known in full detail. In this review, the current knowledge on the multifaceted role of SA in ER stress and UPR is summarized in model plants and crops to gain a better understanding of SA-regulated processes at the physiological, biochemical, and molecular levels.

Keywords: binding protein; cell death; pathogenesis-related genes; systemic acquired resistance; tunicamycin

1. Introduction

1.1. ER Stress

The eukaryotic endoplasmic reticulum (ER) has multiple cellular functions, such as protein synthesis, assembly, folding, and export. The lumen of the ER is a specific environment, which contains a high concentration of Ca^{2+}, playing a role in various cell signaling events. In addition, the lumen of the ER is also an oxidative environment, which regulates the formation of disulphide bonds and proper folding of proteins. Finally, the newly synthesized and correctly folded proteins are loaded for transfer from the ER into the cytosol [1,2]. To maintain the balance between protein folding and transport and the capacity of ER, many Ca^{2+}-dependent molecular chaperones cooperate in the ER, such as calreticulin (CRT) and calnexin (CNX). The binding protein (BiP; glucose-regulated protein 78, Grp78), Grp94, protein disulfide isomerase (PDI), and peptidyl propyl isomerase (PPI) are also central players in protein folding quality control [2–6]. Several abiotic (e.g., high light, high temperature, drought, salt, osmotic and heavy metal stress) and biotic stresses (e.g., bacterial and fungal pathogens, viruses) can induce ER stress in plants [7–10]. Namely, disturbances in ER homeostasis under stress conditions, including those of cellular redox regulation, cause ER stress by the accumulation of unfolded and misfolded proteins that triggers an evolutionarily conserved response, termed the unfolded protein response (UPR). UPR is a protective response to maintain the cellular homeostasis by regulating the expression of a variety of genes (e.g., chaperones) and by reducing protein loading to the ER and enhancing ER-associated protein degradation (ERAD). These processes improve the protein folding capacity and remove the unfolded or misfolded proteins from the ER [10–13]. Programmed cell death (PCD) and autophagy are also associated with ERAD response under prolonged and chronic stress effects [14–17]. Recently, it has been suggested that the plant hormone salicylic acid (SA) induces UPR in plants, but the underlying mechanisms are not completely known yet [10]. To test the potential role of SA in UPR, exogenous application of several chemicals like tunicamycin (Tm, the inhibitor of N-glycosylation of secreted glycoproteins), dithiothreitol (DTT, the inhibitor of the formation of disulphide bonds during protein folding), and azetidine-2-carboxylic acid (AZC, a proline analogue that can interfere with the formation of native protein structure) have been used under laboratory conditions [18]. At the same time, it has been found that after the treatment with Tm caused a four-fold increase in the SA content of *Arabidopsis* [19].

In this review, the current knowledge on the multifaceted role of SA in ER stress and unfolded protein responses will be summarized in model plants and crops to gain a better understanding of SA-regulated processes at the physiological, biochemical, and molecular levels. This knowledge can add a new aspect to the understanding of plant ER stress and UPR signaling and its crosstalk with plant immune responses.

1.2. Basic Properties of UPR

The cytoprotective UPR is initiated by ER-resident stress sensors located in the ER membrane (Figure 1). One of them, the inositol-requiring enzyme 1 (IRE1)-mediated unconventional splicing of basic leucine zipper (bZIP) 60, is the most conserved in eukaryotes [18,20,21]. However, the activation mechanism of plant IRE1 has not been shown in full detail. It was well demonstrated in yeast and animals that the sensor domain of IRE1 binds to the ER-luminal BiP while the full-length bZIP60 is anchored in the ER membrane under normal conditions. The accumulation of unfolded proteins leads to BiP dissociation from IRE1. The released IRE1 is firstly dimerized and then oligomerized after the binding by its luminal domain to the hydrophobic domain of the unfolded proteins. In *Arabidopsis*, two isoforms of IRE1, *IRE1a* and *IRE1b*, are found. It was observed that IRE1b but not IRE1a expressed heterologously in yeast cells showed the oligomerization structure and clustering, indicating the possible conserved step of IRE1 activation in plants, respectively [22–25]. The activated RNAse function of IRE1 results in the splicing of bZIP60 mRNA and bulk degradation of selected mRNAs through regulated IRE1-dependent decay (RIDD) in animals, yeast, and plants. Spliced bZIP60 mRNA is translated to an active transcription factor (TF) and the active bZIP60 protein is translocated to the nucleus and upregulates *UPR* genes containing unfolded protein response element (UPRE) and ER stress element (ERSE) in their promoters [10,20–25].

The activation of ER membrane-anchored TF bZIP28 and the plant B-cell lymphoma2 (Bcl-2)-associated athanogene 7 (BAG7) protein is another mode to control ER stress in plants. Both proteins are anchored to the ER membrane by interactions with BiP under unstressed conditions. Like IRE1, bZIP28 is also activated through the stress-induced accumulation of unfolded proteins in the ER lumen. In response to ER stress, BiP dissociates from bZIP28 and the released bZIP28 translocates from ER to the Golgi through the coat protein II (COPII) vesicles, where it is proteolytically cleaved by site-2TF protease (S2P) but not by site-1 protease (S1P). The cleaved form of bZIP28 translocates into the nucleus and binds to ERSE to activate the *UPR* gene expression [26–30]. BAG7 is also released from the ER membrane by an unknown protease, then it is sumoylated and enters the nucleus, where it interacts with WRKY29 transcription factor and regulates the expression of various chaperone proteins to mitigate ER stress [31]. Another ER membrane-associated transcription factor is *bZIP17*, which is closely related to *bZIP28*. It was found that *bZIP17* could be activated by salt stress in *Arabidopsis* in a manner similar to *bZIP28*. Basically, AtbZIP17 is inserted into the ER lumen. Under stress condition, it is transported firstly to the Golgi apparatus where it is cleaved by the Golgi-localized AtS1P protease, and the N-terminus of AtbZIP17 enters the nucleus to activate stress-responsive genes [12,21].

Another TF, the plant-specific NACs (no apical meristem (NAM), *Arabidopsis* transcription activation factor (ATAF), cup-shaped cotyledon (CUC)) have recently been identified as an important regulators of ER stress responses [8,32,33]. In total, 117 *NAC* genes have been found in the *Arabidopsis* genome, which participate in several developmental and stress-induced processes [10,34]. NAC062 (localized to the plasma membrane) and NAC089 (localized to the ER membrane) undergo proteolytic cleavage under ER stress and translocate to the nucleus to promote the transcription of *UPR* or

PCD genes. NAC089 is dependent on both IRE1/bZIP60 and bZIP28 pathways and plays a role in PCD [32,33]. NAC062 and NAC103 are also controlled by IRE1/bZIP60, inducing the expression of defense genes under stress conditions [32,35].

Finally, protein kinase RNA-like ER kinase (PERK)-mediated translational inhibition was well characterized in mammals, but no PERK homologues have been identified in plant genomes until now [8,10].

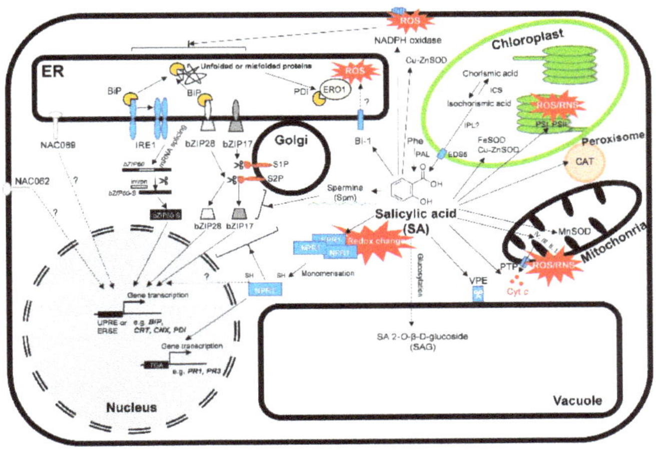

Figure 1. A schematic illustration of unfolded protein response (UPR) and the effects of salicylic acid (SA) under stress condition in plants. The accumulation of unfolded proteins in the ER leads to the conformational changes and activated RNAse function of IRE1 (inositol-requiring enzyme 1), which mediates an unconventional splicing bZIP60 transcription factor mRNA. Spliced bZIP60 mRNA is translated to an active transcription factor and translocated to the nucleus and upregulates *UPR* genes containing unfolded protein response element (UPRE) and ER stress element (ERSE) in their promoters. bZIP28 and bZIP17 are activated by Golgi proteases (S2P and S1P). Then, the cleaved forms of bZIPs translocate into the nucleus and binds to ERSE to activate the *UPR* gene expression. NAC062 and NAC089 also undergo proteolytic cleavage and translocate to the nucleus to promote the transcription of *UPR* or cell death genes. SA has multifaceted roles in the regulation of defense or cell death processes in plants. SA is synthesized by phenylalanine ammonia-lyase (PAL) from L-phenylalanine (Phe) or in the isochorismate (IC) pathway by isochorismate synthase (ICS). Then through the activity of isochorismate pyruvate lyase (IPL) in the chloroplast, it is translocated to the cytosol by EDS5. SA induces high production of reactive oxygen (ROS) and nitrogen species (RNS) in chloroplast and mitochondria and activates NADPH oxidase, respectively. Simultaneously, SA activates various antioxidant enzymes, such as superoxide dismutases (SOD). SA induces cytochrome *c* (Cyt *c*) release from the mitochondrial inner membrane by the permeability transition pore (PTP), decreases the transcript levels of Bax inhibitor-1 (BI-1), and increases the expression of vacuolar processing enzymes (VPEs) inducing cell death. Moreover, SA has a significant effect on polyamine levels (e.g. that of spermine), influencing ER stress in plants. SA changes the redox homeostasis and induces the reduction and monomerization of NPR1, which is translocated to the nucleus where it binds to specific TGA transcriptions factors, inducing the expression of SA-induced defensive response genes (PRs). SA can be inactivated and stored as SA O-β-glucoside (SAG) in the vacuole. *Detailed description and references are in the text.*

2. SA as an ER Stress Signaling Regulator in Plants

The phenolic compound salicylic acid (SA) plays a crucial role in plant defense signaling upon various abiotic and biotic stressors [36,37]. It is required for the establishment of both local and systemic acquired resistance (SAR) after pathogen attack. The elevated concentration of SA under stress conditions induces the accumulation of reactive oxygen species (ROS), leading to oxidized proteins and cell death in the infected tissues [38]. Besides, SA induces expression and accumulation of pathogenesis-related (PR) proteins, which requires optimal coordination and regulation of protein secretory machinery to ensure folding, modification, and transport of PR proteins [39,40]. Thus, SA plays a dominant role in ER stress signaling and regulating UPR under stress conditions [41], but the mode of its action is not known in full detail. In addition, there are contrasting findings from different experiments in the case of SA-mediated UPR. It has to be mentioned, however, that the experimental setups cannot be excluded because it is well known that the action of SA is highly dependent, e.g., on its applied or internal concentrations, on the duration and the mode of the application, on the investigated plant species and organs as well as on the light intensity and daytime of SA treatment [42]. Furthermore, the crosstalk between SA and other plant hormones (e.g., ethylene and jasmonic acid) can overwrite the outcome of defense signals and plays a role in the regulation of UPR [43,44]. From this aspect, a physiological approach is also necessary to draw a more complex picture of the role of SA in ER stress and UPR. In this section, the SA-mediated ER stress signaling is summarized to understand the multifaceted role of SA in this process.

Jelitto-Van Dooren et al. [39] postulated firstly the relationship between ER stress and SA-mediated defense responses and described a spatiotemporal change. They observed that plant cell wall-degrading enzymes (CDEs) secreted by the bacterial pathogen *Erwinia carotovora* induced the expression of the β-1,3-glucanase (*PR3*) gene 4 h after incubation and reached a maximum after 8 h. Nevertheless, *BiP*, *PDI*, and *CRT* transcripts accumulated more rapidly, reaching a maximum after 2 h of CDE incubation both in locally treated tobacco leaves as well as in untreated/systemic distal leaves with the same timing and intensity. However, this CDE-induced *BiP* expression was not dependent on SA based on the use of an SA-insensitive mutant of *Arabidopsis* (*sai1*) and could be regulated by other phytohormones, such as ethylene or jasmonic acid. The authors concluded that *BiP* gene expression during plant–pathogen interactions is required to allow efficient PR protein synthesis because more ER chaperones are required for the synthesis, folding, and transport of defense-related proteins [39]. Later, Wang et al. [40] found that the SA-induced various components of ER stress and UPR during the development of SAR are regulated by the SA-induced master regulator protein NPR1 (nonexpressor of pathogenesis-related (*PR*) genes 1) in *Arabidopsis*. Based on microarray experiments, genes of the *Sec61* translocon complex, which provides a channel for proteins to cross the ER membrane, and a signal recognition particle (SRP) receptor were upregulated. In addition, chaperones, such as *BiP2*, *GRP94*, as well as co-chaperones, including defender against apoptotic death 1 (*DAD1*), *CNXs*, *CRTs*, and *PDIs*, were upregulated in an NPR1-dependent manner. The authors suggested that SA primes the ER capacity to assist in the production, folding, and transport of defense proteins, such as PR1. Consistent with this hypothesis, the expression of *BiP2* was induced before the accumulation of *PR1* [40]. It is well known that SA induces the reduction and monomerization of NPR1, which is translocated to the nucleus and induces the expression of *PRs* through interaction with the TGA TFs at the promoters of *PR* genes [43,45]. However, in other experiments, Wang et al. [40] observed that genes encoding ER-resident proteins are not upregulated by TGA TFs but by TL1-binding transcription factor 1 (TBF1). TBF1, in response to infections, plays a role in the rapid reprogramming of transcription from growth to defense responses [46]. The TGA family of bZIP TFs takes part in the regulation of these defense responses of plants. Unfortunately, only the function of clade I TGA factors, which are independent of NPR1 [47], were investigated upon ER stress [48]. The potential role of TGA clade II and III will be elucidated in the future. At the same time, *tga1-1 tga4-1* mutant seedlings showed increased sensitivity to Tm, which was associated with the upregulation of ER-resident genes encoding *BiP1/2* and *BiP3* chaperones, suggesting that the loss of clade I TGA factors does not impair the IRE1/bZIP60 branch of UPR signaling but impairs ER-based protein folding and/or secretion in an NPR1-independent manner [48].

It has also been found that SA (0.5 mM) and Tm (5 mg mL^{-1}) induced not only *BiP2* but also *BiP3* transcript levels, but Tm induced the expression of both selected chaperon coding sequences more significantly compared to SA in *Arabidopsis* [49]. Interestingly, SA did not induce *BiP3* expression in a *bZIP60* knockout mutant or in an *ire1a ire1b* double mutant, and the transcript levels of *PR1* and *BiP2* also did not change after SA treatment. In addition, *bZIP60s* and *BiP3* were not induced in *NahG*, an SA-deficient transgenic plants. These observations confirmed that SA induced the activation of the IRE1–bZIP60 pathway and thus *BiP3* expression [49]. Surprisingly, it has also been demonstrated that bZIP60-dependent induction of *UPR* genes (*BiP2* and *BiP3*) by SA is independent of NPR1 by the use of *npr1-1* mutants, where levels of *PR1* transcripts did not increase after 5- or 10-h-long SA treatments [49]. Furthermore, it has also been revealed that SA activates not only *bZIP60* but also *bZIP28* independently of NPR1 after 2 h, but bZIP28 levels decreased after 10 h, suggesting that bZIP28 is activated earlier than bZIP60 under these experimental conditions [49]. There were no differences in the induction profiles of *BiP2*, *BiP3*, and *CNX1* in a T-DNA insertion mutant, *hsfb1-1*, suggesting HsfB1-independent (the major molecular switch for the plant growth-to-defense transition) regulation of UPR by SA under these experimental conditions [49]. It has to be noted that the daytime of SA application and light intensity is not known in these works, but it is well known that the effect of SA depends on these external and internal factors [50,51]. Despite this finding, the spliced form of bZIP60 has been observed at 30 min and the maximum after 2 h upon 0.5 mM SA treatment, but it decreased after 5 h of SA application in 7-day-old *Arabidopsis* seedlings [52]. This change in bZIP60 activation suggests that it is a dynamic process because it has been found that the wash-out of SA led to a complete loss of the spliced form of bZIP60 and the re-addition of SA led to an increase of the spliced bZIP60 form again [52]. Surprisingly, the result of Parra-Rojas et al. [52] suggests that the effects of SA on the splicing of bZIP60 is somehow linked to the function of bZIP17 because the level of spliced bZIP60 was higher in *bZIP17* mutants.

Recently, it has been confirmed that Tm-induced ER stress is regulated by NPR1 because the transcriptional role of bZIP28 and bZIP60 in ER stress responses is antagonized by NPR1 [53]. Moreover, the authors suggested that this action could be independent of the role of NPR1 in SA-mediated defense, because *npr1* mutants displayed enhanced resistance to chronic ER stress in the root growth of *Arabidopsis* and Tm treatment did not cause the accumulation of a free and conjugated form of SA. Furthermore, the transcript levels of TBF1-dependent SA-induced genes (*TGA3*, *PAD4*, and *CRT3*) did not change in Col-0 plants, but the transcript levels of *CNX1*, *BiP2*, and *PDI* showed enhanced induction in *npr1* mutants compared with the wild-type plants [53]. Moreover, the authors also demonstrated that Tm-induced ER stress caused a more negative redox potential of the cytosol similar to earlier observations in the case of SA treatment (0.5 mM) and induced the translocation of NPR1 from the cytosol to the nucleus, where NPR1 interacts with bZIP28 and bZIP60 and suppresses the transcriptional activity of these TFs during UPR [53]. Changes in the redox state of cells under stress conditions could be a significant cellular event. Basically, SA accumulation alters the redox potential in the cytosol, resulting in a conformational change of NPR1 from an oligomeric form to a monomeric form and thus causing nuclear translocation and therefore the reprogramming of transcription [45]. Changes in the redox status of cells upon Tm could be interesting because Tm eliminates the N-glycan present in glycoproteins and significantly affects the folding assisted by ER quality control. At the same time, DTT, similar to SA, alters also the redox balance of the cell [18]. However, accumulation of ROS leading to oxidized proteins can also induce UPR after the Tm treatment [13]. ROS generation by ER luminal oxidoreductase 1 (ERO1), the mitochondria-, and/or plastid-originated ROS [54] and NADPH-oxidase activity-dependent ROS [55] suggest a potential link between ER and other organelles in the oxidative processes. In this relation, SA could be an important signaling compound because SA has a significant effect on ROS production in a time- and concentration-dependent manner in these cell compartments [56–58]. However, the direct effects of SA on ERO1 and the relationship between ER and other organelles, which generate ROS, is not known.

The role of bZIP28 and bZIP60 has also been confirmed in SA-mediated ER stress signaling with the interaction of CPR5 (constitutive expresser of pathogenesis-related genes-5), a plant-specific master regulator of growth and defense, which represses the accumulation of SA [59]. In the case of elevated SA in *cpr5* mutants, the IRE1–bZIP60 arm of ER stress is required for the growth inhibition of *Arabidopsis* seedlings. The expression of *BiP3* was also enhanced in *cpr5*, but it was significantly reduced in a *cpr5 bzip28 bzip60* triple mutant [59]. Moreover, it has also been shown that CPR5 plays a role in the UPR induced by Tm treatment after 12 days. However, CPR5 is a negative modulator of the UPR by modulating the bZIP60/bZIP28 arms of ER stress dependently on endogenous SA under stress conditions. In addition, it has also been demonstrated that there is a physical interaction between bZIP60, bZIP28, and CPR5 at the protein level. It can be concluded that CPR5 is a positive modulator of growth under normal conditions, but it acts by antagonizing SA-dependent growth inhibition through UPR modulation under stress condition [59].

There is a strong connection between other ER stress signaling elements and SA. The *Arabidopsis* genome encodes two *IRE1s* (*IRE1a* and *IRE1b*) with different physiological roles. Moreno et al. [60] observed that 4-h-long SA treatment (0.5 mM) induced the expression of both *IRE1a* and *IRE1b* genes. The use of several *ire1a* and *ire1b* mutant and transgenic plants demonstrated that IRE1a plays a predominant role in the secretion of PR proteins upon SA treatment. Mutants of *ire1a* showed enhanced susceptibility to *Pseudomonas syringae* pv. *maculicola* and these plants were not able to establish SAR, whereas *ire1b* mutants were unaffected in these responses. At the same time, IRE1b played a major role in a bZIP60 processing event after Tm treatments. The authors demonstrated that SA-dependent induction of *BiP1/2*, *CRT2*, and *UTr1* was abolished in plants lacking both members of functional IRE1, but the expression of *BiP1/2*, *UTr1*, as well as *PR1* did not change after 3 h in *bzip60* mutants, suggesting bZIP60-independent functions in plant immunity and the potential role of other TFs in this process [60].

Mechanisms of the defense responses can be different in *Arabidopsis* and in another plant species, such as rice (*Oryza sativa* L.). Firstly, 0.1 mM SA-induced activation of OsbZIP74 (also known as OsbZIP50)—an important ER stress regulator in a monocot plant, rice—was observed within 1 h in root cells [61]. In contrast to *Arabidopsis*, IRE1 mediates unconventional splicing of OsbZIP50 in rice, thus inducing ER stress-related factors, such as the ER chaperone *BiP* and counterparts of ER stress signaling, *OsbZIP39* and *OsbZIP60* [62]. At the same time, the endogenous level of SA is much higher in rice than in *Arabidopsis* [63], suggesting the potential concentration-dependent role of SA in ER stress response. OsWRKY45, which is absent from *Arabidopsis*, is an SA-regulated TF and plays a role in the activation of defense response genes upon pathogen infection [64]. Treatment with Tm induced the expression of *OsWRKY45* after 4 h in rice, which was suppressed by chemical chaperon 4-phenylbutyric acid (4-PBA). This induction of *OsWRKY45* was OsbZIP50 dependent upon Tm treatment but it did not depend on OsbZIP50 in the case of application of 0.5 mM SA. Interestingly, co-treatment with Tm and SA was additive to the expression of *OsWRKY45* and *PR1a*, but the transcript levels of *OsBiP1-5* and *OsHSP70* were suppressed by the addition of SA to Tm-treated rice plants. Based on these results, it has been concluded that OsWRKY45 induces the expression of these target genes, which is the priming effect before the activation of SA-activated defense responses. Moreover, it has also been found that ER stress induced by DTT and Tm downregulates the expression of some *PR* genes in an OsIRE1-dependent manner, which can be a protective mechanism by lowering the secretory burden on the ER under stress conditions [65]. Simultaneously, exogenous application of 0.5 mM SA can overwrite the Tm-induced UPR in *Arabidopsis thaliana*. Co-treatment with SA and Tm or DTT significantly decreased transcription levels of *AtBiP3* and *AtbZIP60* after 3 h in root tissues similarly to 4-PBA treatments. These results confirmed that this UPR-suppressive effect of this concentration of SA can be conserved between rice (a monocot) and *Arabidopsis* (a dicot) plants [66]. However, an investigation of the concentration- or time-dependent effects of SA could provide further data to understand the relationship between ER stress and SA-induced defense responses in crops. Interestingly, Tm + SA treatment similarly decreased the expression of *BiP3* after 2 days in roots but not in leaves based on histochemical gene expression analysis of *Arabidopsis* seedlings [66]. These results suggest the potential organ-dependent effects of SA in the regulation of ER stress and UPR, which could be analyzed in the future. An investigation of the changes in different organs and the potential interaction between organs could be important research aims because organ-dependent changes in the level of splicing of bZIP60 have been observed earlier upon heat stress in *Arabidopsis* [52].

Proteolytic activation of a plasma membrane-tethered NAC (NAM/ATAF1/2/CUC2) TF NTL6 is induced by cold stress but not by exogenously applied SA (0.1 mM) in *Arabidopsis*. NTL6 can directly bind to a conserved sequence in the promoters of cold-responsive *PR* genes and induce the expression of *PR1*, *PR2*, and *PR5* under cold stress independently of NPR1/TGA-mediated SA signaling [67]. An analysis of the role of various NAC TFs in SA-dependent and induced defense will provide new research topics in the future.

3. SA-Regulated Chaperons: Survival or Death

SA plays an important role in relaying the pathogen signal to activate defense reactions, such as the synthesis of PR proteins and accumulation of ROS, in the development of hypersensitive reaction (HR) or SAR [38]. Since SA is an important signaling molecule in these defense reactions of plants, its effect on UPR is a major topic in plant science. UPR is dependent on molecular chaperones, which are the key components responsible for protein folding, assembly, translocation, and degradation under normal and stress conditions [68]. However, BiPs have diverse functions; among them, the best-known function is their molecular chaperone activity, but they have a central role in ER stress and UPR, which is essential in plant developmental and immunity processes [6]. At the same time, several findings suggest that BiP induction was independent of *PR* gene induction and SA at the early stage of plant–pathogen interaction, because chaperons are required to support PR protein synthesis in the later phase of the infection [39]. Other authors observed that SA plays a dominant role in the induction of several chaperone-coding genes, such as *BiP2* and *BiP3* in *Arabidopsis* [49] or in the upregulation of *BiP*, *CNX*, and *PDI* in soybean plants [69]. The extremely high concentration of SA (5 mM) also induced the expression of both *BiP* and *PDI* in tobacco leaves [70]. It can be concluded that the selected and applied concentration of SA or the internal concentration of SA in the different plant species (e.g., in rice) [71] could determine the outcome of the stress responses of plants and result in different scientific results. The high concentration of SA induces cell death in plants (e.g., at 1 mM in tomato), but simultaneously, defense responses can also be activated [72]. Thus, the protective mechanisms are dependent on the strength and duration of the stress. Based on these observations, mild and prolonged chronic ER stress have been distinguished [14]. Prolonged and/or chronic ER stress is associated with the generation of ROS and cell death-promoting Ca^{2+} signaling, but the potential relationships with other organelles (e.g., mitochondria, chloroplast, and vacuole) still require more in-depth studies [14]. Investigation of these organelles upon SA could be crucial to understand the role of SA in ER stress and UPR [73]. Thus, the concentration- and time-dependent effects of SA could be essential to survive or to induce cell death. In the case of biotic stress, SA accumulation and high levels of PR1 and BiP proteins have been reported many days after *Pseudomonas syringae* infection during SAR development [74]. In contrast, cell death-inducing concentration of DTT increased the transcript levels of *BiPs*, *GRP94*, *CNX*, and *PDIs* genes but decreased the expression of *PR* genes in wheat seedlings 2 days after treatment [75]. It is also very important that the expression of *PDI* and *BiP* genes is highly dependent on plant tissues under untreated conditions [70], which can also be determined by SA-mediated signaling. However, the dual function of BiP in modulating development and HR has also been reported in soybean and tobacco plants [69]. In soybean transgenic lines (35S::BIP4 and 35S::BiP2), the overexpression of functional BiP and downregulation of the antioxidant system, protein degradation, and cell death-associated genes but upregulation of defense and immune system-related genes, such as *PR* and lignin biosynthetic process genes, can be seen. Interestingly, these lines contained more SA compared to wild-type plants. *BiP*-overexpressing lines displayed delayed leaf senescence under normal conditions based on changes in photosynthetic pigment concentrations. During senescence, UPR was activated, but the expression of *BiP*, *CNX*, *PDI*, and *IRE1* homologs were lower in BiP-overexpressing lines compared to the wild type, suggesting a feedback mechanism that involves the monitoring of BiP protein levels. Although *BiP* overexpression downregulated cell death-associated genes, inoculating soybean seedlings with *Pseudomonas syringae* pv. *tomato* triggered a rapid cell death response within 12 h, which was accompanied by elevated H_2O_2 levels and robust expression of *PR1*, *PR5*, and cysteine protease genes. In contrast to senescence, *BiP*-overexpressing lines showed a similar increase in the expression of *GmNAC81*, a vacuolar processing enzyme (VPE) homolog gene, and SA-mediated *PR* genes, like in case of wild-type plants after *Pseudomonas spp.* infection. Moreover, H_2O_2 production and HR were more pronounced in BiP-enhanced tobacco leaves, and BiP suppression attenuated the HR and SA-responsive *PR1* and chitinase genes were less triggered by nonhost–pathogen interactions. This observation confirmed that BiP antagonistically modulates the SA-mediated induction of *UPR* and *PR* genes, which is coordinated with the induction of the cell death response [69]. Based on these findings, the investigation of the duration and timing of BiP accumulation, the duration of UPR, and long-term effects of SA could also be an interesting research field. In addition, activation of UPR may be regulated differently during the day and night [52] and may also be regulated by circadian rhythms like SA-regulated *PR1* expression and redox balance is [51], which has not been investigated yet. Other studies also demonstrated the role of VPE in ER stress and cell death [76], which controls tonoplast rupture, confirming the potential relationship between ER and other compartments. A lethal concentration of SA induced the expression of *SlVPE1* and the antiapoptotic Bax inhibitor-1 (*SlBI-1*) in tomato roots within three hours after exogenous 1 mM SA treatment, but in the case of sublethal treatment (0.1 mM), transcript levels of *SlVPE1* and *SlBI-1* did not change [77]. This observation may imply the potential role of SA in the coordination of ER stress and proteolysis under PCD [16]. However, BI-1 is involved in the inhibition of PCD in *Arabidopsis* by decreasing ER stress-induced ROS production or by regulating Ca^{2+} homeostasis [78,79]. Not only can the vacuolar membrane be destroyed by SA during HR and PCD, but other membrane structures can also be involved, such as membranes of chloroplasts or mitochondria [57,58]. Thus, compositional changes in the ER membrane, such as in the phospholipid content and distribution upon SA treatment, can be also important signaling events to promote ER stress [49,80]. Polyamines (PA), such as spermine (Spm), could be significant candidates for the activation of UPR. Namely, it was found that Spm induces UPR by activating the splicing of the bZIP60 transcript mediated by IRE1 [81]. It is also well-known that SA in a concentration- and time-dependent manner regulates PA metabolism in plants [82], but the potential relationship between SA and UPR under mild and chronic ER stress is not known.

Not only BiP and PDI but also CRT play a role in plant immunity [83]. SA accumulation was significantly increased in *Arabidopsis* overexpressing *CRT2*, which was associated with the activation of the transcription of *PR1,2* and *5* genes but displayed reduced resistance to virulent *Pseudomonas syringae* pv. *tomato* DC3000 [84]. Based on this observation, CRT2 can act as a self-modulator, which plays a role in the fine-tuning of the SA-dependent immunity triggered by its Ca^{2+}-buffering activity, and may prevent runaway defense responses through the N-terminal domain required for chaperone activity [84]. In contrast, the role of CRT3 is associated to ethylene, because *PR1* expression did not change in CRT3a-silenced tobacco in disease resistance against the oomycete pathogen *Phytophthora infestans* [85]. These results also suggest that the physiological responses to infection are highly dependent on phytohormone interactions and SA and ET/JA levels [43]. Thus, investigation of SA together with other defense-related phytohormones in UPR could be an important future challenge.

4. Concluding Remarks and Future Perspectives

Under various abiotic and biotic stresses, protein synthesis and folding in ER can be inhibited or damaged, leading to the accumulation of misfolded or unfolded proteins in the lumen of ER, thus promoting ER stress and UPR. In plants, different ER stress signaling pathways have been identified, which investigated the ER membrane-bound stress sensors IRE1 and bZIP28 or NAC TFs. Under ER stress, the IRE1-RIDD pathway was induced to cleave mRNAs attached to the ER membrane, thus preventing further protein synthesis. Activated bZIP60 TF is translocated to the nucleus and it upregulates *UPR* genes, such as various chaperones. If UPR is incapable of decreasing ER stress, autophagy and PCD can be induced.

Based on the reviews of the existing literature, there is a link between ER stress responses and SA in plants. However, future studies are needed to reveal how SA modulates the sensing and signaling of ER stress. The time-, concentration-, species-, organ-, and cell-dependent role of SA requires more in-depth studies. The following questions have to be answered:

What is the role of SA in the switch from life to PCD during ER stress? What is the relationship and crosstalk between ER and other organelles in this process? How is SA involved in the co-operation with other phytohormones in cell fate determination upon ER stress? What terminates UPR and inactivates IRE1? How is chaperone synthesis regulated by phytohormones?

Understanding ER stress and defense activation represents an important future challenge. A deeper knowledge of the role of phytohormones in ER stress and UPR can help to design novel strategies for ER stress and plant protection management in agricultural research.

Author Contributions: Writing—original draft preparation, P.P.; writing—review and editing, Z.C., I.T. and A.Ö.

Funding: This work was supported by the grants from the National Research, Development and Innovation Office of Hungary—NKFIH (Grant no. OTKA PD112855 and NKFIH FK 124871) and by the UNKP-18-3-I-SZTE-17 and the UNKP-19-4-SZTE-86 New National Excellence Program of the Ministry of Human Capacities and the University of Szeged Open Access Fund (4416). Péter Poór was supported by the János Bolyai Research Scholarship of the Hungarian Academy of Sciences.

Conflicts of Interest: The authors declare no conflict of interest.

References

1. Martínez, I.M.; Chrispeels, M.J. Genomic analysis of the unfolded protein response in Arabidopsis shows its connection to important cellular processes. *Plant Cell* **2003**, *15*, 561–576. [CrossRef]
2. Xu, C.; Bailly-Maitre, B.; Reed, J.C. Endoplasmic reticulum stress: Cell life and death decisions. *J. Clin. Invest.* **2005**, *115*, 2656–2664. [CrossRef] [PubMed]
3. Vitale, A.; Boston, R.S. Endoplasmic reticulum quality control and the unfolded protein response: Insights from plants. *Traffic* **2008**, *9*, 1581–1588. [CrossRef]
4. Saijo, Y.; Tintor, N.; Lu, X.; Rauf, P.; Pajerowska-Mukhtar, K.; Häweker, H.; Schulze-Lefert, P. Receptor quality control in the endoplasmic reticulum for plant innate immunity. *EMBO J.* **2009**, *28*, 3439–3449. [CrossRef] [PubMed]
5. Del Bem, L.E.V. The evolutionary history of calreticulin and calnexin genes in green plants. *Genetica* **2011**, *139*, 255–259. [CrossRef] [PubMed]
6. Gupta, D.; Tuteja, N. Chaperones and foldases in endoplasmic reticulum stress signaling in plants. *Plant Signaling Behav.* **2011**, *6*, 232–236. [CrossRef] [PubMed]
7. Liu, J.X.; Howell, S.H. Endoplasmic reticulum protein quality control and its relationship to environmental stress responses in plants. *Plant Cell* **2010**, *22*, 2930–2942. [CrossRef]
8. Kørner, C.; Du, X.; Vollmer, M.; Pajerowska-Mukhtar, K. Endoplasmic reticulum stress signaling in plant immunity—at the crossroad of life and death. *Int. J. Mol. Sci.* **2015**, *16*, 26582–26598. [CrossRef]
9. Verchot, J. How does the stressed out ER find relief during virus infection? *Curr. Opin. Plant. Biol.* **2016**, *17*, 74–79. [CrossRef]
10. Nawkar, G.M.; Lee, E.S.; Shelake, R.M.; Park, J.H.; Ryu, S.W.; Kang, C.H.; Lee, S.Y. Activation of the transducers of unfolded protein response in plants. *Front. Plant Sci.* **2018**, *9*, 214. [CrossRef]
11. Liu, Y.; Li, J. Endoplasmic reticulum-mediated protein quality control in *Arabidopsis*. *Front. Plant Sci.* **2014**, *5*, 162. [CrossRef] [PubMed]
12. Chakraborty, R.; Baek, J.H.; Bae, E.Y.; Kim, W.Y.; Lee, S.Y.; Kim, M.G. Comparison and contrast of plant, yeast, and mammalian ER stress and UPR. *Appl. Biol. Chem.* **2016**, *59*, 337–347. [CrossRef]
13. Ozgur, R.; Uzilday, B.; Iwata, Y.; Koizumi, N.; Turkan, I. Interplay between the unfolded protein response and reactive oxygen species: A dynamic duo. *J. Exp. Bot.* **2018**, *69*, 3333–3345. [CrossRef] [PubMed]
14. Schäfer, P.; Eichmann, R. The endoplasmic reticulum in plant immunity and cell death. *Front. Plant Sci.* **2012**, *3*, 200.

15. Pu, Y.; Bassham, D.C. Links between ER stress and autophagy in plants. *Plant Signal. Behav.* **2013**, *8*, e24297. [CrossRef] [PubMed]
16. Cai, Y.; Yu, J.; Gallois, P. Endoplasmic reticulum stress-induced PCD and caspase-like activities involved. *Front. Plant Sci.* **2014**, *5*, 41. [CrossRef] [PubMed]
17. Wan, S.; Jiang, L. Endoplasmic reticulum (ER) stress and the unfolded protein response (UPR) in plants. *Protoplasma* **2016**, *253*, 753–764. [CrossRef] [PubMed]
18. Howell, S.H. Endoplasmic reticulum stress responses in plants. *Annu. Rev. Plant Biol.* **2013**, *64*, 477–499. [CrossRef]
19. Wang, W.; Li, X.; Zhu, M.; Tang, X.; Wang, Z.; Guo, K.; Li, X. *Arabidopsis* GAAP1 to GAAP3 Play Redundant Role in Cell Death Inhibition by Suppressing the Upregulation of Salicylic Acid Pathway Under Endoplasmic Reticulum Stress. *Front. Plant Sci.* **2019**, *10*, 1032. [CrossRef]
20. Iwata, Y.; Koizumi, N. Plant transducers of the endoplasmic reticulum unfolded protein response. *Trends Plant Sci.* **2012**, *17*, 720–727. [CrossRef]
21. Liu, J.X.; Howell, S.H. Managing the protein folding demands in the endoplasmic reticulum of plants. *New Phytol.* **2016**, *211*, 418–428. [CrossRef] [PubMed]
22. Koizumi, N.; Martinez, I.M.; Kimata, Y.; Kohno, K.; Sano, H.; Chrispeels, M.J. Molecular characterization of two *Arabidopsis* Ire1 homologs, endoplasmic reticulum-located transmembrane protein kinases. *Plant Physiol.* **2001**, *127*, 949–962. [CrossRef] [PubMed]
23. Mishiba, K.I.; Nagashima, Y.; Suzuki, E.; Hayashi, N.; Ogata, Y.; Shimada, Y.; Koizumi, N. Defects in IRE1 enhance cell death and fail to degrade mRNAs encoding secretory pathway proteins in the *Arabidopsis* unfolded protein response. *PNAS* **2013**, *110*, 5713–5718. [CrossRef] [PubMed]
24. Zhang, L.; Chen, H.; Brandizzi, F.; Verchot, J.; Wang, A. The UPR branch IRE1-bZIP60 in plants plays an essential role in viral infection and is complementary to the only UPR pathway in yeast. *PLoS Genetics* **2015**, *11*, e1005164. [CrossRef] [PubMed]
25. Zhang, L.; Zhang, C.; Wang, A. Divergence and conservation of the major UPR branch IRE1-bZIP signaling pathway across eukaryotes. *Sci. Rep.* **2016**, *6*, 27362. [CrossRef] [PubMed]
26. Liu, J.X.; Srivastava, R.; Che, P.; Howell, S.H. An endoplasmic reticulum stress response in *Arabidopsis* is mediated by proteolytic processing and nuclear relocation of a membrane-associated transcription factor, bZIP28. *Plant Cell* **2007**, *19*, 4111–4119. [CrossRef] [PubMed]
27. Srivastava, R.; Chen, Y.; Deng, Y.; Brandizzi, F.; Howell, S.H. Elements proximal to and within the transmembrane domain mediate the organelle-to-organelle movement of bZIP28 under ER stress conditions. *Plant J.* **2012**, *70*, 1033–1042. [CrossRef] [PubMed]
28. Srivastava, R.; Deng, Y.; Shah, S.; Rao, A.G.; Howell, S.H. BINDING PROTEIN is a master regulator of the endoplasmic reticulum stress sensor/transducer bZIP28 in *Arabidopsis*. *Plant Cell* **2013**, *25*, 1416–1429. [CrossRef] [PubMed]
29. Srivastava, R.; Deng, Y.; Howell, S.H. Stress sensing in plants by an ER stress sensor/transducer, bZIP28. *Front. Plant Sci.* **2014**, *5*, 59. [CrossRef]
30. Iwata, Y.; Ashida, M.; Hasegawa, C.; Tabara, K.; Mishiba, K.I.; Koizumi, N. Activation of the *Arabidopsis* membrane-bound transcription factor bZIP 28 is mediated by site-2 protease, but not site-1 protease. *Plant J.* **2017**, *91*, 408–415. [CrossRef]
31. Li, Y.; Williams, B.; Dickman, M. *Arabidopsis* B-cell lymphoma2 (Bcl-2)-associated athanogene 7 (BAG7)-mediated heat tolerance requires translocation, sumoylation and binding to WRKY29. *New Phytol.* **2017**, *214*, 695–705. [CrossRef] [PubMed]
32. Yang, Z.T.; Lu, S.J.; Wang, M.J.; Bi, D.L.; Sun, L.; Zhou, S.F.; Liu, J.X. A plasma membrane-tethered transcription factor, NAC 062/ANAC 062/NTL 6, mediates the unfolded protein response in Arabidopsis. *Plant J.* **2014**, *79*, 1033–1043. [CrossRef] [PubMed]
33. Yang, Z.T.; Wang, M.J.; Sun, L.; Lu, S.J.; Bi, D.L.; Sun, L.; Liu, J.X. The membrane-associated transcription factor NAC089 controls ER-stress-induced programmed cell death in plants. *PLoS Genetics* **2014**, *10*, e1004243. [CrossRef] [PubMed]
34. Puranik, S.; Sahu, P.P.; Srivastava, P.S.; Prasad, M. NAC proteins: Regulation and role in stress tolerance. *Trends Plant Sci.* **2012**, *17*, 369–381. [CrossRef] [PubMed]

35. Sun, L.; Yang, Z.T.; Song, Z.T.; Wang, M.J.; Sun, L.; Lu, S.J.; Liu, J.X. The plant-specific transcription factor gene NAC 103 is induced by b ZIP 60 through a new cis-regulatory element to modulate the unfolded protein response in *Arabidopsis*. *Plant J.* **2013**, *76*, 274–286. [PubMed]
36. Loake, G.; Grant, M. Salicylic acid in plant defence—the players and protagonists. *Curr. Opin. Plant. Biol.* **2007**, *10*, 466–472. [CrossRef] [PubMed]
37. Hayat, Q.; Hayat, S.; Irfan, M.; Ahmad, A. Effect of exogenous salicylic acid under changing environment: A review. *Environ. Exp. Bot.* **2010**, *68*, 14–25. [CrossRef]
38. Vlot, A.C.; Dempsey, D.M.A.; Klessig, D.F. Salicylic acid, a multifaceted hormone to combat disease. *Annu. Rev. Phytopathol.* **2009**, *47*, 177–206. [CrossRef]
39. Jelitto-Van Dooren, E.P.; Vidal, S.; Denecke, J. Anticipating endoplasmic reticulum stress: A novel early response before pathogenesis-related gene induction. *Plant Cell* **1999**, *11*, 1935–1943. [CrossRef]
40. Wang, D.; Weaver, N.D.; Kesarwani, M.; Dong, X. Induction of protein secretory pathway is required for systemic acquired resistance. *Science* **2005**, *308*, 1036–1040. [CrossRef]
41. Park, C.J.; Park, J.M. Endoplasmic reticulum plays a critical role in integrating signals generated by both biotic and abiotic stress in plants. *Front. Plant Sci.* **2019**, *10*, 399. [CrossRef] [PubMed]
42. Rivas-San Vicente, M.; Plasencia, J. Salicylic acid beyond defence: Its role in plant growth and development. *J. Exp. Bot.* **2011**, *62*, 3321–3338. [CrossRef] [PubMed]
43. Pieterse, C.M.; Van der Does, D.; Zamioudis, C.; Leon-Reyes, A.; Van Wees, S.C. Hormonal modulation of plant immunity. *Annu. Rev. Cell Dev. Bi.* **2012**, *28*, 489–521. [CrossRef] [PubMed]
44. Boatwright, J.L.; Pajerowska-Mukhtar, K. Salicylic acid: An old hormone up to new tricks. *Mol. Plant Pathol.* **2013**, *14*, 623–634. [CrossRef]
45. Mou, Z.; Fan, W.; Dong, X. Inducers of plant systemic acquired resistance regulate NPR1 function through redox changes. *Cell* **2003**, *113*, 935–944. [CrossRef]
46. Pajerowska-Mukhtar, K.M.; Wang, W.; Tada, Y.; Oka, N.; Tucker, C.L.; Fonseca, J.P.; Dong, X. The HSF-like transcription factor TBF1 is a major molecular switch for plant growth-to-defense transition. *Current Biology* **2012**, *22*, 103–112. [CrossRef]
47. Shearer, H.L.; Cheng, Y.T.; Wang, L.; Liu, J.; Boyle, P.; Després, C.; Fobert, P.R. *Arabidopsis* clade I TGA transcription factors regulate plant defenses in an NPR1-independent fashion. *Mol. Plant Microbe Interact.* **2012**, *25*, 1459–1468. [CrossRef]
48. Wang, L.; Fobert, P.R. *Arabidopsis* clade I TGA factors regulate apoplastic defences against the bacterial pathogen *Pseudomonas syringae* through endoplasmic reticulum-based processes. *PLoS ONE* **2013**, *8*, e77378. [CrossRef]
49. Nagashima, Y.; Iwata, Y.; Ashida, M.; Mishiba, K.I.; Koizumi, N. Exogenous salicylic acid activates two signaling arms of the unfolded protein response in *Arabidopsis*. *Plant Cell Physiol.* **2014**, *55*, 1772–1778. [CrossRef]
50. Liu, P.P.; von Dahl, C.C.; Klessig, D.F. The extent to which methyl salicylate is required for signaling systemic acquired resistance is dependent on exposure to light after infection. *Plant Physiol.* **2011**, *157*, 2216–2226. [CrossRef]
51. Karapetyan, S.; Dong, X. Redox and the circadian clock in plant immunity: A balancing act. *Free Radic. Biol. Med.* **2018**, *119*, 56–61. [CrossRef] [PubMed]
52. Parra-Rojas, J.; Moreno, A.A.; Mitina, I.; Orellana, A. The dynamic of the splicing of bZIP60 and the proteins encoded by the spliced and unspliced mRNAs reveals some unique features during the activation of UPR in *Arabidopsis thaliana*. *PLoS ONE* **2015**, *10*, e0122936. [CrossRef] [PubMed]
53. Lai, Y.S.; Renna, L.; Yarema, J.; Ruberti, C.; He, S.Y.; Brandizzi, F. Salicylic acid-independent role of NPR1 is required for protection from proteotoxic stress in the plant endoplasmic reticulum. *PNAS* **2018**, *115*, E5203–E5212. [CrossRef] [PubMed]
54. Ozgur, R.; Uzilday, B.; Sekmen, A.H.; Turkan, I. The effects of induced production of reactive oxygen species in organelles on endoplasmic reticulum stress and on the unfolded protein response in *Arabidopsis*. *Ann. Bot.* **2015**, *116*, 541–553. [CrossRef] [PubMed]
55. Angelos, E.; Brandizzi, F. NADPH oxidase activity is required for ER stress survival in plants. *Plant J.* **2018**, *96*, 1106–1120. [CrossRef]

56. Poór, P.; Takács, Z.; Bela, K.; Czékus, Z.; Szalai, G.; Tari, I. Prolonged dark period modulates the oxidative burst and enzymatic antioxidant systems in the leaves of salicylic acid-treated tomato. *J. Plant Physiol.* **2017**, *213*, 216–226. [CrossRef]
57. Poór, P.; Patyi, G.; Takács, Z.; Szekeres, A.; Bódi, N.; Bagyánszki, M.; Tari, I. Salicylic acid-induced ROS production by mitochondrial electron transport chain depends on the activity of mitochondrial hexokinases in tomato (*Solanum lycopersicum* L.). *J. Plant Res.* **2019**, *132*, 273–283. [CrossRef]
58. Poór, P.; Borbély, P.G.; Bódi, N.; Bagyánszki, M.; Görgényi Miklósné Tari, I. Effects of salicylic acid on photosynthetic activity and chloroplast morphology under light and prolonged darkness. *Photosynthetica* **2019**, *57*, 367–376. [CrossRef]
59. Meng, Z.; Ruberti, C.; Gong, Z.; Brandizzi, F. CPR5 modulates salicylic acid and the unfolded protein response to manage tradeoffs between plant growth and stress responses. *Plant J.* **2017**, *89*, 486–501. [CrossRef]
60. Moreno, A.A.; Mukhtar, M.S.; Blanco, F.; Boatwright, J.L.; Moreno, I.; Jordan, M.R.; Pajerowska-Mukhtar, K.M. IRE1/bZIP60-mediated unfolded protein response plays distinct roles in plant immunity and abiotic stress responses. *PLoS ONE* **2012**, *7*, e31944. [CrossRef]
61. Lu, S.J.; Yang, Z.T.; Sun, L.; Sun, L.; Song, Z.T.; Liu, J.X. Conservation of IRE1-regulated bZIP74 mRNA unconventional splicing in rice (*Oryza sativa* L.) involved in ER stress responses. *Mol. Plant* **2012**, *5*, 504–514. [CrossRef] [PubMed]
62. Hayashi, S.; Wakasa, Y.; Takaiwa, F. Functional integration between defence and IRE1-mediated ER stress response in rice. *Sci. Rep.* **2012**, *2*, 670. [CrossRef] [PubMed]
63. Yang, Y.; Qi, M.; Mei, C. Endogenous salicylic acid protects rice plants from oxidative damage caused by aging as well as biotic and abiotic stress. *Plant J.* **2004**, *40*, 909–919. [CrossRef]
64. Shimono, M.; Sugano, S.; Nakayama, A.; Jiang, C.J.; Ono, K.; Toki, S.; Takatsuji, H. Rice WRKY45 plays a crucial role in benzothiadiazole-inducible blast resistance. *Plant Cell* **2007**, *19*, 2064–2076. [CrossRef] [PubMed]
65. Hayashi, S.; Wakasa, Y.; Takahashi, H.; Kawakatsu, T.; Takaiwa, F. Signal transduction by IRE1-mediated splicing of bZIP50 and other stress sensors in the endoplasmic reticulum stress response of rice. *Plant J.* **2012**, *69*, 946–956. [CrossRef] [PubMed]
66. Hayashi, S.; Takaiwa, F. Visualization of endoplasmic reticulum stressed cells for forward genetic studies in plants. *J. Plant Physiol.* **2015**, *180*, 61–66. [CrossRef] [PubMed]
67. Seo, P.J.; Kim, M.J.; Park, J.Y.; Kim, S.Y.; Jeon, J.; Lee, Y.H.; Park, C.M. Cold activation of a plasma membrane-tethered NAC transcription factor induces a pathogen resistance response in *Arabidopsis*. *Plant J.* **2010**, *61*, 661–671. [CrossRef]
68. Park, C.J.; Seo, Y.S. Heat shock proteins: A review of the molecular chaperones for plant immunity. *Plant Pathol. J.* **2015**, *31*, 323. [CrossRef]
69. Carvalho, H.H.; Silva, P.A.; Mendes, G.C.; Brustolini, O.J.; Pimenta, M.R.; Gouveia, B.C.; Fontes, E.P. The endoplasmic reticulum binding protein BiP displays dual function in modulating cell death events. *Plant Physiol.* **2014**, *164*, 654–670. [CrossRef]
70. Denecke, J.; Carlsson, L.E.; Vidal, S.; Höglund, A.S.; Ek, B.; van Zeijl, M.J.; Palva, E.T. The tobacco homolog of mammalian calreticulin is present in protein complexes in vivo. *Plant Cell* **1995**, *7*, 391–406.
71. De Vleesschauwer, D.; Xu, J.; Höfte, M. Making sense of hormone-mediated defense networking: From rice to *Arabidopsis*. *Front. Plant Sci.* **2014**, *5*, 611. [CrossRef] [PubMed]
72. Poór, P.; Kovács, J.; Szopkó, D.; Tari, I. Ethylene signaling in salt stress-and salicylic acid-induced programmed cell death in tomato suspension cells. *Protoplasma* **2013**, *250*, 273–284. [CrossRef]
73. Bobik, K.; Burch-Smith, T.M. Chloroplast signaling within, between and beyond cells. *Front. Plant Sci.* **2015**, *6*, 781. [CrossRef] [PubMed]
74. Sticher, L.; Metraux, J.P. Inhibitors of N-linked glycosylation induce systemic acquired resistance in cucumber. *Physiol. Mol. Plant P.* **2000**, *56*, 245–252. [CrossRef]
75. Yu, X.; Wang, T.; Zhu, M.; Zhang, L.; Zhang, F.; Jing, E.; Lin, T. Transcriptome and physiological analyses for revealing genes involved in wheat response to endoplasmic reticulum stress. *BMC Plant Biol.* **2019**, *19*, 193. [CrossRef] [PubMed]
76. Qiang, X.; Zechmann, B.; Reitz, M.U.; Kogel, K.H.; Schäfer, P. The mutualistic fungus Piriformospora indica colonizes *Arabidopsis* roots by inducing an endoplasmic reticulum stress–triggered caspase-dependent cell death. *Plant Cell* **2012**, *24*, 794–809. [CrossRef]

77. Kovács, J.; Poór, P.; Szepesi, Á.; Tari, I. Salicylic acid induced cysteine protease activity during programmed cell death in tomato plants. *Acta Biol. Hung.* **2016**, *67*, 148–158. [CrossRef]
78. Watanabe, N.; Lam, E. BAX inhibitor-1 modulates endoplasmic reticulum stress-mediated programmed cell death in *Arabidopsis*. *J. Biol. Chem.* **2008**, *283*, 3200–3210. [CrossRef]
79. Watanabe, N.; Lam, E. Bax inhibitor-1, a conserved cell death suppressor, is a key molecular switch downstream from a variety of biotic and abiotic stress signals in plants. *Int. J. Mol. Sci.* **2009**, *10*, 3149–3167. [CrossRef]
80. Krinke, O.; Ruelland, E.; Valentová, O.; Vergnolle, C.; Renou, J.P.; Taconnat, L.; Zachowski, A. Phosphatidylinositol 4-kinase activation is an early response to salicylic acid in Arabidopsis suspension cells. *Plant Physiol.* **2007**, *144*, 1347–1359. [CrossRef]
81. Sagor, G.H.M.; Chawla, P.; Kim, D.W.; Berberich, T.; Kojima, S.; Niitsu, M.; Kusano, T. The polyamine spermine induces the unfolded protein response via the MAPK cascade in *Arabidopsis*. *Front. Plant Sci.* **2015**, *6*, 687. [CrossRef] [PubMed]
82. Takács, Z.; Poór, P.; Tari, I. Comparison of polyamine metabolism in tomato plants exposed to different concentrations of salicylic acid under light or dark conditions. *Plant Physiol. Biochem.* **2016**, *108*, 266–278. [CrossRef] [PubMed]
83. Li, J.; Zhao-Hui, C.; Batoux, M.; Nekrasov, V.; Roux, M.; Chinchilla, D.; Jones, J.D. Specific ER quality control components required for biogenesis of the plant innate immune receptor EFR. *PNAS* **2009**, *106*, 15973–15978. [CrossRef] [PubMed]
84. Qiu, Y.; Xi, J.; Du, L.; Roje, S.; Poovaiah, B.W. A dual regulatory role of *Arabidopsis* calreticulin-2 in plant innate immunity. *Plant J.* **2012**, *69*, 489–500. [CrossRef] [PubMed]
85. Matsukawa, M.; Shibata, Y.; Ohtsu, M.; Mizutani, A.; Mori, H.; Wang, P.; Takemoto, D. Nicotiana benthamiana calreticulin 3a is required for the ethylene-mediated production of phytoalexins and disease resistance against oomycete pathogen *Phytophthora infestans*. *Mol. Plant Microbe Interact.* **2013**, *26*, 880–892. [CrossRef]

International Journal of
Molecular Sciences

MDPI

Review

The Ecology of Salicylic Acid Signaling: Primary, Secondary and Tertiary Effects with Applications in Agriculture

Camila C. Filgueiras *,†, Adalvan D. Martins, Ramom V. Pereira and Denis S. Willett †

Department of Entomology, College of Agriculture and Life Sciences, Cornell AgriTech, Cornell University, Geneva, NY 14456, USA; adm285@cornell.edu (A.D.M.); rvp34@cornell.edu (R.V.P.); deniswillett@cornell.edu (D.S.W.)

* Correspondence: camilafilgueiras@cornell.edu; Tel.:+1-315-787-2319

† These authors contributed equally to this work.

Received: 26 October 2019; Accepted: 19 November 2019; Published: 21 November 2019

Abstract: The salicylic acid pathway is one of the primary plant defense pathways, is ubiquitous in vascular plants, and plays a role in rapid adaptions to dynamic abiotic and biotic stress. Its prominence and ubiquity make it uniquely suited for understanding how biochemistry within plants can mediate ecological consequences. Induction of the salicylic acid pathway has primary effects on the plant in which it is induced resulting in genetic, metabolomic, and physiologic changes as the plant adapts to challenges. These primary effects can in turn have secondary consequences for herbivores and pathogens attacking the plant. These secondary effects can both directly influence plant attackers and mediate indirect interactions between herbivores and pathogens. Additionally, stimulation of salicylic acid related defenses can affect natural enemies, predators and parasitoids, which can recruit to plant signals with consequences for herbivore populations and plant herbivory aboveground and belowground. These primary, secondary, and tertiary ecological consequences of salicylic acid signaling hold great promise for application in agricultural systems in developing sustainable high-yielding management practices that adapt to changing abiotic and biotic environments.

Keywords: aboveground belowground; plant defense in agriculture; natural enemies; indirect interactions; indirect effects; plant mediated interactions

1. Introduction

Agricultural productivity is a global priority [1]. With a growing population and increasingly dynamic climate, there has been an intense focus on genetic improvement of food crops for human consumption [2]. Much of this effort has been focused on directly increasing yield under diverse abiotic and biotic conditions. While yields have increased substantially, especially after the green revolution, yield improvements have been stagnating in critical areas [2,3].

As a consequence of these genetic improvement efforts, modern cultivars have lost the adaptability [4] and defenses inherent to their wild ancestors [5–8]. While this increased productivity, in many cases the modern cultivars can be more susceptible to attack by pests and pathogens [5–8]. Plants respond to these attack in different ways, defending themselves both directly through physical and chemical defenses against herbivores and pathogens, and indirectly by, for example, recruiting natural enemies of herbivores [9].

These direct and indirect defenses are regulated through biochemical pathways that rely on plant hormones to mediate physiological changes that aid in plant defense [10]. These changes can be genetic involving alterations to transcription and translation, metabolomic affecting synthesis of secondary metabolites, and volatilomic inducing release of volatile signals. While there are a few principal plant defense pathways primarily responsible for defense against pests and pathogens, such as the jasmonic acid (JA) pathway, the salicylic acid (SA) pathway garnered substantial interest for its role in regulating defenses, its inducibility, and potential applications for applied agricultural management in the field [11,12].

Because of these reasons, the SA pathway is the focus of this review. It is important to keep in mind, however, that these systems of plant defense do not occur in isolation. There is ample evidence of cross-talk between plant defense pathways with the JA pathway, for example [13–15]. These interactions are usually reciprocally antagonistic; for example, SA can transcriptionally control JA signaling [16,17].

Keeping in mind the potential for cross-talk, this review will focus primarily on the ecology of the SA pathway. Much work has been done elucidating the individual steps in synthesis, mechanisms of induction, and biochemical pathways that form the SA pathway. This review will touch on many of those points, but with a focus on how those pathways and reactions effect communication with the plant itself and with other organisms. The ecology of the SA pathway—how the SA pathway in a given plant mediates interactions between and with other organisms—is just beginning to be understood. The goal of this review is to provide a basis for future work that aims to explore this space more fully.

To that end, this review will be structured with separate sections focusing on the primary, secondary, and tertiary effects of inducing the SA pathway. The objective of these sections is to highlight primary effects of SA on the plant, the secondary effects of SA on pests and pathogens, and the tertiary effects of SA on natural enemies (Figure 1). In each of those sections, this review will lay the groundwork for what has been done in the area while pointing out opportunities for further work into the ecology of these different effects. The review will close with a discussion of a relatively new advancements and an exciting area of active research: use of the SA pathway for applied control of agricultural pests and pathogens with a discussion of costs and benefits of this approach for plants and managers in applied agricultural systems.

Figure 1. Primary, Secondary and Tertiary Effects of SA Signaling. Primary effects encompass plant specific effects. Secondary effects involve direct and indirect interactions with pathogens and herbivores. Tertiary effects comprise interactions with natural enemies.

2. Primary Effects of SA

Salicylic acid, as a molecule, has remarkable properties in multiple fields [18]. Independent of its rich and storied medical history and modern relationship with aspirin, salicylic acid is a common, nearly ubiquitous, phenolic secondary metabolite of plants [18–20]. The chemical properties of the molecule make it readily soluble inside plant tissue [20] and easily transported in its methylated form [21,22]. Its methylated form, methyl salicylate (MeSA), is readily volatilized and, in addition to its role as a cues for other organims, is used in oil of wintergreen [23].

2.1. Production and Processing

Production of salicylic acid occurs in plant plastids where the end product in the shikimic acid pathway, chorismic acid can be further processed into either isochorismic acid or prephenic acid then L-phenylalanine and trans-cinnamic acid [19]. These two parallel pathways each rely on separate enzymes, isochorismate synthase (ICS) and phenylalanine ammonia lyase (PAL, responsible for conversion of L-phenylalanine to trans-cinnamic acid) respectively [19,24]. Genes responsible for production of these enzymes and their homologs were identified in several plant species [19,24]. Further modification of salicylic acid to its methylated form can be mediated by BA/SA carboxyl methyltransferase 1 (BSMT1) identified through work in *Arabidopsis* [24].

Whether production of salicylic acid by either the ICS or PAL pathways has ecological consequences for induction of the pathway or downstream effects on other organisms remains unclear. Previous work showed that the ICS pathway may be primarily responsible for production of SA [25], but both pathways were implicated in plant responses to abiotic stressors such as UV exposure and biotic stressors such as pathogen infection [24]. Much of the work exploring these pathways was done in *Arabidopsis* model systems [24,26]. To our knowledge, and lacking from recent reviews, no work has been done exploring differential effects of pathogen and pest stimulation of the IC and PAL pathways for salicylic acid biosynthesis either in *Arabidopsis* or other plant systems [19,24–27].

Following production of SA, the plant hormone can be modified in several different ways that affects its solubility, mobility, and use as a signal. Among other steps that may occur post-synthesis, SA can be glucosylated, methylated, and conjugated with amino acids [19,24,25]. Glucosylatation of SA via glucosyltransferases convert SA to a relatively non-toxic deactivated form that can can be stored long-term in cell vacuoles [24]. Methylation of SA to form MeSA creates a highly mobile signal with a host of ecological effects [19,24,25]. Amino acid conjugation is an active area of research and is likely involved in degradation of SA [24]. These three forms of SA modification-storage, transport, and degradation-among others are important means by which which plants regulate levels of SA post-production and mediate effects on plant physiology.

Regulation of SA is almost universally important for plant physiology, but not universally uniform; plants vary substantially in basal levels of SA [28,29]. Multiple order of magnitude differences between species were recorded, even within the same family [28,29]. Basal levels of SA in *Arabidopsis* range from 0.250 µg to 1 µg g^{-1} FW [28]. Within a given plant levels of SA can also vary widely. Shoots and roots can have wildly different amounts of SA both basally and as a result of the physiologic responses of different plant tissues [28,30].

2.2. Plant Response

Downstream of production, SA can have a multitude of effects on plant physiology in all parts of the plant. SA affects plant thermogenesis, stomatal dynamics, seed germination, cell growth, vegetative growth, flowering, photosynthesis, responses to abiotic stresses and defensive responses against pests and pathogens [19,29,31–38].

In mediating responses to abiotic stress, SA increases the efficiency of the antioxidant system in plants [32]. SA can lower levels of reactive oxygen species preventing cell damage from free radicals and promoting intracellular redox homeostasis [32,39]. This increased tolerance to oxidative stress also plays a role in mediating interactions with other organisms; reactive oxygen species are involved in cell death responses and generation of long-term resistance to pests and pathogens [32,40]. Of particular importance to the ecology of the SA pathway, induction of the SA pathway can result in localized and systemic defense responses within the plant.

When plants are attacked by microbes and pathogens, a series of plant defense responses can be triggered. Pattern recognition receptors in the plant can detect pathogen and microbe specific patterns that can, in turn, stimulate pattern triggered immunity (PTI) [41–43]. If PTI is suppressed by pathogen effectors, plants can rely on an additional level of defense in effector triggered immunity (ETI) [37,42,43]. These defenses can result in programmed cell death at the site of infection regulated by specific plant resistance genes and termed the hypersensitive response [43,44]. SA was implicated in mediating both PTI and ETI responses in monocots and dicots and is a necessary component of systemic acquired resistance [37,43,45,46].

Systemic acquired resistance (SAR), the ability of plants to develop long-term resistance to micro-organisms even in parts of the plant not initially attacked, is a key aspect of plant defense and dependent upon accumulation of SA [45,46]. SA can increase amounts of pathogenesis-related (PR) proteins with anti-microbial properties through systemic changes to transcriptional programming via interaction with transcriptional cofactors of Non-expresser of PR genes (*NPR*) [37,47–51]. Importantly, while SA is an important and necessary component of SAR, it is not the mobile signal for induction. Methyl salicylate, azelaic acid, pipecolic acid, and its derivative N-hydroxypipecolic acid among others, all play a role as mobile, and some cases, volatile signals for systemic acquired resistance [52–56].

2.3. Induction

While induction of the SA pathway by pathogens can result in SAR, activation of the SA pathway can be induced exogenously through application of elicitors and plays important roles in regulating responses to other organisms including attack by other plants, and by insect herbivores.

Ever since early recognition of SA as a plant signal, exogenous application of SA has been used to induce adaptive responses in plants to both abiotic and biotic stress [32,39,57]. Induction of SA pathway with exogenous elicitors was successfully conducted using a variety of compounds including SA proper, MeSA, and Benzothiadiazole (BTH, benzo(1,2,3) thiadiazole-7-carbothioic acid S-methyl ester, a synthetic analog of SA), and S-methylmethionine salicylate, among others [58]. Exogenous application of these elicitors through spraying or seed treatment can often mimic adaptations to abiotic and biotic stress, particularly defense responses triggered by pathogenic micro-organisms, plants, and insects [12,58–61].

Plants attacked by other parasitic plants can exhibit pathogenically similar responses with regulation by SA [62]. *Striga* infection in *Sorghum* can elicit hypersensitive responses at sites of attack [63]. Similarly, Dodder infection in alfalfa can induce expression of PR genes for defense [64]. Dodder attacks on tomato induced hypersensitive-like responses and elevated SA levels along with other plant hormones involved in defense [65]. This is still an active area of research; the mechanisms, elicitors, and ramifications of SA induction by plant parasites is still being explored [62].

Given the prominent role of SA in regulating plant responses to other forms of attack, it should come as little surprise that feeding by insect herbivores can also induce SA mediated effects for plant defense. While SA is relatively better explored in the context of pathogen defense, insects-particularly sucking insects-can also elicit responses. Aphid feeding by the green peach aphid (*Myzus persicae*) on *Arabidopsis* induces SA related gene expression involved in pathogen defense [66]. Feeding by the Russian wheat aphid (*Diuraphis noxia* Mordvilko) on wheat induces SA accumulation which plays a role in resistance [67]. Likewise, tomato resistance to potato aphid mediated by the Mi-1 gene relies on SA and is a case study for similar plant defense responses to plant pathogens and sucking insects [68]. Such responses can even alter activation of other plant defense pathways; silverleaf whitefly (*Bemisia tabaci* type B) feeding on *Arabidopsis* induces the SA related plant defenses while suppressing the JA defense pathway [69].

Additionally, there is some evidence that SA may be involved in defense against chewing insects. While many studies of plant defense pathway induction by chewing insects point to the role of the JA pathway [70] in mediating chewing herbivore responses, SA was implicated in plant defenses of tomato in response to feeding by larvae of the cotton bollworm (*Helicoverpa armigera* Hubner) [71].

Induction of the SA acid pathway in response to attack by plants, pathogens, and pests is not a given, however. Other factors can play a role. Endogenous levels of other plant hormones can be crucial in determining whether and how SA related responses develop within a plant [17,72]. Ethylene, for example, not only can interact with SA responses, but also affect JA-SA crosstalk [72,73]. Circadian rhythms and presence of light can be important in regulating responses affecting both the attacker and how the plant responds to pathogens and herbivores [74–76].

3. Secondary Effects of SA

3.1. Direct Effects

Induction of the SA pathway has a direct role in mediating interactions with and between pathogens and herbivores. As was touched on briefly, activation of SA related plant defenses can result in production of PR proteins and induction of SAR with negative consequences for infecting pathogens [26,37,45,49]. Viral, bacterial, fungal, and oomycete pathogens can all be negatively impacted through induction of SA associated resistance in both monocots and dicots [77,78]. Specifically, SA plays an important role in resistance to tobacco mosaic virus in tobacco [46], resistance to *Pseudomonas* in *Arabidopsis* [79], resistance to *Alternaria* fungus in potato [80], rice blast fungus in rice [81], and defense against *Phytophthora infestans* in potato [82]. In addition to halting pathogen infection, induction of SAR may render a plant inaccessible for future pathogen attack by altering patterns and distributions of pathogen infection locally. These effects can even cascade to alter community structures of microorganisms [83].

Community effects of SA mediated plant defenses are not only limited to effects at the micro-organism community level, but also extend to other plants. Volatile communication between plants can cause cascading effects of SA induction in plant communities by triggering SA related defenses in neighboring plants [84,85]. This phenomenon was best studied in willow and sagebrush systems [86,87] and relies on transmission of plant volatiles such as MeSA among others aboveground [84,88]. Communication can also take place belowground [88]; recent work has indicated a role for SA signaling, among other plant defense pathways, in belowground plant-plant communication [89–91].

Similarly, SA induction can have negative consequences for the fitness of attacking herbivores. As a defensive molecule, phenolic compounds such as SA can act as deterrents and be toxic to insect herbivores [92]. In willows, for example, SA and related compounds play a role in reducing performance of generalist chewing insects such as *Manduca* and *Operophtera* [92,93]. Also in willows, SA mediates resistance development against a gall midge that produces a characteristic hypersensitive response resulting in reduced insect larval survival [94].

SA induction can have additional direct effects against insect herbivores. In *Arabidopsis*, egg deposition and larval feeding by *Pieris brassicae* can interact to induce higher SA levels [95]. SA can accumulate at sites of oviposition [96] and larval feeding can exacerbate this effect [95]. At the same time, there is enhanced expression of PR genes [95]. Larvae feeding on egg-induced plants perform poorly and gain less weight [95]. Importantly, *Arabidopsis* mutants deficient in SA pathway components do not show the same larval effects [95]. SA not only affects larval performance of *P. brassicae* in *Arabidopsis*, but also potentially affects oviposition behavior; MeSA tends to deter oviposition by *P. brassicae* when either dispensed exogenously or expressed constitutively at high levels [97]. Similar effects on oviposition and performance were observed to some extent in other systems, but either been tested solely through exogenous application or not explored to the same extent [98,99].

Fitness effects of SA induction are not always negative for the offending herbivore, however. In tobacco plants infected with the tobacco mosaic virus, over-expression of PAL increases SAR to the tobacco mosaic virus while under-expression weakens it [100]. The inverse is true for insect herbivory; plants with a weak SAR response were better able to fend off herbivory while plants with a strong SAR response were not able to defend as well against herbivory by *Heliothis virescens* larvae [100]. This phenomenon was observed to some extent in other pathosystems and is mediated by cross-talk and trade-offs between plant defense pathways [100,101]. As mentioned in the introduction, the SA pathway does not act in isolation but can and often interacts with other plant signalling systems, such as the JA pathway.

3.2. Interactions

The preceding example is one prominent case of the SA pathway mediating indirect interactions between plant pathogens and insect herbivores. The ecological consequences of these indirect interactions are only beginning to be understood, but exist in numerous study systems. Interactions can go both ways; in the above example, SA was shown to play a role in pathogen resistance affecting herbivore feeding but herbivore feeding can also impact pathogen infection. Feeding by *Helicoverpa zea* larvae on tomato leaves caused a reduction in infection by the bacterial pathogen *Pseudomonas syringae* [102]. While SA has not always been implicated in these interactions, the role of plant defense pathways in mediating many forms of indirect interactions was excellently reviewed [77] and can occur even across the aboveground-belowground divide [103]. The mechanisms mediating these interactions are still being explored, but sequence of infection or attack is incredibly important as are the identities of the pathogens, plants, and herbivores involved in the interaction [104–106].

3.3. Temporal Considerations

The role of the SA pathway in mediating ecological effects and interactions between other organisms has an inter-generational temporal component as well. There are epigenetic effects of SA mediated plant defense signaling [19]. DNA methylation and histone modification can play a role in mediating plant defenses through regulation of plant defense genes and affecting SAR [19,107,108]. There is also limited evidence that some of these epigenetic modifications can be heritable in both *Arabidopsis* and bean [107,109–111]. The ecological ramifications of epigenetic effects of SA remain to be explored but one can imagine a situation where induction of the SA pathway not only has a priming effect on the plant during its lifetime perhaps resulting in SAR, but also has inter-generational effects that affect pathogen and herbivore populations long term.

4. Tertiary Effects of SA

Just as induction of the SA pathway has primary metabolic consequences for the plant and secondary consequences for other organisms such as pests and pathogens, induction of the SA pathway can have tertiary effects on additional trophic levels affecting behavior and recruitment of natural enemies both aboveground and belowground.

4.1. Aboveground Natural Enemies

Since early work showing the role of herbivore induced plant volatiles in recruiting natural enemies aboveground [112], the role of SA has been explored in influencing natural enemies of plant herbivores. Natural enemies can be predators or parasites of plant herbivores and range from specialist parasitic wasps to generalist predatory beetles. Natural enemies of insect herbivores, particularly parasitic wasps, are known to respond to a variety of cues released from plants and insects in order to locate their future hosts and can learn to respond to a variety of dynamic and ephemeral cues [113,114].

One prominent cue that can be directly linked to induction of the SA pathway is the volatile methyl salicylate (MeSA). MeSA is a phenolic signal produced from SA and is involved in plant-plant communication as a mobile and volatile signal for systemic acquired resistance [23,53]. MeSA in its role as a plant defense signal is likely conserved [115]; many plants release MeSA as a component of herbivore induced plant volatile blends in response to feeding by insect herbivores (Table 1).

Table 1. MeSA released from listed plant species in a blend of herbivore released plant volatiles after feeding by listed herbivore species.

Plant			Herbivore	
Family	**Species**	**Common Name**	**Species**	**Citation**
Fabacae	*Phaseolus lunatus*	Spider Mite	*Tetranychus urticae* Koch	[116]
Fabacae	*Phaseolus lunatus*	Spider Mite	*Tetranychus urticae* Koch	[117]
Solanaceae	*Solanum lycopersicum*	Spider Mite	*Tetranychus urticae* Koch	[118]
Cucurbitaceae	*Cucumis sativus*	Spider Mite	*Tetranychus urticae* Koch	[119]
Brassicaceae	*Brassica oleracea capitata L.*	Garden Whites	*Pieris spp.*	[120]
Rosaceae	*Pyrus communis*	Pear Psyllid	*Psylla pyricola* Forster	[121]
Cannabaceae	*Humulus iupulus* L.	Damson Hop Aphid	*Phorodon humuli* Forster	[122]
Rosaceae	*Prunus padus*	Bird Cherry–Oat Aphid	*Rhopalosiphum padi* L.	[123]
Solanaceae	*Nicotiana attenuata* Torr. ex Wats	Five-Spotted Hawkmoth	*Manduca quinquemaculata*	[124]
Solanaceae	*Nicotiana attenuata* Torr. ex Wats	Leaf Bug	*Dicyphus minimus*	[124]
Solanaceae	*Nicotiana attenuata* Torr. ex Wats	Tobacco Flea Beetle	*Epitrix hirtipennis*	[124]
Fabaceae	*Lotus japonicus*	Spider Mite	*Tetranychus urticae*	[117]

Predators and parasites of these insect herbivores can also perceive and respond to MeSA. Indeed, MeSA has been shown to be attractive to a wide range of insect natural enemies ranging from micro-hymenoptera to lacewings in laboratory and field studies (Table 2). In many cases, recruitment of these natural enemies in response to release of MeSA can occur over relatively large distances and reduce pest populations. It bears noting, however, that despite the apparent broad use of MeSA as a beneficial plant signal resulting in the attraction of predators and parasites to reduce herbivore feeding, this is not universally true. In trials comparing *Arabidopsis thaliana* plants compromised in the production of MeSA with wild-type plants, MeSA compromised plants were more attractive to parasitoids, natural enemies of biotic stressors, than their wild-type counterparts releasing MeSA [125].

Table 2. Natural Enemies Recruited by MeSA.

Plant		Herbivore		Natural Enemy				
Family	**Species**	**Common**	**Species**	**Common**	**Species**	**Impact on Herbivore**	**Setting**	**Citation**
Cannabaceae	*Humulus lupulus*	Damson-Hop aphid	*Phorodon humuli* [122]	Green Lacewing	*Chrysopa nigricornis*	Not Reported	Field	[126]
Fabaceae	*Phaseolus lunatus*	Spider Mite	*Tetranychus urticae*		*Phytoseiulus persimilis*	Not Reported	Lab	[127]
Fabaceae	*Glycine max* L.	Soybean Aphid	*Aphis glycines*	Green Lacewing	*Chrysopa nigricornis*	Reduced	Field/Lab	[128]
Fabaceae	*Glycine max* L.	Soybean Aphid	*Aphis glycines*	Syrphid Flies		Reduced	Field/Lab	[128]
Cannabaceae	*Humulus lupulus*			Ladybeetle	*Stethorus punctum picipes*	Not Reported	Field	[129]
Cannabaceae	*Humulus lupulus*			Minute Pirate Bug	*Orius tristicolor*	Not Reported	Field	[129]
Cannabaceae	*Humulus lupulus*			Bigeyed Bug	*Geocoris pallens*	Not Reported	Field	[129]
Cannabaceae	*Humulus lupulus*			Syrphidae		Not Reported	Field	[129]
Cannabaceae	*Humulus lupulus*			Empididae		Not Reported	Field	[129]
Cannabaceae	*Humulus lupulus*			Sarcophagidae		Not Reported	Field	[129]
Cannabaceae	*Humulus lupulus*			Agromyzidae		Not Reported	Field	[129]
Cannabaceae	*Humulus lupulus*			Micro-hymenoptera		Not Reported	Field	[129]
Vitaceae	*Vitis labrusca*			Green Lacewing	*Chrysopa nigricornis*	Not Reported	Field	[130]
Vitaceae	*Vitis labrusca*			Lacewing	*Hemerobius sp.*	Not Reported	Field	[130]
Vitaceae	*Vitis labrusca*				*Deraeocoris brevis*	Not Reported	Field	[130]
Vitaceae	*Vitis labrusca*				*Stethorus punctum picipes*	Not Reported	Field	[130]
Vitaceae	*Vitis labrusca*			Minute Pirate Bug	*Orius tristicolor*	Not Reported	Field	[130]
Rosaceae	*Fragaria sp.*	Aphididae, Thripidae, Cicadellidae		Minute Pirate Bug	*Orius tristicolor*	No Effect [1]	Field	[131]
Rosaceae	*Fragaria sp.*	Aphididae, Thripidae, Cicadellidae		Chrysopidae		No Effect [1]	Field	[131]
Poaceae	*Zea mays*	Cucurbit Beetle	*Diabrotica speciosa*	Entomopathogenic Nematode	*Heterorhabditis amazonensis*	Not Reported	Lab	[132]
Rutaceae	*Citrus paradisi x Poncirus trifoliata*	Citrus Root Weevil	*Diaprepes abbreviatus*	Entomopathogenic Nematode	*Steinernema diaprepesi*	Not Reported	Lab	[133]

[1] Most effects were found to be not significant, although a marginal decrease was observed for one year for leafhoppers [131].

4.2. Belowground Natural Enemies

While the SA pathway plays a significant role in regulating plant defenses and mediating interactions with herbivores and natural enemies aboveground, its role in belowground indirect defenses is just beginning to be understood. Plants inhabit two distinct environments; just as the shoots and leaves of plants aboveground can benefit from recruitment of natural enemies to reduce aboveground herbivore pressures, so too can plant roots. Belowground herbivory is an important factor affecting plant performance, if relatively unexplored [134]. New methodologies, technologies, and approaches have been opening up the frontier of belowground plant defense interactions in recent years [135,136] with discoveries that highlight the importance of SA in belowground interactions with natural enemies.

In addition to adapting to two distinct environments, plants must also contend with an embedded corollary; natural enemies belowground are inherently different than those aboveground. Parasitic wasps tend to be less effective against belowground herbivores. Instead, natural enemies belowground can include entomopathogenic nematodes and soil-dwelling mites. Similar to aboveground systems, belowground feeding by root herbivores can induce release of volatiles that recruit predatory mites [137] and entomopathogenic nematodes [138–140]. These entomopathogenic nematodes effect the death of their insect hosts with the aid of symbiotic bacteria [141]. Release of herbivore induced plant volatiles belowground has been shown to reduce herbivory [142,143] and can increase probability of pest insect mortality by approximately 90% [144].

The role of the SA pathway in mediating these changes is just beginning to be explored. Exogenous application of MeSA in citrus and corn plants can cause recruitment of entomopathogenic nematodes [132,133]. In citrus, this relationship has been explored further; exogenous application of MeSA can induce release of the terpene volatile limonene which is attractive to entomopathogenic nematodes [133]. Effects of SA induction on natural enemies belowground could potentially have far reaching consequences. Trials examining distances of recruitment suggest that release and diffusion of these volatile signals can attract beneficial natural enemies from distances as great as 60 cm in sandy soil types [145].

4.3. Connecting Aboveground and Belowground with SA

This work suggests a specific role for SA in not only connecting belowground and aboveground plant systems, but also in mediating tertiary effects between aboveground organisms and belowground natural enemies and vice versa. Indeed, recent work showed that belowground feeding by insect larvae can induce release of volatiles aboveground attractive to parasitoids of the adult insect [146]. While no plant defense pathway was implicated in that work, the ability of induction of the SA pathway to effect release of terpene volatiles and recruit entomopathogenic nematodes belowground suggests a broad role for the SA pathway facilitating tertiary effects by communicating with and connecting natural enemies below and aboveground.

5. Applications of SA Induction for Control of Pests and Pathogens

Induction of the SA pathway has clear ramifications for plants, pathogens, herbivores, and natural enemies above and belowground with ecological consequences radiating from SA mediated interactions at multiple trophic levels. Observation of these types of effects naturally leads to contemplation of possible application in agriculture. Using the SA pathway in agriculture to manage pests and pathogens has generally followed three approaches: (1) exogenous application of compounds that induce the SA pathway for direct plant defense; (2) genetic modification of plants to alter plant defense expression; (3) exogenous application of SA related volatile compounds (e.g., MeSA) to attract natural enemies that control insect pests.

5.1. Exogenous Induction

Early investigations into using exogenous applications of SA inducing compounds to elicit plant defenses pathways, induce resistance, and augment SA signaling opened the way for consideration of novel management strategies for control of agricultural pests and pathogens. Exogenous application can take many forms involving a wide range of elicitors and synthetic SA analogs that can result in SAR and enhanced resistance to pathogen load [147,148]. In addition to the examples cited above, exogenous application of MeSA to *Nicotiana benthamiana* increased resistance to challenges by the bacterial pathogen *Pseudomonas syringae* [149]. Repeated application of MeSA strengthened this response [149] suggesting that exogenous application in field settings could hold potential for managing pathogen resistance. Field wide applications of exogenous elicitors may not even be necessary; recent work showed that positive feedback loops involving SAR and monoterpene communication could potentially propagate SAR at the population level using plant–plant communication to magnify spatial effects [150]. Effects of exogenous elicitors could also be magnified through time across generations as epigenetic effects of exogenous elicitors have been documented and could potentially engender inter-generational defense with fewer applications [151].

This approach could not only have appreciable effects on plant disease, but also increase yield. In some crop systems, exogenous application of elicitors can have beneficial effects for crop management comparable to pesticide controls [11]. In several important agricultural crops including monocots and dicots, exogenous induction of SAR against bacterial and fungal pathogens in the field has beneficial effects in reducing disease severity in some cases even exceeding benefits seen by pesticide controls [11]. Importantly for consideration of utility in agricultural settings, disease reduction as a result of SAR induction was in some cases associated with increases in yield [11].

Beneficial outcomes on disease management from exogenous application of SA elicitors are not guaranteed, however. Results can be dependent on the identities of the plant and pathogen species [11]. Fusarium wilt (*Fusarium oxysporum f.sp. cucumerinum*) in cucumber, for example, does not respond to exogenous induction of SAR [152]. Multiple elicitors in peanut also failed to control fungal late leaf spot, even creating undesirable effects and augmenting fungal growth [153]. These undesirable outcomes could reflect a lack of understanding of elicitor mechanisms. Recent work has shown that the SA synthetic analog BTH does not confer resistance to *Rhizoctonia solani*, the causal agent of sheath blight disease in the grain species *Brachypodium distachyon* because it induces genes related to JA signaling [154]. In this same system, exogenous application of SA does confer resistance [154].

Non-beneficial agricultural outcomes of exogenous induction of SA defenses are not limited to lack of pathogen control, however. As discussed above, stimulation of systemic acquired resistance may be beneficial in reducing pathogen load, but may result in crop plants becoming more susceptible to herbivory [155]. There are well documented examples of trade-offs between pathogen and herbivore resistance [101]. Trade-offs also encompass other physiological effects; if a crop plant is allocating more energy and resources to defense, less may be allocated to production. Plant defense pathways and growth regulation are inextricably entwined [156]. In the case of SA, the shikimic acid pathway is a starting point for biosynthesis and also critically important in amino acid production [19]. Optimizing defense and yield trade-offs will continue to be a consideration in applying knowledge of SA defenses in agriculture and may have to be considered on a case by case basis. With certain crops under pathogen pressure, but not insect pressure, exogenous stimulation of plant defenses could increase yields, while in other situations, yield may be suppressed by exogenous applications of elicitors either due to no effect on pathogen resistance, negative effect on herbivore defenses, negative effect on yield investments, or some combination of all three.

5.2. Genetic Approaches to Using SA in Agriculture

An ideal solution to address the challenges and trade-offs listed above would be crops that have defenses turned off in situations where pest and pathogen pressure is absent that then are strongly activated in situations where pest and pathogen pressure are prevalent [156]. While this characteristic may have been present to some extent in wild progenitors, domestication of agricultural crops of economic importance can have substantial effects on plant secondary chemistry with potential consequences for the ecology of pest and pathogen interactions [157,158]. Over-expression and constitutive expression of specific genes involved in plant defense pathways can create enhanced and broad spectrum resistance to pathogens [159]. While these approaches might be effective under situations of intense pathogen pressure, they suffer from the same trade-offs and drawbacks discussed in the exogenous induction section; constant allocation of plant resources to defense through genetically modified constitutive expression likely has negative consequences for crop yield.

Recent work shows promise in obviating those drawbacks by developing a switch that would balance defense and production trade-offs [156]. Induction of plant immunity through PTI can result in global translational reprogamming that occurs rapidly following pathogen infection [160]. The genetic elements responsible for that reprogramming can be packaged and inserted into *Arabidopsis* and rice resulting in resistance to agriculturally relevant pathogens across generations while balancing fitness and yield costs [161]. These results hold considerable promise for development and engineering of high-yielding plant varieties adaptive to and successful across wide ranges of pathogen pressure.

5.3. Natural Enemy Attraction

While engineering plants that selectively activate SA defense pathways in areas of high pathogen pressure holds considerable promise for defense against plant pathogens, herbivore pressure can also substantially impact yields. The SA pathway has limited efficacy for defense against herbivore pests of agricultural pests. As discussed above, SA can play a role in direct defense against sucking insects such as aphids [66–69]. A potentially more promising approach for control of insect herbivores through SA related defenses in agricultural systems is via attraction of natural enemies. As mentioned above, release of volatiles related to SA defenses can recruit natural enemies above and belowground and reduce herbivore populations [133] (Table 2). This attraction can be accomplished either through deployment of lures releasing volatiles such as MeSA or through exogenous induction of SA in plants. Meta-analysis of this attraction showed large and relatively invariant effects of attraction across a wide range of predator and parasitoid taxa [12].

The efficacy of this approach on reducing pest populations could be highly variable however and merits further exploration into appropriate means of implementation. Attraction of large numbers of predators and parasitoids may have non-target effects and does not necessarily result in improved pest control; attracted natural enemies may not necessarily be effective in controlling the offending pest species for several reasons including phenology (pest life stage plays an important role in susceptibility), relative population densities, and competing hosts [162]. Additionally, long term release of attractive signals such as MeSA could diminish in efficacy over time, particularly if used prophylatically in the absence of abundant pest populations. Natural enemies responding to a volatile cues in the absence of host resources will learn to avoid this deceptive signal potentially to the chagrin of agricultural producers hoping for continuous natural enemy protection.

5.4. Opportunities

Knowledge of SA related defense signaling and potential applications to agricultural challenges burgeoned in recent years. While there is substantial work that remains to be done in understanding basic mechanisms behind primary, secondary, and tertiary ecological effects of SA signaling, what work that has been done points to numerous opportunities for developing methods that enhance sustainable production of important agricultural crops through efficient and efficacious management of abiotic and biotic challenges. These contributions could engender a second green revolution: a plant defense-based revolution-leveraging intelligent adaptations to abiotic and biotic challenges to preserve and increase sustainable crop yields and feed a hungry planet.

Author Contributions: Conceptualization, D.S.W. and C.C.F.; literature review, C.C.F., A.D.M., R.V.P. and D.S.W.; Writing—original draft preparation, D.S.W. and C.C.F.; Writing—review and editing, C.C.F., A.D.M., R.V.P., and D.S.W.

Funding: This research received no external funding

Acknowledgments: Bo Holladay provided illustrations (yadalloh@gmail.com). Bo Holladay and Lois Schertz Willett provided invaluable feedback on early versions of this manuscript.

Conflicts of Interest: The authors declare no conflict of interest.

Abbreviations

SA	Salicylic Acid
JA	Jasmonic Acid
PAL	Phenylalanine ammonia lysase
ICS	Isochorismate synthase
MeSA	Methyl Salicylate
SAR	Systemic Acquired Resistance
PTI	Pattern Triggered Immunity
PR	Pathogenesis related
NPR	Non-expressor of PR
ETI	Effector Triggered Immunity
BTH	Benzo(1,2,3)thiadiazole-7-carbothioic acid S-methyl ester

References

1. Bruinsma, J. *World Agriculture: Towards 2015/2030: An FAO Study*; Routledge: London, UK, 2017.
2. Ray, D.K.; Mueller, N.D.; West, P.C.; Foley, J.A. Yield trends are insufficient to double global crop production by 2050. *PLoS ONE* **2013**, *8*, e66428. [CrossRef] [PubMed]
3. Ray, D.K.; Ramankutty, N.; Mueller, N.D.; West, P.C.; Foley, J.A. Recent patterns of crop yield growth and stagnation. *Nat. Commun.* **2012**, *3*, 1293. [CrossRef] [PubMed]
4. Lobell, D.B.; Roberts, M.J.; Schlenker, W.; Braun, N.; Little, B.B.; Rejesus, R.M.; Hammer, G.L. Greater sensitivity to drought accompanies maize yield increase in the US Midwest. *Science* **2014**, *344*, 516–519. [CrossRef] [PubMed]
5. Moreira, X.; Abdala-Roberts, L.; Gols, R.; Francisco, M. Plant domestication decreases both constitutive and induced chemical defences by direct selection against defensive traits. *Sci. Rep.* **2018**, *8*, 12678. [CrossRef] [PubMed]
6. Paudel, S.; Lin, P.A.; Foolad, M.R.; Ali, J.G.; Rajotte, E.G.; Felton, G.W. Induced Plant Defenses Against Herbivory in Cultivated and Wild Tomato. *J. Chem. Ecol.* **2019**, *45*, 693–707. [CrossRef]
7. Whitehead, S.R.; Turcotte, M.M.; Poveda, K. Domestication impacts on plant–herbivore interactions: A meta-analysis. *Philos. Trans. R. Soc. B Biol. Sci.* **2017**, *372*, 20160034. [CrossRef]
8. Chen, Y.H.; Gols, R.; Stratton, C.A.; Brevik, K.A.; Benrey, B. Complex tritrophic interactions in response to crop domestication: Predictions from the wild. *Entomol. Exp. Appl.* **2015**, *157*, 40–59. [CrossRef]
9. Aljbory, Z.; Chen, M.S. Indirect plant defense against insect herbivores: A review. *Insect Sci.* **2018**, *25*, 2–23. [CrossRef]
10. Berens, M.L.; Berry, H.M.; Mine, A.; Argueso, C.T.; Tsuda, K. Evolution of hormone signaling networks in plant defense. *Annu. Rev. Phytopathol.* **2017**, *55*, 401–425. [CrossRef]
11. Vallad, G.E.; Goodman, R.M. Systemic acquired resistance and induced systemic resistance in conventional agriculture. *Crop Sci.* **2004**, *44*, 1920–1934. [CrossRef]
12. Rodriguez-Saona, C.; Kaplan, I.; Braasch, J.; Chinnasamy, D.; Williams, L. Field responses of predaceous arthropods to methyl salicylate: A meta-analysis and case study in cranberries. *Biol. Control* **2011**, *59*, 294–303. [CrossRef]
13. Koornneef, A.; Pieterse, C.M. Cross talk in defense signaling. *Plant Physiol.* **2008**, *146*, 839–844. [CrossRef] [PubMed]
14. Kunkel, B.N.; Brooks, D.M. Cross talk between signaling pathways in pathogen defense. *Curr. Opin. Plant Biol.* **2002**, *5*, 325–331. [CrossRef]

15. Vos, I.A.; Moritz, L.; Pieterse, C.M.; Van Wees, S. Impact of hormonal crosstalk on plant resistance and fitness under multi-attacker conditions. *Front. Plant Sci.* **2015**, *6*, 639. [CrossRef]
16. Caarls, L.; Pieterse, C.M.; Van Wees, S. How salicylic acid takes transcriptional control over jasmonic acid signaling. *Front. Plant Sci.* **2015**, *6*, 170. [CrossRef]
17. Thaler, J.S.; Humphrey, P.T.; Whiteman, N.K. Evolution of jasmonate and salicylate signal crosstalk. *Trends Plant Sci.* **2012**, *17*, 260–270. [CrossRef]
18. Mahdi, J.; Mahdi, A.; Mahdi, A.; Bowen, I. The historical analysis of aspirin discovery, its relation to the willow tree and antiproliferative and anticancer potential. *Cell Prolif.* **2006**, *39*, 147–155. [CrossRef]
19. Kumar, D.; Haq, I.; Chapagai, D.; Tripathi, D.; Donald, D.; Hossain, M.; Devaiah, S. Hormone signaling: Current perspectives on the roles of salicylic acid and its derivatives in plants. In *The Formation, Structure and Activity of Phytochemicals*; Springer: Berlin/Heidelberg, Germany, 2015; pp. 115–136.
20. Raskin, I. Role of salicylic acid in plants. *Annu. Rev. Plant Biol.* **1992**, *43*, 439–463. [CrossRef]
21. Heil, M.; Ton, J. Long-distance signalling in plant defence. *Trends Plant Sci.* **2008**, *13*, 264–272. [CrossRef]
22. Shah, J. Plants under attack: Systemic signals in defence. *Curr. Opin. Plant Biol.* **2009**, *12*, 459–464. [CrossRef]
23. Shulaev, V.; Silverman, P.; Raskin, I. Airborne signalling by methyl salicylate in plant pathogen resistance. *Nature* **1997**, *385*, 718. [CrossRef]
24. D'Maris Amick Dempsey, A.C.; Vlot, M.C.W.; Daniel, F.K. Salicylic acid biosynthesis and metabolism. *Arab. Book Am. Soc. Plant Biol.* **2011**, *9*, e0156.
25. Chen, Z.; Zheng, Z.; Huang, J.; Lai, Z.; Fan, B. Biosynthesis of salicylic acid in plants. *Plant Signal. Behav.* **2009**, *4*, 493–496. [CrossRef] [PubMed]
26. Dempsey, D.A.; Klessig, D.F. How does the multifaceted plant hormone salicylic acid combat disease in plants and are similar mechanisms utilized in humans? *BMC Biol.* **2017**, *15*, 23. [CrossRef] [PubMed]
27. Kumar, D.; Chapagai, D.; Dean, P.; Davenport, M. Biotic and abiotic stress signaling mediated by salicylic acid. In *Elucidation of Abiotic Stress Signaling in Plants*; Springer: Berlin/Heidelberg, Germany, 2015; pp. 329–346.
28. Rivas-San Vicente, M.; Plasencia, J. Salicylic acid beyond defence: Its role in plant growth and development. *J. Exp. Bot.* **2011**, *62*, 3321–3338. [CrossRef]
29. Raskin, I.; Skubatz, H.; Tang, W.; Meeuse, B.J. Salicylic acid levels in thermogenic and non-thermogenic plants. *Ann. Bot.* **1990**, *66*, 369–373. [CrossRef]
30. Chen, Z.; Iyer, S.; Caplan, A.; Klessig, D.F.; Fan, B. Differential accumulation of salicylic acid and salicylic acid-sensitive catalase in different rice tissues. *Plant Physiol.* **1997**, *114*, 193–201. [CrossRef]
31. Raskin, I.; Ehmann, A.; Melander, W.R.; Meeuse, B.J. Salicylic acid: A natural inducer of heat production in Arum lilies. *Science* **1987**, *237*, 1601–1602. [CrossRef]
32. Hayat, Q.; Hayat, S.; Irfan, M.; Ahmad, A. Effect of exogenous salicylic acid under changing environment: A review. *Environ. Exp. Bot.* **2010**, *68*, 14–25. [CrossRef]
33. Manthe, B.; Schulz, M.; Schnabl, H. Effects of salicylic acid on growth and stomatal movements ofVicia faba L.: Evidence for salicylic acid metabolization. *J. Chem. Ecol.* **1992**, *18*, 1525–1539. [CrossRef]
34. Rate, D.N.; Cuenca, J.V.; Bowman, G.R.; Guttman, D.S.; Greenberg, J.T. The gain-of-function Arabidopsis acd6 mutant reveals novel regulation and function of the salicylic acid signaling pathway in controlling cell death, defenses, and cell growth. *Plant Cell* **1999**, *11*, 1695–1708. [CrossRef] [PubMed]
35. Janda, T.; Gondor, O.K.; Yordanova, R.; Szalai, G.; Pál, M. Salicylic acid and photosynthesis: Signalling and effects. *Acta Physiol. Plant.* **2014**, *36*, 2537–2546. [CrossRef]
36. Rajjou, L.; Belghazi, M.; Huguet, R.; Robin, C.; Moreau, A.; Job, C.; Job, D. Proteomic investigation of the effect of salicylic acid on Arabidopsis seed germination and establishment of early defense mechanisms. *Plant Physiol.* **2006**, *141*, 910–923. [CrossRef] [PubMed]
37. Klessig, D.F.; Choi, H.W.; Dempsey, D.A. Systemic acquired resistance and salicylic acid: Past, present, and future. *Mol. Plant-Microbe Interact.* **2018**, *31*, 871–888. [CrossRef] [PubMed]
38. Vanacker, H.; Lu, H.; Rate, D.N.; Greenberg, J.T. A role for salicylic acid and NPR1 in regulating cell growth in Arabidopsis. *Plant J.* **2001**, *28*, 209–216. [CrossRef] [PubMed]
39. Kadioglu, A.; Saruhan, N.; Sağlam, A.; Terzi, R.; Acet, T. Exogenous salicylic acid alleviates effects of long term drought stress and delays leaf rolling by inducing antioxidant system. *Plant Growth Regul.* **2011**, *64*, 27–37. [CrossRef]
40. Lu, H.; Greenberg, J.T.; Holuigue, L. Salicylic acid signaling networks. *Front. Plant Sci.* **2016**, *7*, 238. [CrossRef]

41. Bigeard, J.; Colcombet, J.; Hirt, H. Signaling mechanisms in pattern-triggered immunity (PTI). *Mol. Plant* **2015**, *8*, 521–539. [CrossRef]
42. Tsuda, K.; Katagiri, F. Comparing signaling mechanisms engaged in pattern-triggered and effector-triggered immunity. *Curr. Opin. Plant Biol.* **2010**, *13*, 459–465. [CrossRef]
43. Maruri-López, I.; Aviles-Baltazar, N.Y.; Buchala, A.; Serrano, M. Intra and extracellular journey of the phytohormone salicylic acid. *Front. Plant Sci.* **2019**, *10*. [CrossRef]
44. Heath, M.C. Hypersensitive response-related death. In *Programmed Cell Death in Higher Plants*; Springer: Berlin/Heidelberg, Germany, 2000; pp. 77–90.
45. Durrant, W.E.; Dong, X. Systemic acquired resistance. *Annu. Rev. Phytopathol.* **2004**, *42*, 185–209. [CrossRef] [PubMed]
46. Gaffney, T.; Friedrich, L.; Vernooij, B.; Negrotto, D.; Nye, G.; Uknes, S.; Ward, E.; Kessmann, H.; Ryals, J. Requirement of salicylic acid for the induction of systemic acquired resistance. *Science* **1993**, *261*, 754–756. [CrossRef] [PubMed]
47. Innes, R. The positives and negatives of NPR: A unifying model for salicylic acid signaling in plants. *Cell* **2018**, *173*, 1314–1315. [CrossRef] [PubMed]
48. Sudisha, J.; Sharathchandra, R.; Amruthesh, K.; Kumar, A.; Shetty, H.S. Pathogenesis related proteins in plant defense response. In *Plant Defence: Biological Control*; Springer: Berlin/Heidelberg, Germany, 2012; pp. 379–403.
49. Dangl, J.L.; Jones, J.D. Plant pathogens and integrated defence responses to infection. *Nature* **2001**, *411*, 826. [CrossRef] [PubMed]
50. Fu, Z.Q.; Yan, S.; Saleh, A.; Wang, W.; Ruble, J.; Oka, N.; Mohan, R.; Spoel, S.H.; Tada, Y.; Zheng, N.; et al. NPR3 and NPR4 are receptors for the immune signal salicylic acid in plants. *Nature* **2012**, *486*, 228. [CrossRef] [PubMed]
51. Ali, A.; Shah, L.; Rahman, S.; Riaz, M.W.; Yahya, M.; Xu, Y.J.; Liu, F.; Si, W.; Jiang, H.; Cheng, B. Plant defense mechanism and current understanding of salicylic acid and NPRs in activating SAR. *Physiol. Mol. Plant Pathol.* **2018**, *104*, 15–22. [CrossRef]
52. Dempsey, D.A.; Klessig, D.F. SOS–too many signals for systemic acquired resistance? *Trends Plant Sci.* **2012**, *17*, 538–545. [CrossRef]
53. Park, S.W.; Kaimoyo, E.; Kumar, D.; Mosher, S.; Klessig, D.F. Methyl salicylate is a critical mobile signal for plant systemic acquired resistance. *Science* **2007**, *318*, 113–116. [CrossRef]
54. Bernsdorff, F.; Döring, A.C.; Gruner, K.; Schuck, S.; Bräutigam, A.; Zeier, J. Pipecolic acid orchestrates plant systemic acquired resistance and defense priming via salicylic acid-dependent and-independent pathways. *Plant Cell* **2016**, *28*, 102–129. [CrossRef]
55. Shah, J.; Chaturvedi, R.; Chowdhury, Z.; Venables, B.; Petros, R.A. Signaling by small metabolites in systemic acquired resistance. *Plant J.* **2014**, *79*, 645–658. [CrossRef]
56. Hartmann, M.; Zeier, J. N-Hydroxypipecolic acid and salicylic acid: A metabolic duo for systemic acquired resistance. *Curr. Opin. Plant Biol.* **2019**, *50*, 44–57. [CrossRef] [PubMed]
57. Doherty, H.M.; Selvendran, R.R.; Bowles, D.J. The wound response of tomato plants can be inhibited by aspirin and related hydroxy-benzoic acids. *Physiol. Mol. Plant Pathol.* **1988**, *33*, 377–384. [CrossRef]
58. Bektas, Y.; Eulgem, T. Synthetic plant defense elicitors. *Front. Plant Sci.* **2015**, *5*, 804. [CrossRef] [PubMed]
59. Farooq, M.; Basra, S.; Wahid, A.; Ahmad, N.; Saleem, B. Improving the drought tolerance in rice (*Oryza sativa* L.) by exogenous application of salicylic acid. *J. Agron. Crop Sci.* **2009**, *195*, 237–246. [CrossRef]
60. Yang, C.; Hu, L.; Ali, B.; Islam, F.; Bai, Q.; Yun, X.; Yoneyama, K.; Zhou, W. Seed treatment with salicylic acid invokes defence mechanism of Helianthus annuus against Orobanche cumana. *Ann. Appl. Biol.* **2016**, *169*, 408–422. [CrossRef]
61. Ansari, A.; Razmjoo, J.; Karimmojeni, H. Mycorrhizal colonization and seed treatment with salicylic acid to improve physiological traits and tolerance of flaxseed (*Linum usitatissimum* L.) plants grown under drought stress. *Acta Physiol. Plant.* **2016**, *38*, 34. [CrossRef]
62. Smith, J.L.; De Moraes, C.M.; Mescher, M.C. Jasmonate-and salicylate-mediated plant defense responses to insect herbivores, pathogens and parasitic plants. *Pest Manag. Sci. Former. Pestic. Sci.* **2009**, *65*, 497–503. [CrossRef]
63. Mohamed, A.; Ellicott, A.; Housley, T.; Ejeta, G. Hypersensitive response to Striga infection in sorghum. *Crop Sci.* **2003**, *43*, 1320–1324. [CrossRef]

64. Borsics, T.; Lados, M. Dodder infection induces the expression of a pathogenesis-related gene of the family PR-10 in alfalfa. *J. Exp. Bot.* **2002**, *53*, 1831–1832. [CrossRef]
65. Runyon, J.B.; Mescher, M.C.; Felton, G.W.; De Moraes, C.M. Parasitism by Cuscuta pentagona sequentially induces JA and SA defence pathways in tomato. *Plant Cell Environ.* **2010**, *33*, 290–303. [CrossRef]
66. Moran, P.J.; Thompson, G.A. Molecular responses to aphid feeding in Arabidopsis in relation to plant defense pathways. *Plant Physiol.* **2001**, *125*, 1074–1085. [CrossRef] [PubMed]
67. Mohase, L.; van der Westhuizen, A.J. Salicylic acid is involved in resistance responses in the Russian wheat aphid-wheat interaction. *J. Plant Physiol.* **2002**, *159*, 585–590. [CrossRef]
68. Li, Q.; Xie, Q.G.; Smith-Becker, J.; Navarre, D.A.; Kaloshian, I. Mi-1-mediated aphid resistance involves salicylic acid and mitogen-activated protein kinase signaling cascades. *Mol. Plant-Microbe Interact.* **2006**, *19*, 655–664. [CrossRef] [PubMed]
69. Zarate, S.I.; Kempema, L.A.; Walling, L.L. Silverleaf whitefly induces salicylic acid defenses and suppresses effectual jasmonic acid defenses. *Plant Physiol.* **2007**, *143*, 866–875. [CrossRef] [PubMed]
70. Gatehouse, J.A. Plant resistance towards insect herbivores: A dynamic interaction. *New Phytol.* **2002**, *156*, 145–169. [CrossRef]
71. Peng, J.; Deng, X.; Jia, S.; Huang, J.; Miao, X.; Huang, Y. Role of salicylic acid in tomato defense against cotton bollworm, Helicoverpa armigera Hubner. *Z. Für Naturforschung* **2004**, *59*, 856–862. [CrossRef]
72. Li, N.; Han, X.; Feng, D.; Yuan, D.; Huang, L.J. Signaling crosstalk between salicylic acid and ethylene/jasmonate in plant defense: Do we understand what they are whispering? *Int. J. Mol. Sci.* **2019**, *20*, 671. [CrossRef]
73. Broekgaarden, C.; Caarls, L.; Vos, I.A.; Pieterse, C.M.; Van Wees, S.C. Ethylene: Traffic controller on hormonal crossroads to defense. *Plant Physiol.* **2015**, *169*, 2371–2379. [CrossRef]
74. Karapetyan, S.; Dong, X. Redox and the circadian clock in plant immunity: A balancing act. *Free Radic. Biol. Med.* **2018**, *119*, 56–61. [CrossRef]
75. Suárez-Vidal, E.; López-Goldar, X.; Sampedro, L.; Zas, R. Effect of light availability on the interaction between maritime pine and the pine weevil: Light drives insect feeding behavior but also the defensive capabilities of the host. *Front. Plant Sci.* **2017**, *8*, 1452. [CrossRef]
76. Poór, P.; Takács, Z.; Bela, K.; Czékus, Z.; Szalai, G.; Tari, I. Prolonged dark period modulates the oxidative burst and enzymatic antioxidant systems in the leaves of salicylic acid-treated tomato. *J. Plant Physiol.* **2017**, *213*, 216–226. [CrossRef] [PubMed]
77. Stout, M.J.; Thaler, J.S.; Thomma, B.P. Plant-mediated interactions between pathogenic microorganisms and herbivorous arthropods. *Annu. Rev. Entomol.* **2006**, *51*, 663–689. [CrossRef] [PubMed]
78. Chaturvedi, R.; Shah, J. Salicylic acid in plant disease resistance. In *Salicylic Acid: A Plant Hormone*; Springer: Berlin/Heidelberg, Germany, 2007; pp. 335–370.
79. Yang, L.; Li, B.; Zheng, X.y.; Li, J.; Yang, M.; Dong, X.; He, G.; An, C.; Deng, X.W. Salicylic acid biosynthesis is enhanced and contributes to increased biotrophic pathogen resistance in Arabidopsis hybrids. *Nat. Commun.* **2015**, *6*, 7309. [CrossRef] [PubMed]
80. Coquoz, J.; Buchala, A.; Meuwly, P.; Metraux, J. Arachidonic acid induces local but not systemic synthesis of salicylic acid and confers systemic resistance in potato plants to Phytophthora infestans and Alternaria solani. *Phytopathology* **1995**, *85*, 1219–1224. [CrossRef]
81. Silverman, P.; Seskar, M.; Kanter, D.; Schweizer, P.; Metraux, J.P.; Raskin, I. Salicylic acid in rice (biosynthesis, conjugation, and possible role). *Plant Physiol.* **1995**, *108*, 633–639. [CrossRef]
82. Halim, V.A.; Eschen-Lippold, L.; Altmann, S.; Birschwilks, M.; Scheel, D.; Rosahl, S. Salicylic acid is important for basal defense of Solanum tuberosum against Phytophthora infestans. *Mol. Plant Microbe Interact.* **2007**, *20*, 1346–1352. [CrossRef]
83. Hein, J.W.; Wolfe, G.V.; Blee, K.A. Comparison of rhizosphere bacterial communities in Arabidopsis thaliana mutants for systemic acquired resistance. *Microb. Ecol.* **2008**, *55*, 333–343. [CrossRef]
84. Baldwin, I.T.; Halitschke, R.; Paschold, A.; Von Dahl, C.C.; Preston, C.A. Volatile signaling in plant-plant interactions: "talking trees" in the genomics era. *Science* **2006**, *311*, 812–815. [CrossRef]
85. Baldwin, I.T.; Kessler, A.; Halitschke, R. Volatile signaling in plant–plant–herbivore interactions: What is ewblock? *Curr. Opin. Plant Biol.* **2002**, *5*, 351–354. [CrossRef]
86. Pearse, I.S.; Hughes, K.; Shiojiri, K.; Ishizaki, S.; Karban, R. Interplant volatile signaling in willows: Revisiting the original talking trees. *Oecologia* **2013**, *172*, 869–875. [CrossRef]

87. Karban, R.; Huntzinger, M.; McCall, A.C. The specificity of eavesdropping on sagebrush by other plants. *Ecology* **2004**, *85*, 1846–1852. [CrossRef]
88. Holopainen, J.K.; Blande, J.D. Molecular plant volatile communication. In *Sensing in Nature*; Springer: Berlin/Heidelberg, Germany, 2012; pp. 17–31.
89. You, L.X.; Wang, P.; Kong, C.H. The levels of jasmonic acid and salicylic acid in a rice-barnyardgrass coexistence system and their relation to rice allelochemicals. *Biochem. Syst. Ecol.* **2011**, *39*, 491–497. [CrossRef]
90. Schreiner, M.; Krumbein, A.; Knorr, D.; Smetanska, I. Enhanced glucosinolates in root exudates of Brassica rapa ssp. rapa mediated by salicylic acid and methyl jasmonate. *J. Agric. Food Chem.* **2011**, *59*, 1400–1405. [CrossRef]
91. Kong, C.H.; Zhang, S.Z.; Li, Y.H.; Xia, Z.C.; Yang, X.F.; Meiners, S.J.; Wang, P. Plant neighbor detection and allelochemical response are driven by root-secreted signaling chemicals. *Nat. Commun.* **2018**, *9*, 3867. [CrossRef] [PubMed]
92. Pasteels, J.M.; Rowell-Rahier, M. The chemical ecology of herbivory on willows. *Proc. R. Soc.* **1992**, *98*, 63–73. [CrossRef]
93. Ruuhola, T. *Dynamics of Salicylates in Willows and Its Relation to Herbivory*; University of Joensuu: Joensuu, Finland, 2001.
94. Ollerstam, O.; Larsson, S. Salicylic acid mediates resistance in the willow Salix viminalis against the gall midge Dasineura marginemtorquens. *J. Chem. Ecol.* **2003**, *29*, 163–174. [CrossRef]
95. Lortzing, V.; Oberländer, J.; Lortzing, T.; Tohge, T.; Steppuhn, A.; Kunze, R.; Hilker, M. Insect egg deposition renders plant defence against hatching larvae more effective in a salicylic acid-dependent manner. *Plant Cell Environ.* **2019**, *42*, 1019–1032. [CrossRef]
96. Bruessow, F.; Gouhier-Darimont, C.; Buchala, A.; Metraux, J.P.; Reymond, P. Insect eggs suppress plant defence against chewing herbivores. *Plant J.* **2010**, *62*, 876–885. [CrossRef]
97. Groux, R.; Hilfiker, O.; Gouhier-Darimont, C.; Peñaflor, M.F.G.V.; Erb, M.; Reymond, P. Role of methyl salicylate on oviposition deterrence in Arabidopsis thaliana. *J. Chem. Ecol.* **2014**, *40*, 754–759. [CrossRef]
98. Ulland, S.; Ian, E.; Mozuraitis, R.; Borg-Karlson, A.K.; Meadow, R.; Mustaparta, H. Methyl salicylate, identified as primary odorant of a specific receptor neuron type, inhibits oviposition by the moth *Mamestra brassicae* L. (Lepidoptera, Noctuidae). *Chem. Senses* **2007**, *33*, 35–46. [CrossRef]
99. Favaro, R.; Resende, J.T.V.; Gabriel, A.; Zeist, A.R.; Cordeiro, E.C.N.; Favaro Júnior, J.L. Salicylic acid: Resistance inducer to two-spotted spider mite in strawberry crop. *Hortic. Bras.* **2019**, *37*, 60–64. [CrossRef]
100. Felton, G.; Korth, K.; Bi, J.; Wesley, S.; Huhman, D.; Mathews, M.; Murphy, J.; Lamb, C.; Dixon, R. Inverse relationship between systemic resistance of plants to microorganisms and to insect herbivory. *Curr. Biol.* **1999**, *9*, 317–320. [CrossRef]
101. Felton, G.W.; Korth, K.L. Trade-offs between pathogen and herbivore resistance. *Curr. Opin. Plant Biol.* **2000**, *3*, 309–314. [CrossRef]
102. Stout, M.; Fidantsef, A.; Duffey, S.; Bostock, R. Signal interactions in pathogen and insect attack: Systemic plant-mediated interactions between pathogens and herbivores of the tomato, Lycopersicon esculentum. *Physiol. Mol. Plant Pathol.* **1999**, *54*, 115–130. [CrossRef]
103. Johnson, S.N.; Clark, K.E.; Hartley, S.E.; Jones, T.H.; McKenzie, S.W.; Koricheva, J. Aboveground–belowground herbivore interactions: A meta-analysis. *Ecology* **2012**, *93*, 2208–2215. [CrossRef]
104. Wang, M.; Biere, A.; Van der Putten, W.H.; Bezemer, T.M. Sequential effects of root and foliar herbivory on aboveground and belowground induced plant defense responses and insect performance. *Oecologia* **2014**, *175*, 187–198. [CrossRef]
105. Erb, M.; Robert, C.A.; Hibbard, B.E.; Turlings, T.C. Sequence of arrival determines plant-mediated interactions between herbivores. *J. Ecol.* **2011**, *99*, 7–15. [CrossRef]
106. Huang, W.; Robert, C.A.; Hervé, M.R.; Hu, L.; Bont, Z.; Erb, M. A mechanism for sequence specificity in plant-mediated interactions between herbivores. *New Phytol.* **2017**, *214*, 169–179. [CrossRef]
107. Ramirez-Prado, J.S.; Abulfaraj, A.A.; Rayapuram, N.; Benhamed, M.; Hirt, H. Plant immunity: From signaling to epigenetic control of defense. *Trends Plant Sci.* **2018**, *23*, 833–844. [CrossRef]
108. Espinas, N.A.; Saze, H.; Saijo, Y. Epigenetic control of defense signaling and priming in plants. *Front. Plant Sci.* **2016**, *7*, 1201. [CrossRef]

109. Luna, E.; Bruce, T.J.; Roberts, M.R.; Flors, V.; Ton, J. Next-generation systemic acquired resistance. *Plant Physiol.* **2012**, *158*, 844–853. [CrossRef] [PubMed]
110. Slaughter, A.; Daniel, X.; Flors, V.; Luna, E.; Hohn, B.; Mauch-Mani, B. Descendants of primed Arabidopsis plants exhibit resistance to biotic stress. *Plant Physiol.* **2012**, *158*, 835–843. [CrossRef]
111. Ramírez-Carrasco, G.; Martínez-Aguilar, K.; Alvarez-Venegas, R. Transgenerational defense priming for crop protection against plant pathogens: A hypothesis. *Front. Plant Sci.* **2017**, *8*, 696. [CrossRef]
112. De Moraes, C.M.; Lewis, W.; Pare, P.; Alborn, H.; Tumlinson, J. Herbivore-infested plants selectively attract parasitoids. *Nature* **1998**, *393*, 570. [CrossRef]
113. Turlings, T.C.; Wäckers, F.L.; Vet, L.E.; Lewis, W.J.; Tumlinson, J.H. Learning of host-finding cues by hymenopterous parasitoids. In *Insect Learning*; Springer: Berlin/Heidelberg, Germany, 1993; pp. 51–78.
114. Vet, L.E.; Lewis, W.J.; Carde, R.T. Parasitoid foraging and learning. In *Chemical Ecology of Insects 2*; Springer: Berlin/Heidelberg, Germany, 1995; pp. 65–101.
115. Vlot, A.C.; Klessig, D.F.; Park, S.W. Systemic acquired resistance: The elusive signals. *Curr. Opin. Plant Biol.* **2008**, *11*, 436–442. [CrossRef]
116. Dicke, M.; Sabelis, M.W.; Takabayashi, J.; Bruin, J.; Posthumus, M.A. Plant strategies of manipulating predatorprey interactions through allelochemicals: Prospects for application in pest control. *J. Chem. Ecol.* **1990**, *16*, 3091–3118. [CrossRef]
117. Ozawa, R.; Shimoda, T.; Kawaguchi, M.; Arimura, G.i.; Horiuchi, J.i.; Nishioka, T.; Takabayashi, J. Lotus japonicus infested with herbivorous mites emits volatile compounds that attract predatory mites. *J. Plant Res.* **2000**, *113*, 427–433. [CrossRef]
118. Dicke, M.; Takabayashi, J.; Posthumus, M.A.; Schütte, C.; Krips, O.E. Plant—Phytoseiid interactions mediated by herbivore-induced plant volatiles: Variation in production of cues and in responses of predatory mites. *Exp. Appl. Acarol.* **1998**, *22*, 311–333. [CrossRef]
119. Agrawal, A.A.; Janssen, A.; Bruin, J.; Posthumus, M.A.; Sabelis, M.W. An ecological cost of plant defence: Attractiveness of bitter cucumber plants to natural enemies of herbivores. *Ecol. Lett.* **2002**, *5*, 377–385. [CrossRef]
120. Geervliet, J.B.; Posthumus, M.A.; Vet, L.E.; Dicke, M. Comparative analysis of headspace volatiles from different caterpillar-infested or uninfested food plants of Pieris species. *J. Chem. Ecol.* **1997**, *23*, 2935–2954. [CrossRef]
121. Scutareanu, P.; Drukker, B.; Bruin, J.; Posthumus, M.A.; Sabelis, M.W. Volatiles from Psylla-infested pear trees and their possible involvement in attraction of anthocorid predators. *J. Chem. Ecol.* **1997**, *23*, 2241–2260. [CrossRef]
122. Campbell, C.; Pettersson, J.; Pickett, J.; Wadhams, L.; Woodcock, C. Spring migration of damson-hop aphid, Phorodon humuli (Homoptera, Aphididae), and summer host plant-derived semiochemicals released on feeding. *J. Chem. Ecol.* **1993**, *19*, 1569–1576. [CrossRef] [PubMed]
123. Glinwood, R.; Pettersson, J. Host choice and host leaving in Rhopalosiphum padi (Hemiptera: Aphididae) emigrants and repellency of aphid colonies on the winter host. *Bull. Entomol. Res.* **2000**, *90*, 57–61. [CrossRef]
124. Kessler, A.; Baldwin, I.T. Defensive function of herbivore-induced plant volatile emissions in nature. *Science* **2001**, *291*, 2141–2144. [CrossRef]
125. Snoeren, T.A.; Mumm, R.; Poelman, E.H.; Yang, Y.; Pichersky, E.; Dicke, M. The herbivore-induced plant volatile methyl salicylate negatively affects attraction of the parasitoid Diadegma semiclausum. *J. Chem. Ecol.* **2010**, *36*, 479–489. [CrossRef]
126. James, D.G. Field evaluation of herbivore-induced plant volatiles as attractants for beneficial insects: Methyl salicylate and the green lacewing, Chrysopa nigricornis. *J. Chem. Ecol.* **2003**, *29*, 1601–1609. [CrossRef]
127. De Boer, J.G.; Dicke, M. The role of methyl salicylate in prey searching behavior of the predatory mite Phytoseiulus persimilis. *J. Chem. Ecol.* **2004**, *30*, 255–271. [CrossRef]
128. Mallinger, R.E.; Hogg, D.B.; Gratton, C. Methyl salicylate attracts natural enemies and reduces populations of soybean aphids (Hemiptera: Aphididae) in soybean agroecosystems. *J. Econ. Entomol.* **2011**, *104*, 115–124. [CrossRef]
129. James, D.G. Further field evaluation of synthetic herbivore-induced plan volatiles as attractants for beneficial insects. *J. Chem. Ecol.* **2005**, *31*, 481–495. [CrossRef]
130. James, D.G.; Price, T.S. Field-testing of methyl salicylate for recruitment and retention of beneficial insects in grapes and hops. *J. Chem. Ecol.* **2004**, *30*, 1613–1628. [CrossRef]

131. Lee, J.C. Effect of methyl salicylate-based lures on beneficial and pest arthropods in strawberry. *Environ. Entomol.* **2010**, *39*, 653–660. [CrossRef]
132. Filgueiras, C.C.; Willett, D.S.; Pereira, R.V.; Junior, A.M.; Pareja, M.; Duncan, L.W. Eliciting maize defense pathways aboveground attracts belowground biocontrol agents. *Sci. Rep.* **2016**, *6*, 36484. [CrossRef] [PubMed]
133. Filgueiras, C.C.; Willett, D.S.; Junior, A.M.; Pareja, M.; El Borai, F.; Dickson, D.W.; Stelinski, L.L.; Duncan, L.W. Stimulation of the salicylic acid pathway aboveground recruits entomopathogenic nematodes belowground. *PLoS ONE* **2016**, *11*, e0154712. [CrossRef] [PubMed]
134. van Dam, N.M. Belowground herbivory and plant defenses. *Annu. Rev. Ecol. Evol. Syst.* **2009**, *40*, 373–391. [CrossRef]
135. Rasmann, S.; Hiltpold, I.; Ali, J. The role of root-produced volatile secondary metabolites in mediating soil interactions. In *Advances in Selected Plant Physiology Aspects*; IntechOpen: Rijeka, Croatia, 2012.
136. van Dam, N.M.; Bouwmeester, H.J. Metabolomics in the rhizosphere: Tapping into belowground chemical communication. *Trends Plant Sci.* **2016**, *21*, 256–265. [CrossRef]
137. Aratchige, N.; Lesna, I.; Sabelis, M. Below-ground plant parts emit herbivore-induced volatiles: Olfactory responses of a predatory mite to tulip bulbs infested by rust mites. *Exp. Appl. Acarol.* **2004**, *33*, 21–30. [CrossRef]
138. Ali, J.G.; Alborn, H.T.; Stelinski, L.L. Subterranean herbivore-induced volatiles released by citrus roots upon feeding by Diaprepes abbreviatus recruit entomopathogenic nematodes. *J. Chem. Ecol.* **2010**, *36*, 361–368. [CrossRef]
139. Van Tol, R.W.; Van Der Sommen, A.T.; Boff, M.I.; Van Bezooijen, J.; Sabelis, M.W.; Smits, P.H. Plants protect their roots by alerting the enemies of grubs. *Ecol. Lett.* **2001**, *4*, 292–294. [CrossRef]
140. Rasmann, S.; Köllner, T.G.; Degenhardt, J.; Hiltpold, I.; Toepfer, S.; Kuhlmann, U.; Gershenzon, J.; Turlings, T.C. Recruitment of entomopathogenic nematodes by insect-damaged maize roots. *Nature* **2005**, *434*, 732. [CrossRef]
141. Kaya, H.K.; Gaugler, R. Entomopathogenic nematodes. *Annu. Rev. Entomol.* **1993**, *38*, 181–206. [CrossRef]
142. Ali, J.G.; Alborn, H.T.; Campos-Herrera, R.; Kaplan, F.; Duncan, L.W.; Rodriguez-Saona, C.; Koppenhöfer, A.M.; Stelinski, L.L. Subterranean, herbivore-induced plant volatile increases biological control activity of multiple beneficial nematode species in distinct habitats. *PLoS ONE* **2012**, *7*, e38146. [CrossRef] [PubMed]
143. Degenhardt, J.; Hiltpold, I.; Köllner, T.G.; Frey, M.; Gierl, A.; Gershenzon, J.; Hibbard, B.E.; Ellersieck, M.R.; Turlings, T.C. Restoring a maize root signal that attracts insect-killing nematodes to control a major pest. *Proc. Natl. Acad. Sci. USA* **2009**, *106*, 13213–13218. [CrossRef] [PubMed]
144. Willett, D.S.; Alborn, H.T.; Stelinski, L.L. Multitrophic effects of belowground parasitoid learning. *Sci. Rep.* **2017**, *7*, 2067. [CrossRef] [PubMed]
145. Filgueiras, C.C.; Willett, D.S.; Pereira, R.V.; Sabino, P.H.d.S.; Junior, A.M.; Pareja, M.; Dickson, D.W. Parameters affecting plant defense pathway mediated recruitment of entomopathogenic nematodes. *Biocontrol Sci. Technol.* **2017**, *27*, 833–843. [CrossRef]
146. Neveu, N.; Grandgirard, J.; Nenon, J.; Cortesero, A. Systemic release of herbivore-induced plant volatiles by turnips infested by concealed root-feeding larvae Delia radicum L. *J. Chem. Ecol.* **2002**, *28*, 1717–1732. [CrossRef]
147. Faize, L.; Faize, M. Functional analogues of salicylic acid and their use in crop protection. *Agronomy* **2018**, *8*, 5. [CrossRef]
148. Tripathi, D.; Raikhy, G.; Kumar, D. Chemical elicitors of systemic acquired resistance-salicylic acid and its functional analogs. *Curr. Plant Biol.* **2019**, *17*, 48–59. [CrossRef]
149. Song, G.C.; Ryu, C.M. Evidence for Volatile Memory in Plants: Boosting Defence Priming through the Recurrent Application of Plant Volatiles. *Mol. Cells* **2018**, *41*, 724.
150. Wenig, M.; Ghirardo, A.; Sales, J.H.; Pabst, E.S.; Breitenbach, H.H.; Antritter, F.; Weber, B.; Lange, B.; Lenk, M.; Cameron, R.K.; et al. Systemic acquired resistance networks amplify airborne defense cues. *Nat. Commun.* **2019**, *10*, 3813. [CrossRef]
151. Jaskiewicz, M.; Conrath, U.; Peterhänsel, C. Chromatin modification acts as a memory for systemic acquired resistance in the plant stress response. *EMBO Rep.* **2011**, *12*, 50–55. [CrossRef]

152. Ishii, H.; Tomita, Y.; Horio, T.; Narusaka, Y.; Nakazawa, Y.; Nishimura, K.; Iwamoto, S. Induced resistance of acibenzolar-S-methyl (CGA 245704) to cucumber and Japanese pear diseases. *Eur. J. Plant Pathol.* **1999**, *105*, 77–85. [CrossRef]
153. Zhang, S.; Reddy, M.; Kokalis-Burelle, N.; Wells, L.W.; Nightengale, S.P.; Kloepper, J.W. Lack of induced systemic resistance in peanut to late leaf spot disease by plant growth-promoting rhizobacteria and chemical elicitors. *Plant Dis.* **2001**, *85*, 879–884. [CrossRef]
154. Kouzai, Y.; Noutoshi, Y.; Inoue, K.; Shimizu, M.; Onda, Y.; Mochida, K. Benzothiadiazole, a plant defense inducer, negatively regulates sheath blight resistance in Brachypodium distachyon. *Sci. Rep.* **2018**, *8*, 17358. [CrossRef]
155. Thaler, J.S.; Fidantsef, A.L.; Duffey, S.S.; Bostock, R.M. Trade-offs in plant defense against pathogens and herbivores: A field demonstration of chemical elicitors of induced resistance. *J. Chem. Ecol.* **1999**, *25*, 1597–1609. [CrossRef]
156. Ning, Y.; Liu, W.; Wang, G.L. Balancing immunity and yield in crop plants. *Trends Plant Sci.* **2017**, *22*, 1069–1079. [CrossRef]
157. Milla, R.; Osborne, C.P.; Turcotte, M.M.; Violle, C. Plant domestication through an ecological lens. *Trends Ecol. Evol.* **2015**, *30*, 463–469. [CrossRef]
158. Meyer, R.S.; DuVal, A.E.; Jensen, H.R. Patterns and processes in crop domestication: An historical review and quantitative analysis of 203 global food crops. *New Phytol.* **2012**, *196*, 29–48. [CrossRef]
159. Li, Y.; Huang, F.; Lu, Y.; Shi, Y.; Zhang, M.; Fan, J.; Wang, W. Mechanism of plant–microbe interaction and its utilization in disease-resistance breeding for modern agriculture. *Physiol. Mol. Plant Pathol.* **2013**, *83*, 51–58. [CrossRef]
160. Xu, G.; Greene, G.H.; Yoo, H.; Liu, L.; Marqués, J.; Motley, J.; Dong, X. Global translational reprogramming is a fundamental layer of immune regulation in plants. *Nature* **2017**, *545*, 487. [CrossRef]
161. Xu, G.; Yuan, M.; Ai, C.; Liu, L.; Zhuang, E.; Karapetyan, S.; Wang, S.; Dong, X. uORF-mediated translation allows engineered plant disease resistance without fitness costs. *Nature* **2017**, *545*, 491. [CrossRef]
162. Hajek, A.E.; Eilenberg, J. *Natural Enemies: An Introduction to Biological Control*; Cambridge University Press: Cambridge, UK, 2018.

International Journal of
Molecular Sciences

MDPI

Article

Involvement of Salicylic Acid in Anthracnose Infection in Tea Plants Revealed by Transcriptome Profiling

Yun-Long Shi [1], Yue-Yue Sheng [1], Zhuo-Yu Cai [1], Rui Yang [1], Qing-Sheng Li [1], Xu-Min Li [1], Da Li [1], Xiao-Yuan Guo [1], Jian-Liang Lu [1], Jian-Hui Ye [1], Kai-Rong Wang [2], Long-Jie Zhang [2], Yue-Rong Liang [1,]* and Xin-Qiang Zheng [1,]*

1 Tea Research Institute, Zhejiang University, # 866 Yuhangtang Road, Hangzhou 310058, China; 11516051@zju.edu.cn (Y.-L.S.); 21816156@zju.edu.cn (Y.-Y.S.); 21716160@zju.edu.cn (Z.-Y.C.); kexiaonao@163.com (R.Y.); qsli@zju.edu.cn (Q.-S.L.); 21616096@zju.edu.cn (X.-M.L.); 11816014@zju.edu.cn (D.L.); GRACE_XY@126.com (X.-Y.G.); jllu@zju.edu.cn (J.-L.L.); jianhuiye@zju.edu.cn (J.-H.Y.)

2 Ningbo Huangjinyun Tea Science and Technology Co. Ltd., Yuyao 315412, China; wkrtea321hjytea@163.com (K.-R.W.); zhanglongjie0701@163.com (L.-J.Z.)

* Correspondence: yrliang@zju.edu.cn (Y.-R.L.); xqzheng@zju.edu.cn (X.-Q.Z.); Tel.: +86-571-8898-2704 (Y.-R.L. & X.-Q.Z.)

Received: 9 April 2019; Accepted: 16 May 2019; Published: 17 May 2019

Abstract: Anthracnose is a major leaf disease in tea plant induced by *Colletotrichum*, which has led to substantial losses in yield and quality of tea. The molecular mechanism with regards to responses or resistance to anthracnose in tea remains unclear. A de novo transcriptome assembly dataset was generated from healthy and anthracnose-infected leaves on tea cultivars "Longjing-43" (LJ43) and "Zhenong-139" (ZN139), with 381.52 million pair-end reads, encompassing 47.78 billion bases. The unigenes were annotated versus Gene Ontology (GO), Kyoto Encyclopedia of Genes and Genomes (KEGG), National Center for Biotechnology Information (NCBI) non-redundant protein sequences (Nr), evolutionary genealogy of genes: Non-supervised Orthologous Groups (eggNOG) and Swiss-prot. The number of differential expression genes (DEGs) detected between healthy and infected leaves was 1621 in LJ43 and 3089 in ZN139. The GO and KEGG enrichment analysis revealed that the DEGs were highly enriched in catalytic activity, oxidation-reduction, cell-wall reinforcement, plant hormone signal transduction and plant-pathogen interaction. Further studies by quantitative reverse transcriptase polymerase chain reaction (qRT-PCR) and high-performance liquid chromatography (HPLC) showed that expression of genes involved in endogenous salicylic acid biosynthesis and also accumulation of foliar salicylic acid are involved in the response of tea plant to anthracnose infection. This study firstly provided novel insight in salicylic acid acting as a key compound in the responses of tea plant to anthracnose disease. The transcriptome dataset in this study will facilitate to profile gene expression and metabolic networks associated with tea plant immunity against anthracnose.

Keywords: anthracnose; tea plant; salicylic acid; *Camellia sinensis*; plant immunity; RNA sequencing; PR1; plant-pathogen interaction

---

1. Introduction

Anthracnose is a key leaf disease in the tea plant (*Camellia sinensis* (L.) O. Kuntze) induced by *Colletotrichums* [1,2], which has caused a great loss in tea yield and quality owing to its strong infectivity and widespread distribution [3]. Although there have been hundreds of *Colletotrichum* isolates detected in tea plant [4,5], the dominant species or major pathogen on the tea plant are considered to be

Colletotrichum gloeosporioidesis, *C. camelliae* and *C. fructicola* [3–5], which grow and spread quickly in humid and hot summer and autumn seasons, leading to gray sunken or shrunk necrotic lesions on tea leaves and twigs [6,7]. The loss of tea yield induced by anthracnose was estimated to range from 5% to 20% or even more, depending on the cultivar planted [8,9]. Although plant activators [10], antagonistic bacteria [11] and agronomic techniques such as leaf trimming and rational fertilization have been used to control the anthracnose disease, spraying fungicides is still a major control method in tea fields, resulting in a serious fungicides residue problem. Breeding tea cultivars resistant to anthracnose is considered to be the most effective measure to control anthracnose because there is a great difference in resistance to this disease between various tea cultivars [6,8,9].

Revealing molecular defense responses of tea plant to anthracnose infection and developing a molecular assisted selection (MAS) method are badly needed in the tea plant breeding field. However, little is known about the molecular mechanisms regulating the defense response in tea plants [6] although attempts were made to probe this. Transcriptional analysis and histochemistry revealed that the hypersensitive response (HR) and H_2O_2 play critical roles in tea plant defense response to *C. fructicola* [12], and chemical changes of caffeine is considered to be associated with tea–fungi interaction [13]. Nonpathogenic species of *Colletotrichum* was more vulnerable to catechins and caffeine, and differentiation in secondary metabolites might be an important factor leading to the difference in pathogenicity between cultivars [14]. Different communities of *Colletotrichum* with little variability within internal transcribed spacer (ITS) and glyceraldehyde 3-phosphate dehydrogenase (*GAPDH*) regions in the DNA sequence have different virulence [15].

Generally, it was reported that salicylic acid (SA), a secondary metabolite belonging to phenolic compounds, acts as a plant signal or hormone with a regulatory role in a variety of physiological processes under abiotic and biotic stresses, especially as a module in activating disease resistance [16–18]. The endogenous SA concentration is related to the conditions of stresses. For instance, SA accumulation, which is responsible for the phenylpropanoid synthesis pathway, was observed in wheat leaves after heavy metal cadmium treatment [19]. In tobacco (*Nicotiana tabacum* L. cv. *Xanthinc*), a transient sharp increase in SA induced by ultraviolet UV light, ozone and mosaic virus parallelly led to pathogenesis-related proteins accumulation and enhanced disease resistance [20]. The *Arabidopsis thaliana* mutants of *sid* and *pad4*, which are involved in synthesis of SA and camalexin (a kind of phytoalexin), were defective in SA synthesis, resulting in them being more susceptible to pathogens [21,22]. The defense compounds including pathogenesis-related proteins (PR-1, PR-2, PR-5), camalexin and H_2O_2 could be induced by endogenous SA [22–24]. Transgenic plants with salicylate hydroxylase, which converts SA to other compounds, accumulated almost no SA and lost the resistance to virus [25]. When blocked by an inhibitor of phenylpropanoid (precursor of SA) synthesis, the exogenous SA recovered the ability to plant resistance [23]. All these show that SA indeed plays the key role and an important signaling effect in plant resistance, especially in systemic acquired resistance [25,26]. Due to the fact that there has been no study revealing the relationship of SA to the response of tea plant to diseases, it will be interesting to reveal the responses of SA biosynthesis genes and molecular function of SA in the defense response of tea plant to anthracnose infection.

RNA-seq can be focused on assessing the degree of RNA processing and the types of RNA modification, both of which may play an important role in the disease infection process [27,28]. Information of RNA-seq can be used to profile gene expression levels and to reveal the genes involved in plant defense responses [29,30]. RNA-seq analysis needs no reference genome [31] even though tea plant genome data has been available [32]. Developing a transcriptome assembly dataset generating from anthracnose infected tea leaves will be helpful to reveal the molecular defense responses of tea plant to the anthracnose and also to mine molecular markers for MAS used in resistance breeding.

In the present study, transcriptome sequencing by Trinity was performed on the platform of Illumina HiSeq to construct a de novo transcriptome assembly database generated from anthracnose infected tea leaves of two susceptible tea cultivars "Longjing 43" (LJ43) and "Zhenong 139" (ZN139), in which the unigenes were generated and annotated, and important functional genes and metabolic pathways were also revealed. The data sets will provide references for further study on gene expression profiles, biochemical processes and regulation networks associated with tea plant immunity against anthracnose.

2. Results

2.1. Symptoms of Infected Tea Plants

We inspected the typical anthracnose symptoms in *Colletotrichum* susceptible tea cultivars "Longjing-43" (LJ43) and "Zhenong-139" (ZN139). The infected leaves were observed to be gray sunken or shrunk necrotic lesions (Figure 1a). In the late stage, the pathogens made the leaves partially withered, fragile, or easily broken, compared to the healthy leaf (Figure 1a left). The infected leaves of both cultivars "LJ43" and "ZN139" showed the same symptoms and the major pathogen isolated from the infected leaves was *Colletotrichum gloeosporioides*. A detailed view of anthracnose-infected tea plant occurrence in tea fields as well as sample preparation and experiment scheme are shown in Figure 1.

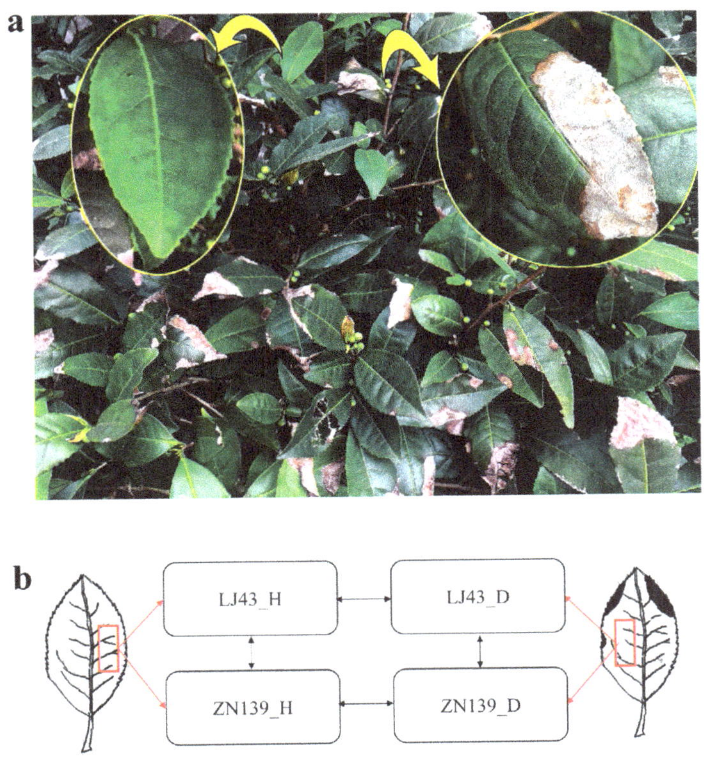

Figure 1. Sample preparation and experiment scheme. (**a**) Anthracnose of tea plant occurrence in tea field. Yellow arrow represents healthy leaf (left) and partially enlarged drawing of anthracnose infected leaf (right). (**b**) A schematic diagram of a healthy leaf (left) and an infected leaf (right). The leaf slices were cut for RNA isolation as described in red boxes. Pairwise comparisons for LJ43 and ZN139 were conducted by arrows. Horizontal arrows mean comparisons between healthy leaves (LJ43_H and ZN139_H) and anthracnose disease infected leaves (LJ43_D & ZN139_D) within cultivar, and the vertical arrow indicates comparisons between cultivars.

2.2. *Transcriptome Profiling*

2.2.1. Sequencing Quality, Assembly Characterization and Functional Annotation

Transcriptome sequencing was performed using a *de novo* assembly (Figure 2) to reveal molecular resistance or defense information in the host tea plant. Based on Illumina sequencing, we generated 381.52 million pair-end reads encompassed 47.78 billion bases, in total, from the two cultivars with two biological replicants. The percentage of N (ambiguous bases), Q20 (reads with mean error rate <1%), Q30 (reads with mean error rate < 0.1%), GC (guanine-cytosine content) were 0.00%, 95%, 89%, 48.5%, respectively. After quality filtering, about 98% clean reads from the raw reads in each sample were reserved for a de novo assembly (Table 1).

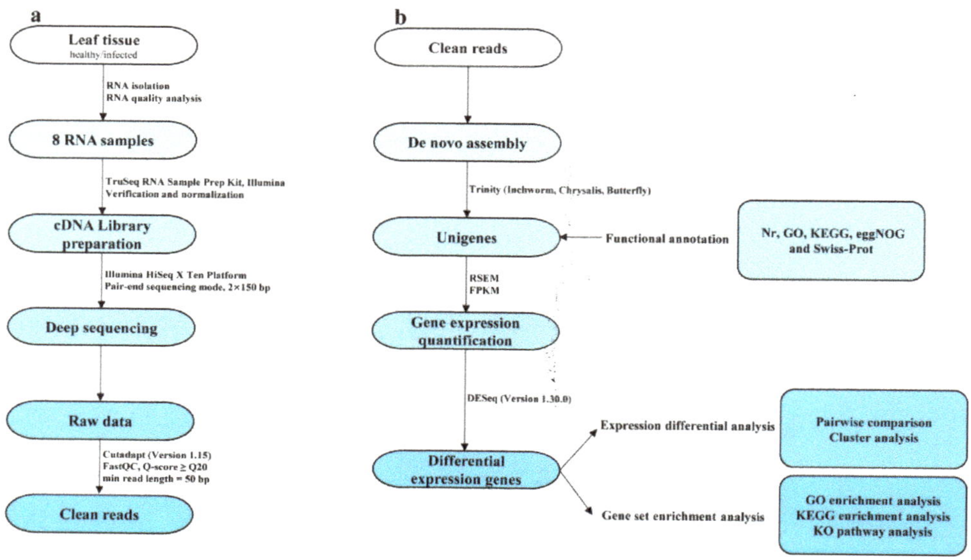

Figure 2. Overview of RNA-seq analysis workflows. (**a**) Acquisition of high-quality sequence (clean reads) from 8 normalized RNA samples. (**b**) The protocol of de novo assembly by Trinity, annotation and downstream analysis for differential expression genes (DEGs).

Table 1. Summary statistics of raw reads and clean reads.

Sample Name	LJ43_H	LJ43_D	ZN139_H	ZN139_D	All Samples
Raw Reads Number (million)	38.86 ± 1.19	42.94 ± 0.03	40.06 ± 3.52	37.40 ± 2.27	318.52
Total Bases (billion bp)	5.83 ± 0.18	6.44 ± 0.00	6.01 ± 0.53	5.61 ± 0.34	47.78
N (%)	0.00 ± 0.00	0.00 ± 0.00	0.00 ± 0.00	0.00 ± 0.00	
Q20 (%)	95.29 ± 0.38	95.43 ± 0.13	95.06 ± 0.24	95.32 ± 0.32	
Q30 (%)	89.38 ± 0.69	89.67 ± 0.27	88.89 ± 0.40	89.47 ± 0.57	
GC (%)	48.72 ± 0.01	49.07 ± 0.13	48.79 ± 0.30	48.41 ± 0.72	
Clean Reads (%)	98.04 ± 0.28	98.11 ± 0.05	97.99 ± 0.17	97.99 ± 0.25	

After assembling, 335,186 contigs and 109,316 unigenes were generated from cultivar LJ43, while 352,038 contigs and 115,953 unigenes were generated from cultivar ZN139. Details of total length, mean length, N50 (length-weighted median, i.e., 50% contig length), N50 sequence No. (the number of sequences longer than N50), N90 (length-weighted 90% contig length) and N90 sequence No. (the number of sequences longer than N90) were listed in Table 2.

Table 2. Information of assembled contigs, unigenes and annotations for two cultivars.

Cultivar		LJ43	ZN139
Contig	Total Length (bp)	102,521,451	107,758,129
	Sequence Number	335,186	352,038
	Mean Length (bp)	306	306
	N50 (bp)	404	405
	N50 Sequence No.	54,939	58,050
	N90 (bp)	150	150
	N90 Sequence No.	251,498	264,172
Unigene	Total Length (bp)	65,163,329	68,536,138
	Sequence Number	109,316	115,953
	Mean Length (bp)	596	591
	N50 (bp)	821	803
	N50 Sequence No.	20,020	21,541
	N90 (bp)	261	260
	N90 Sequence No.	81,051	86,082
Annotated Unigene Number in Database	Nr	45,230 (41.38%)	46,383 (40%)
	GO	24,892 (22.77%)	25,168 (21.71%)
	KEGG	5799 (5.3%)	5873 (5.06%)
	eggNOG	43,025 (39.36%)	44,067 (38%)
	Swiss-Prot	33,895 (31.01%)	34,860 (30.06%)
	In all database	4605 (4.21%)	4677 (4.03%)

All unigene sequences were aligned against the five databases, i.e., GO (Gene Ontology), KEGG (Kyoto Encyclopedia of Genes and Genomes), Nr (NCBI non-redundant protein sequences), eggNOG (evolutionary genealogy of genes: Non-supervised Orthologous Groups) and Swiss-prot (Table 2). In cultivar LJ43, 45,230 unigenes (41.4%) were annotated by Nr, 24,892 unigenes (22.8%) by GO, 5799 unigenes (5.3%) by KEGG, 43,025 unigenes (39.4%) by eggNOG and 33,895 unigenes (31.0%) by Swiss-prot, while in cultivar ZN139, 46,383 unigenes (40.0%) were annotated by Nr, 25,168 unigenes (21.7%) by GO, 5873 unigenes (5.1%) by KEGG, 44,067 unigenes (38.0%) by eggNOG and 34,860 unigenes (30.1%) by Swiss-prot. A total of 4605 unigenes (4.21%) and 4677 unigenes (4.03%) in two cultivars were annotated in all five reference databases.

Homology searching against the Nr database gave the explanations containing species distribution of the top hits, *E*-value distribution and similarity distribution (Figure S1). Importantly, about 20.3% unigenes shared the highest homologies to *Coffea canephora* genes. In the eggNOG analysis, annotated unigenes were classified into 26 items with similar descriptions in both cultivars LJ43 (Figure S2a) and ZN139 (Figure S2b). Based on GO classification, all unigenes were classified to three main categories, i.e., molecular function, cellular component and biological process (Table S1). Catalytic activity (GO: 0003824), binding (GO: 0005488), transporter activity (GO: 0005215) and structural molecule activity (GO: 0005198) were the vital subcategories under the category of molecular function. As for cellular component, cell (GO: 0005623), cell part (GO: 0044464), membrane (GO: 0016020), organelle (GO: 0043226) and membrane part (GO: 0044425) were the primary subcategories. With the respect to biological process category, the top five subcategories were metabolic process (GO: 0008152), cellular process (GO: 0009987), single-organism process (GO: 0044699), biological regulation (GO: 0065007) and localization (GO: 0051179). Moreover, all the annotation results of unigenes for both LJ43 and ZN139 were deposited at the public database Figshare (http://doi.org/10.6084/m9.figshare.7706237.v1, 12 February 2019).

2.2.2. Differential Expression and Enrichment Analysis

Differential expression genes (DEGs) in both cultivars were calculated through anthracnose-infected leaves versus healthy leaves. As shown in Table 3, 1621 DEGs were found in infected leaves of cultivar LJ43, with 1082 up-regulated and 539 down-regulated, while 3089 DEGs were found in infected leaves of cultivar ZN139, with 1527 up-regulated and 1562 down-regulated, compared to the healthy leaves; 755 and 1487 unigenes of the transcripts with GO IDs were mapped by Blast2GO, respectively. Among the DEGs with $p < 0.05$, 7 were annotated by GO, with 4 for biological process and 3 for cellular component in cultivar LJ43 (Figure 3a), while 26 were annotated by GO, with 8 for the biological process, 7 for the cellular component and 11 for the molecular function in cultivar ZN139 (Figure 3b). It was found that catalytic activity, membrane, oxidation-reduction process, cell periphery and carbohydrate metabolic process were strongly responsible to the infection of *Colletotrichum*. Other cellular component categories including the extracellular region, cell wall and external encapsulating structure, as well as the secondary metabolic process were next enriched.

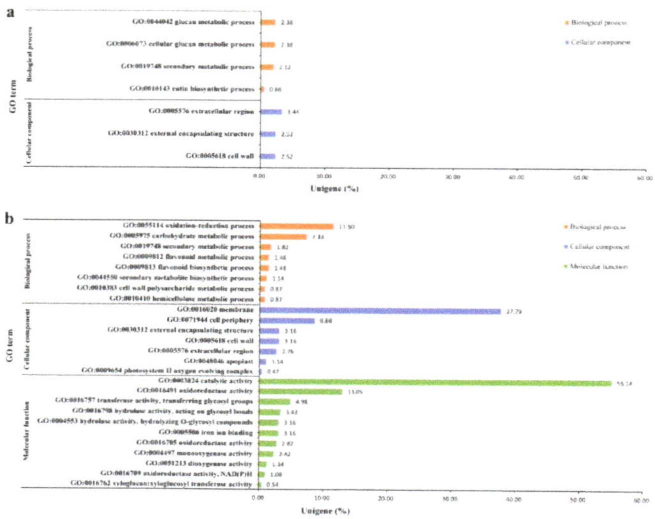

Figure 3. Gene Ontology (GO) enrichment test of cultivars LJ43 (**a**) and ZN139 (**b**). GO terms are divided into three catalogs, i.e., molecular function (green bars), biological processes (orange bars), cellular component (blue bars), which is shown in *Y*-axis. *X*-axis shows the percentage of unigenes in all DEGs aligned to the GO database with Corrected *p*-value (p) < 0.05.

Of course, there are also questions to be explained in the present study. As shown in Figure 3, there was great difference in the GO enrichment between cultivars LJ43 and ZN139, but the exact details leading to the difference remains unknown. It was reported that caffeine and catechins such as (−)-epigallocatechin gallate and (+)-catechin may be involved in the resistance of tea plants to anthracnose [13]. LJ43 has higher level of polyphenols than ZN139 although there are no differences in amino acids and caffeine [33]. Difference in tea polyphenols might be one of the many factors leading to difference in GO enrichment between the two cultivars. Differences in gene abundance might also be partially responsible for the difference in GO enrichment.

As we focused on the top 25 KEGG pathways (Figure 4), it was found that the DEGs were enriched in the pathways involving in plant hormone signal transduction (*ko04075*), plant-pathogen interaction (*ko04626*), starch and sucrose metabolism (*ko00500*) and mineral absorption (*ko04978*) in both cultivars. Besides, flavonoid biosynthesis (*ko00941*), phenylpropanoid biosynthesis (*ko00940*), pentose and glucuronate interconversions (*ko00040*), cutin, suberine and wax (*ko00073*), biosynthesis metabolism of xenobiotics by cytochrome P450 (*ko00980*), and drug metabolism-cytochrome P450 (*ko00982*) also responded to the anthracnose infection (Figure 4).

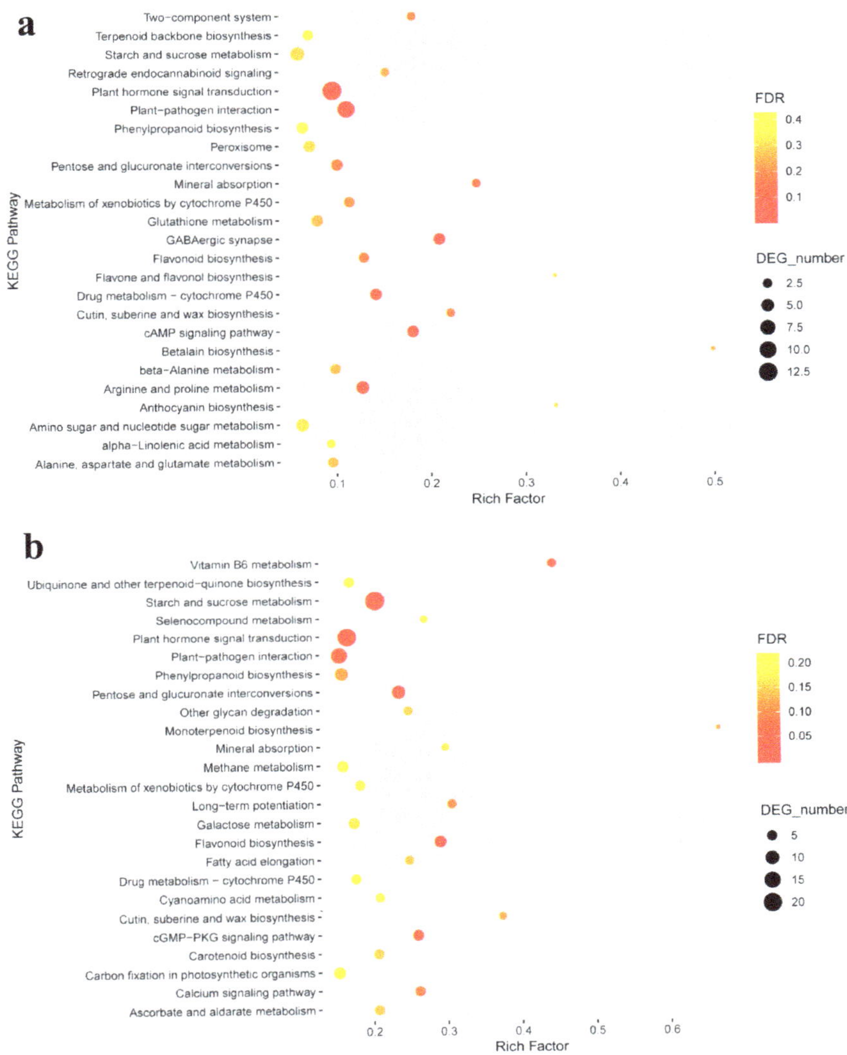

Figure 4. Kyoto Encyclopedia of Genes and Genomes (KEGG) pathway enrichment bubble chart of LJ43 (**a**) and ZN139 (**b**). The Top 25 of pathway enrichment was revealed with low false discovery rate (FDR) (in red colour, significantly) and high FDR (in green colour, insignificantly). *X*-axis shows the percentage of DEGs in total number of genes involved in corresponding KEGG pathway. *Y*-axis expresses KEGG pathway. As well, the size of bubble means the number of DEGs in this item.

Table 3. List of total DEGs in case-control study [1].

Control	Case	Up-Regulated DEGs	Down-Regulated DEGs	Total DEGs
LJ43_H	LJ43_D	1082	539	1621
ZN139_H	ZN139_D	1527	1562	3089

[1] Healthy leaf were identified as control group, and anthracnose infected leaves as case group.

The differential expression and enrichment analysis results were deposited at the public database Figshare (http://doi.org/10.6084/m9.figshare.7706237.v1, 12 February 2019).

2.2.3. Visualization of Two Vital Plant Metabolic Pathways

Pant hormone signal transduction (*ko04075*, Figure 5) and plant-pathogen interaction (*ko04626*, Figure S3) were vital plant metabolic pathways involved in anthracnose infected process. The nomenclature of gene was based on the entry name from KEGG in this paper. The DEGs mapped to these two pathways showed a same trend in both tested tea cultivars. When the two KEGG pathway results were integrated in one figure (Figure 5), it shows that the genes involving in salicylic acid signal transduction pathway, including *NPR1* (nonexpressor of pathogenesis-related gene 1), *TGA* (TGACG motif-binding factor) and pathogenesis-related protein 1 (*PR1*) were upregulated significantly. *PR1* is directly linked to disease resistance. Some plant hormones with signal transduction components, such as auxin (AUX1, AUX/IAA, GH3, SAUR), cytokinine (AHP, A-ARR), gibberellin (GID1, DELLA), jasmonic acid (JAR1) also play an important role in cell enlargement and division, plant growth, induced germination and stress response. As mentioned in pathogen-associated molecular patterns (PAMP) -triggered immunity, RBOH (respiratory burst oxidase), CALM/CLM (calmodulin), WRKY 25/33 and PR1 were found to be upregulated in the anthracnose infected tea leaves. Upstream CNGC (cyclic nucleotide gated channel) involved in calcium ion transferring was downregulated. The changes of gene expression level might be related to HR, cell wall reinforcement, stomatal closure and defense-related gene induction. The expression level of R genes (*RPM1, PBS1, EDS1*) was upregulated, while the downstream HSP90 (heat shock protein 90kDa beta) was downregulated in effector-triggered immunity.

It is reported that system acquired resistance (SAR) is dependent on salicylic acid signaling and systemic expression of *PR* genes. Immune system resistance (ISR) depends on ethylene and jasmonic acid (JA) but is not associated with the expression of the *PR* genes. Both SAR and ISR do result in broad spectrum resistance. Although an antagonistic interaction between SA and JA pathways was revealed, synergistic interactions were also observed [34]. It is considered that there may be a link between JAR1 and SA-related genes. We searched using key words of brassinosteroids, spermine and heat shock transcription factor in the datasets and found that only heat shock transcription factor (HSP90, being down-regulated by infection) contributed to protect from pathogen infection (Figure 5).

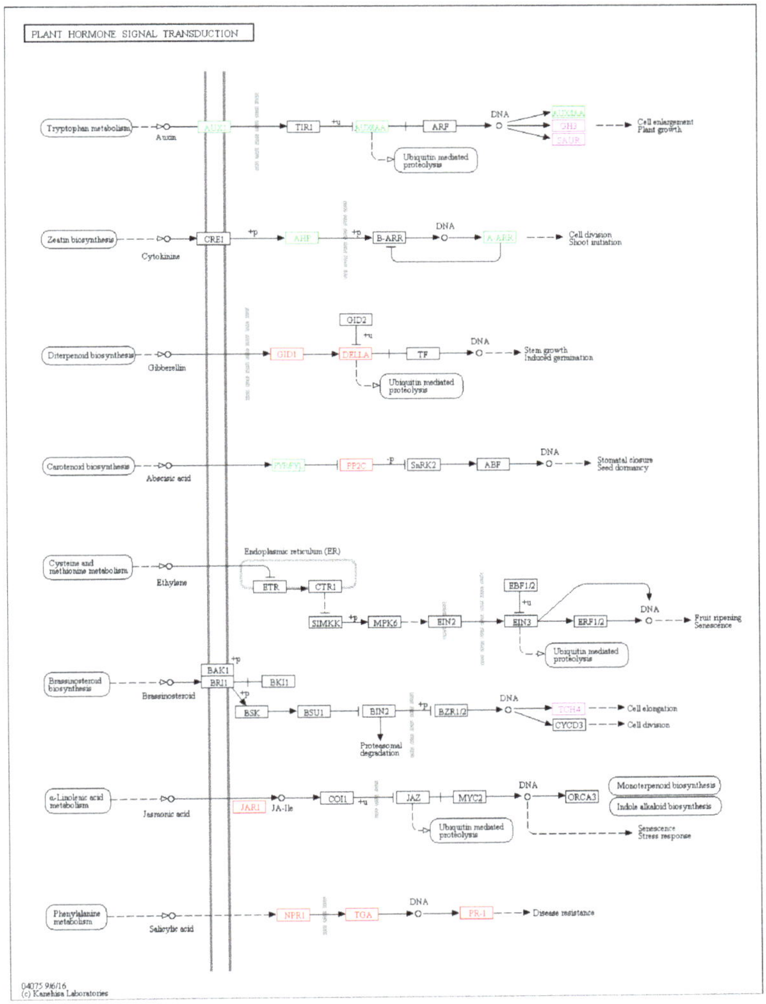

Figure 5. Visualization of plant hormone signal transduction (ko04075) KEGG pathway for two tea cultivars. The DEGs comparisons of anthracnose infected leaves versus healthy leaves was depicted by boxes in different colors, with red colour representing upregulation, green colour representing downregulation and purple colour representing mixture of upregulation and downregulation. Other boxes mean no DEGs mapping to KEGG ontology terms.

2.3. *Quantitative Reverse Transcriptase Polymerase Chain Reaction (qRT-PCR) Validation of Salicylic Acid Signaling Related Genes*

To further verify the sequencing results, expression levels of genes involving in salicylic acid signal transduction pathway in healthy leaves and anthracnose infected leaves of cultivars LJ43 and ZN139 were tested by qRT-PCR (quantitative reverse transcriptase polymerase chain reaction). Genes *ALD1* (AGD2-like defense response gene 1, KEGG Orthology K10206), *TGA2* and *TGA3* (K14431), *PR1* (K13449) in the anthracnose infected leaves showed significant higher expression level than healthy leaves (Figure 6). The expression of *NPR1* (K14508) differentiated between cultivar ZN139 and cultivar LJ43, with upregulation in anthracnose infected leaves of LJ43 but no significant difference between healthy leaves (ZN139_H) and infected leaves (ZN139_D) of cultivar ZN139. *NPR1* is constitutively expressed in many plants and is activated by modification after infection rather than at the transcriptional level. However, our transcriptome dataset shows that the gene encoding predicted regulatory protein NPR3 isoform X1 (LJ69756_g5) was significantly upregulated in the infected leaves, but it was not found in SA pathway. The reason why NPR1 and NPR3 were upregulated during the anthracnose infection remains to be investigated.

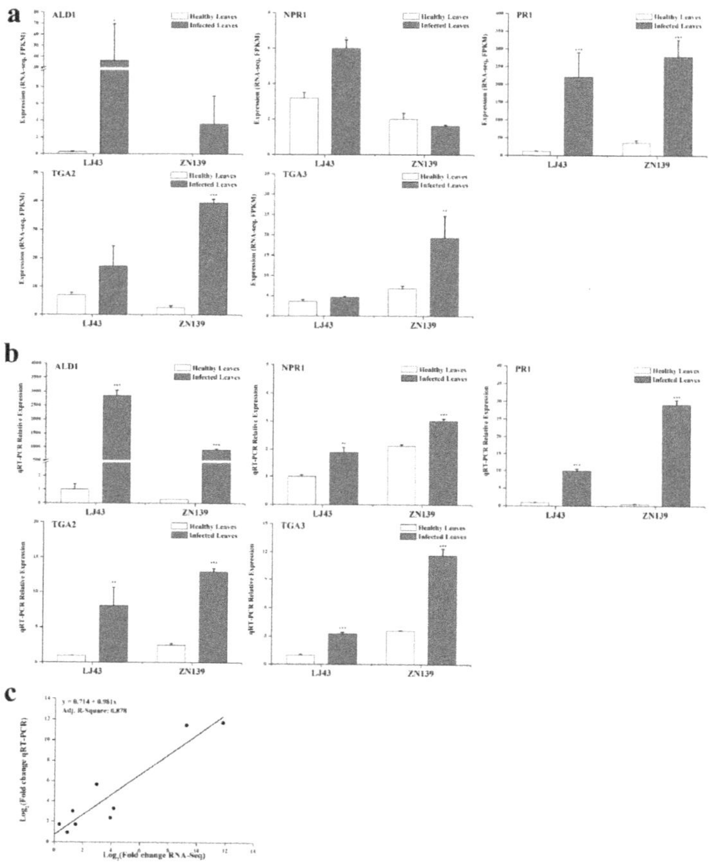

Figure 6. Quantitative reverse transcriptase polymerase chain reaction (qRT-PCR) validation of DEGs and correlation analysis of results between RNA-seq and qRT-PCR. (**a**) FPKM (expected number of Fragments Per Kilobase of transcript sequence per Millions base pairs sequenced) values present RNA-seq expression of 5 DEGs. The sequence of *ALD1* could be found in unigene ID LJ62765_g1 and ZN66762_g1, as *NPR1* in LJ68588_g3 and ZN72573_g8, *TGA2* in LJ58444_g1 and ZN66591_g2, *TGA3* in LJ61836_g1 and ZN62926_g1, *PR1* in LJ62639_g1 and ZN62871_g1. All RNA-seq experiments were performed in two biological replicates (n = 2), statistical significance p of which was generated from DEseq software. (**b**) The relative expression values were firstly normalized using actin as an internal reference, and then made relative to LJ43_H in which the expression value of LJ43_H was fixed as 1. All qRT-PCR experiments were performed in triplicate (n = 3). (**c**) Correlation analysis of results between RNA-seq and qRT-PCR. FPKM values from RNA-seq were compared to relative expression levels detected by qRT-PCR. Data are represented as mean ± standard deviation (SD). Compared with related healthy leaves, * means $p < 0\ 05$, ** means $p < 0\ 01$ and ***means $p < 0\ 001$.

Correlation analysis confirmed that relative expression values generated by the two methods were consistent, with a linear correlation coefficient R-square of 0.878 (Figure 6c), suggesting that endogenous salicylic acid is involved in the response to anthracnose in the tea plant.

2.4. Salicylic Acid Content

Concentrations of free salicylic acid (SA) and bound SA (the inactive storage form 2-O-β-D-glucosylsalicylic acid) [35] were determined by high-performance liquid chromatography (HPLC) (Figure 7) and it showed that levels of free SA, bound SA and total SA in the anthracnose infected leaves were significantly higher than healthy leaves, suggesting that salicylic acid is an important signaling agent responding to anthracnose infection in tea plant.

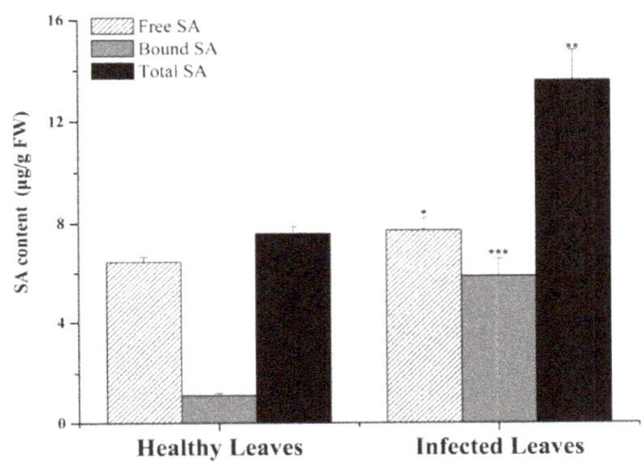

Figure 7. Salicylic acid (SA) content determined by high-performance liquid chromatography (HPLC) in LJ43. All HPLC experiments were performed in triplicate ($n = 3$). Data are represented as means ± SD. Compared with related healthy leaves, * means $p < 0.05$, ** means $p < 0.01$ and *** means $p < 0.001$.

3. Discussion

Profiling plant immune systems and the pathogen molecules as well as their interaction will provide extraordinary insights into molecular recognition and cell biology of plants [36]. Next-generation molecular approaches such as transcriptomics, metabolomics, whole-genome sequencing and proteomics can be used for this purpose [37]. In the present study, a systematic and accurate transcriptome of healthy leaves and anthracnose-infected leaves of two susceptible tea cultivars was profiled, whose dataset will be of significance for further revealing molecular mechanism of tea plant resistance to anthracnose and mining molecular markers for MAS in tea plant breeding.

Host gene expression fluctuates after pathogen infection and the timing of sampling is important for transcriptome study. However, there have been no studies focused on the expression fluctuation of host genes after pathogen infection. Field investigation showed that the latent period after inoculation of anthracnose pathogens was about 2 weeks [38]. Laboratory tests showed that obvious symptom cold be observed and the resistance capacities of various tea cultivars could be identified after two weeks of wound-inoculation of anthracnose pathogens [8]. In Hangzhou, where we sampled the leaves, the rainy season occurs in June, with humid weather during which anthracnose infection usually takes place. The symptom is observed in early July and so we sampled in early July. Also, tea is a self-sterile plant and all tea cultivars we are planting are cross-pollinated hybrids with a genetically complex background. It would be ideal to obtain a resistant mutant from a disease susceptible tea cultivar. Unfortunately, we have not got one. If a gene showed a same response trend to the infection, it is more definite to conclude that the response has the generality among the susceptible cultivars and is more likely to be related to anthracnose infection when two susceptible cultivars were chosen to be tested.

The present transcriptomic profiling revealed that the oxidation-reduction, catalytic activity, cell wall, membrane and carbohydrate metabolic process in GO analysis (Figure 3), as well as the pathways of Ca^{2+} signaling and SA signaling in KEGG analysis, were strongly responsible to anthracnose infection. SA signaling pathways with remarkable downstream defense gene expression are validated by qRT-PCR in this paper.

Catalytic activity and oxidation-reduction commonly exist in many biological processes or reactions. Most plant cells generate reactive oxygen species (ROS) [39], including hydrogen peroxide, singlet oxygen and hydroxyl radical [40] during their interaction with pathogens, for example *Phytophthora infestans* [41] and *Colletotrichum lindemuthianum* [42]. The generation and accumulation of hydrogen peroxide with a membrane-bound NADPH (nicotinamide adenine dinucleotide phosphate) oxidase is a symptom of the oxidative burst [43], which coincides with programmed cell death (PCD) during HR. The ROS intermediates redox signaling and oxidation-reduction in host plant resistance [44]. Previous study also showed that hydrogen peroxide and HR played a crucial role in tea plant defense [12].

The modification and reinforcement of the cell wall (RCW) as well as deposition of papillae containing callose are considered to be a structural barrier, which can block the penetration of fungal pathogens [42,45]. The detection of hydrogen peroxide at subcellular levels showed that the cross-linking between papillae and HR cells help the pathogen arrest by reinforcing the cell wall apposition [46]. A genetic network of cell-wall damage, regulated by lignin biosynthesis, was characterized in *Arabidopsis thaliana (Col)* [47]. Plants may have evolved a system of dynamic cell wall remodeling to prevent diseases [48]. Many researches elucidated that the changing of cell wall is a way to hinder pathogens entering the plant cells. Hence, we speculated that the cell wall and callose, with complex chemical structure and physiological function, have direct antifungal effects on facilitating tea plant defense.

KEGG analysis revealed that many DEGs were associated with several pathways, particularly in plant hormone signal transduction (Figure 5) and plant-pathogen interaction (Figure S3) in both cultivars. As shown in Figure 4 and Figure S3, a defense-related gene, HR and PCD could be induced through pathogen-associated molecular patterns triggered immunity (PTI) and effector-triggered immunity (ETI) [36]. Ca^{2+} is a fateful factor in PTI, resulting in three regulating pathways. One is the ROS-related pathway, in which upstream RBOH with NADPH oxidases [49] is enhanced highly in anthracnose-infected leaves. The generation of ROS acts as a signal to induce HR and RCW. Ca^{2+} has been reported to activate RBOHs, which is directly phosphorylated responsible to PAMPs perception, at the region of cytosolic N-terminal [50,51]. Ectopic and heterologous expression in the potato (*Solanum tuberosum* cv Rishiri) suggests that calcium-dependent protein kinases 5 regulates the ROS production via St RBOHB (RBOH) [52]. Another is the calmodulin-related pathway. The expression of calmodulin may result in RCW and stomatal closure. A study on barley (*mlo* mutations) against mildew showed that plant defense was regulated by Ca^{2+} dependent interaction with calmodulin and mildew resistance locus O protein, a modulator of plant defense and cell death, in early signaling cascades [53]. Calmodulin detection by enzyme-linked immunosorbent assay (ELISA) showed that calmodulin promoted cell wall regeneration and cell division [54]. The third pathway is related to WRKY transcription factor 25/33, whose high expression induces accumulation of downstream defense proteins encoded by gene PR1. The present results demonstrate that Ca^{2+} is a key component during plant-pathogen interactions, like that described in a previous study [55].

As for ETI in Figure S3, specific disease resistance genes (R) recognize the effectors that enable pathogens overcoming PTI and ETI [36]. When tea plant cell comes into contact with effectors delivered by fungus, the expression level of nucleotide binding leucine rich repeat domains (NB-LRR) proteins [56] encoded by R genes (*RPM1, PBS1, EDS1*) increased markedly, resulting in activation of ETI with a threshold of HR, PCD and defense amplification. In contrast, heat shock protein 90 (HSP90) was down-regulated, which was perhaps related to RPM1 function [57]. A study on heat-shock treatment reported that inhibition of HSP90 caused enhancement of disease resistance by released heat shock transcription factor 1 in tomato cultivar *Natsunokoma* [58], which is consistent with our findings.

It is known that SA, JA and ethylene are three key phytohormones responding to biotic stresses in plant. It is suggested that effective defense against biotrophic or hemibiotrophic pathogens is largely due to PCD in the host, and to associated activation of defense responses regulated by the SA-dependent pathway, while necrotrophic pathogens, benefiting from host cell death, are not limited by SA-dependent defenses, but by JA and ethylene-signaling defense responses [59,60]. Figure 5 shows that the remarkable up-regulation of *NPR1, TGA* and *PR1* genes induced by SA signaling finally facilitates the PR proteins associating with disease resistance to *Colletotrichum.* Based on SA signaling, the effect of *Colletotrichum* on tea leaves is likely to be hemibiotrophic. The qRT-PCR testing confirmed that SA signaling was involved in tea resistance to anthracnose, in which notably high expression levels of *NPR1, TGA2, TGA3* and *PR1* in anthracnose infected leaves were detected in both cultivars (Figure 6). The accumulation of the PR1 protein with antifungal activity enhances plant immunity in the vacuole or extracellular space such as cell wall [26,61]. Among these genes, we demonstrated here that PR1 can be used as a marker for SA-dependent gene expression, being consistent with that in *Arabidopsis* [62]. NPR1 is a critical regulator and receptor, directly binding to SA [63,64] and a test on yeast indicated that the TGA family was a link between NPR1 and PR1 expression [65]. However, another study showed that there were more genes in SA signaling, not limited to those we confirmed before, such as a high affinity SA-binding protein (SABP2) [66]. WRKY factors act as repressors or activators in system acquired resistance (SAR) network, which are induced by SA [67,68]. The present HPLC analysis revealed that the levels of free SA (active form), bound SA (inactive form) and total SA in tea leaves were significantly increased by *Colletotrichum* infection (Figure 7). The results are in accordance with those in *Arabidopsis thaliana (npr1* and *NahG)* showing that antimicrobial proteins encoded by the PR1 superfamily were increased with the increase in endogenous SA levels [69], and also are in accordance with those described in two previous studies on tobacco (*Nicotiana tabacum* L. cv. *Xanthinc*) [20]. The increase in various forms of SA suggests that the free SA is directly involved in tea plant resistance to anthracnose infection and the bound SA might be a stored form of SA, with a potential for defense resistance. In addition, the expression of ALD1, which is involved in the lysine metabolic pathway, was extremely enhanced in the anthracnose infected leaves. The previous study on *Arabidopsis* indicated that ALD1 involving in mediation of plant immunity was more similar to SA-triggered immunity [62]. ALD1-triggered pipecolic acid (PA) and N-hydroxypipecolic acid, whose molecular structures are similar to SA, are largely accumulated in the disease infected plant, where they act as a regulator of SAR activated by SA [70]. The cross-links between SA and PA in tea plant defense need to be further investigated. A hypothetical model was clarified in Figure 8 for SA accumulation and related gene expression involving in the SA signaling pathway in tea immunity.

In conclusion, PR1 and endogenous SA, acting as a key compound, clearly play a pivotal role in defense activation of tea immunity to anthracnose. Fully understanding the pathogen molecule pathways and the interaction between *Colletotrichum* and the tea plant will be helpful to develop agricultural measures to control tea plant anthracnose disease and also to mine molecular markers used in MAS. Despite many recent insights into the tea–*Colletotrichum* interactions by several assays, there is still a large gap in knowledge of mechanisms for further study, such as focuses on mapping to the reference genome, transgenic tea verification and field testing on exogenous SA, SA agonist or SA inhibitor. Furthermore, many other related topics mentioned above, such as cross talks among signaling pathways, other plant hormones, metabolic processes, phytoalexins and etc., need to be verified in future work.

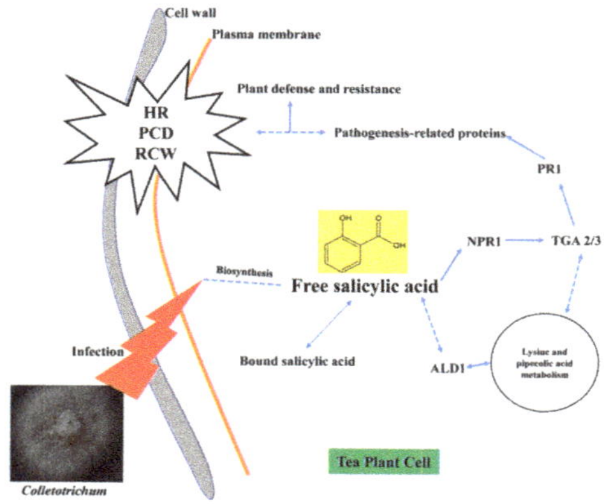

Figure 8. A hypothetical model for SA accumulation and related gene expression in tea plant immunity. Solid lines indicate actual regulations of signaling effect. Dotted lines indicate hypothetical regulations of signaling effect.

4. Materials and Methods

4.1. Plant Material

Tea cultivars "Longjing 43" (LJ43) and "Zhenong 139" (ZN139) which are susceptible to anthracnose were used as plant materials in this study. The tea plants were grown in the Experimental Tea Farm of Zhejiang University (Hangzhou, China, 30.3 N, 120.2 E). Leaf samples (10 leaves each sample) were collected in early July 2017 and the leaf slices were cut as described in Figure 1. The responses of a plant to pathogen infections were induced systemically. The healthy leaves were sampled from the individuals which had not completely infected with any part so as to avoid the effects of systematic responses. The leaves were washed with distilled water before cutting the leaf slices to eliminate contamination of RNA from the pathogens and the leaf slices were frozen in liquid nitrogen immediately and then stored at −80 °C until RNA extraction. The healthy leaf samples from LJ43 and ZN139 were labeled as LJ43_H and ZN139_H, and the anthracnose infected leaf samples from LJ43 and ZN139 were labeled as LJ43_D and ZN139_D, respectively (Figure 1). There were two biological replicates for each treatment in the study of de novo transcriptome assembly, and three biological replicates for the studies of qRT-PCR validation and HPLC analysis of salicylic acid.

4.2. Total RNA Extraction, cDNA Library Construction and Deep Sequencing

Total RNA was extracted using RNAprep Pure Kit for Polysaccharides and Polyphenolics rich plant (Cat. no. DP441, Tiangen Biotech Co., Ltd., Beijing, China) according to the instructions of the kit. Concentration and quality of the RNA were assessed using Agilent 2100 Bioanalyzer (Agilent Technologies Inc., Santa Clara, USA), and those with $1.8 \leq$ OD260 nm/OD280 nm ≤ 2.1 and RIN ≥ 6.5 were used in subsequent tests. RNA integrity was evaluated by agarose gel electrophoresis. Three micrograms of RNA were used as input material for cDNA preparations.

The cDNA library was constructed according to the protocol of TruSeqRNA Sample Prep Kit (Illumina Inco., San Diego, CA, USA). The quality and quantity of libraries were checked by fluorospectrophotometry (Quant-iTPicoGreen dsDNA Assay Kit, Invitrogen, P7589; Quantifluor-ST fluorometer E6090, Promega, CA, USA) and Agilent 2100 Bioanalyzer with Agilent Bioanalyzer High Sensitivity DNA chip Kit (5067-4626, Agilent Technologies Inc., Santa Clara, USA). Each normalized cDNA library (2 nM) was gradually diluted to 4–5 pM and the sequencing was performed on Illumina HiSeq X Ten Platform in 2×150 bp pair-end sequencing mode (Figure 2a) and the original data in FASTQ format (Raw Data) was then generated. The quality assessment of raw data in FASTQ format was done with FastQC [71] (http://www.bioinformatics.babraham.ac.uk/projects/fastqc, 18 January 2018, Q-score ≥ Q20, min read length = 50 bp). The connectors were removed by Cutadapt (Version 1.15) and clean reads were prepared for the further analysis (Figure 2a).

4.3. De Novo Assembly and Functional Annotation

The present tests were carried out in 2017, when the reference genome of tea plant had not been published. Accordingly, de novo assembly was performed to reconstruct the transcriptomes. We used Trinity (Version r20140717, default k-mer = 25 bp) [72,73], consisting of three software modules: Inchworm, Chrysalis and Butterfly, to efficiently process massive clean reads based on robust reconstruction and analysis of De Bruijn Graph (DBG) [72]. Briefly, clean reads with overlap joints were firstly combined to form longer fragments, i.e., contigs. Related pair-end reads and contigs were clustered to form unigenes. Finally, non-redundant unigenes for functional annotation were obtained [73].

The unigene sequences were compared using the local BLASTX [74] program (*E*-value ≤1e-5) against 5 public databases: Nr [75], GO [76], KEGG [77], Swiss-Prot [78] and eggNOG [79]. GO terms were identified on the platform BLAST2GO [80]. KEGG Pathways and KEGG Orthologs were annotated by KAAS [81] (KEGG Automatic Annotation Server, http://www.genome.jp/tools/kaas/, 18 January 2018).

4.4. Differential Expression Genes (DEGs) Analysis

Firstly, the software RSEM [82] was used to compare read count values on each unigene as original expression, and then standardized as FPKM (expected number of fragments per kilobase of transcript sequence per millions of base pairs sequenced). Then DESeq (Version 1.30.0) was used to analyze DEGs [83]. The screen criteria were expression difference multiple |log2 Fold Change| > 1 ($p < 0.05$). The bi-directional clustering analysis of DEGs in different samples was performed using R language. Then we mapped all the DEGs to GO databases to reveal the gene enrichment in each term. Furthermore, the DEGs involved in the KEGG pathways including signaling and metabolic pathways were assigned. All the downstream analysis was described in Figure 2b.

4.5. qRT-PCR Validation

To verify the expression levels of DEGs obtained from RNA-seq, qRT-PCR were carried out following the previous protocol [84]. The primary DEGs involved in salicylic acid signal transduction pathway were chosen for validation and specific primers were designed using the PrimerQuest Tool (https://sg.idtdna.com/PrimerQuest/Home/Index, 10 April 2018–23 April 2018). qRT-PCR was run in the machine StepOne Plus (Applied Biosystems, Foster City, CA, USA). The qRT-PCR test was generated using triplicate samples. Each relative expression level was calculated using the $2^{-\Delta\Delta CT}$ method [85] with *β-actin* gene as internal control.

4.6. Salicylic Acid Extraction and High-Performance Liquid Chromatography (HPLC) Quantification

The extraction of total SA including free SA and bound SA was performed following the published method [35,86]. The cut leaf slices stored at −80 °C were pulverized in a mortar with liquid nitrogen condition using a pestle. About 0.1 g homogenized powders was mixed with 1 mL pre-cooling 90% methanol, stood overnight, and centrifuged at 8000× *g* for 10 min. The supernatants were collected while the sediment was re-suspended in 0.5 mL 90% methanol and further extracted for 2 h. The centrifugation and collection were repeated as before. The combined supernatants were evaporated using a SpeedVac machine at 40 °C to 0.3 mL aqueous solution. 20 μL trichloroacetic acid (1 mg/mL) was added to the aqueous solution and mixed by vortex. Partition with 1 mL ethyl acetate: cyclohexane (1:1, *v*/*v*) was carried out twice. The organic phase with free SA was dried by nitrogen blowing and dissolved in 0.5 mL mobile phase for HPLC test. The aqueous phase containing bound SA was acid-hydrolyze by adding 0.5 mL HCl (8 M) to release free SA. The same steps for obtaining free SA were implemented again. At this moment, the bound SA was determined in the form of free SA. The contents of free SA and bound SA in the samples were then determined by an Agilent

1100 HPLC system with fluorescence detection. Chromatographic separation was performed using a Kromasil C18 reverse phase column (5 μm, 250 mm × 4.6 mm). The mobile phase was a mixture of 1% acetic acid aqueous solution with 1% phosphoric acid: methanol (40:60, *v*/*v*). The flow rate was set at 0.8 mL/min for 50 min in total. The column temperature was set at 35 °C and injection volume was 10 μL. Excitation and emission wavelengths were 294 nm and 426 nm, respectively.

The data of qRT-PCR and HPLC were expressed as mean (mean values of three independent experiments) ± SD (standard deviation). An analysis of variance (ANOVA) test, mainly using least significant difference (LSD), was used for measuring statistical significance by means of SPSS and EXCEL software.

Supplementary Materials: Supplementary materials can be found at http://www.mdpi.com/1422-0067/20/10/2439/s1. Raw data files generated by RNA-seq were deposited in the NCBI Sequence Read Archive SRP162639. The files of unigenes were deposited at GenBank Transcriptome Shotgun Assembly TSA GHEL00000000. The annotation and differential expression analysis results of the unigenes were deposited at Figshare (http://doi.org/10.6084/m9.figshare.7706237.v1, 12 February 2019). Table S1. All subcategories of GO annotations. Table S2. qRT-PCR primers of selected unigenes. Figure S1. Characteristics of homology searching in *Camellia sinensis* against the Nr database. Figure S2. eggNOG functional classification of consensus sequence. Figure S3. Visualization of plant-pathogen interaction (ko04626) KEGG pathway for two tea cultivars.

Author Contributions: Y.-L.S., Y.R.L. and X.-Q.Z conceived the project; Y.-L.S., Y.-Y.S., Z.-Y.C. and X.-Y.G. prepared samples and generated the experiments; R.Y., Q.-S.L., X.-M.L., J.-L.L., J.-H.Y. and D.L. obtained and analyzed the RNA-seq data; Y.-L.S. performed bioinformatics analysis and prepared the manuscript; Z.-Y.C. and R.Y. optimized figures and tables; K.-R.W. and L.-J.Z., field management and disease identification; Y.-R.L. and X.-Q.Z. experimental design, writing and polishing the manuscript. All authors read the final manuscript and approved submission.

Funding: This research was financially supported by grants from the Science Technology Department of Zhejiang Province (Project No. 2016C02053-5) and the Science Technology Bureau of Ningbo City (Project No. 2017C10001).

Acknowledgments: The authors acknowledge the Personal Biotechnology Co., Ltd. (Shanghai, China) for offering the sequencing service.

Conflicts of Interest: All of the authors declare no conflict of interest.

Abbreviations

ALD1	AGD2-like defense response gene 1
DBG	De Bruijn Graph
DEGs	Differential expression genes
eggNOG	Evolutionary genealogy of genes: Non-supervised Orthologous Groups
ETI	Effector-triggered immunity
FDR	False discovery rate
FPKM	Expected number of Fragments Per Kilobase of transcript sequence per Millions base pairs sequenced
GO	Gene Ontology
HPLC	High-performance liquid chromatography
HR	Hypersensitive response
ISR	Immune system resistance
ITS	Internal transcribed spacer
KAAS	KEGG Automatic Annotation Server
KEGG	Kyoto Encyclopedia of Genes and Genome
MAS	Molecular assisted selection
NPR1	Nonexpressor of pathogenesis-related gene 1
Nr	NCBI non-redundant protein sequences
p	*p*-value
PCD	Programmed cell death
PR1	Pathogenesis-related gene 1
PTI	Pathogen-associated molecular patterns triggered immunity
qRT-PCR	Quantitative reverse transcriptase polymerase chain reaction
RCW	Reinforcement of the cell wall
ROS	Reactive oxygen species
SA	Salicylic acid
SAR	System acquired resistance
TGA	Trans-activating TGA Factors

References

1. Crous, P.W.; Gams, W.; Stalpers, J.A.; Robert, V.; Stegehuis, G. MycoBank: An online initiative to launch mycology into the 21st century. *Stud. Mycol.* **2004**, 19–22.
2. O'Connell, R.J.; Thon, M.R.; Hacquard, S.; Amyotte, S.G.; Kleemann, J.; Torres, M.F.; Damm, U.; Buiate, E.A.; Epstein, L.; Alkan, N.; et al. Lifestyle transitions in plant pathogenic *Colletotrichum* fungi deciphered by genome and transcriptome analyses. *Nat. Genet.* **2012**, *44*, 1060. [CrossRef]
3. Cannon, P.F.; Damm, U.; Johnston, P.R.; Weir, B.S. *Colletotrichum*—Current status and future directions. *Stud. Mycol.* **2012**, *73*, 181–213. [CrossRef] [PubMed]
4. Wang, Y.C.; Hao, X.Y.; Wang, L.; Xiao, B.; Wang, X.C.; Yang, Y.J. Diverse *Colletotrichum* species cause anthracnose of tea plants (*Camellia sinensis* (L.) O. Kuntze) in China. *Sci. Rep.* **2016**, *6*, 35287. [CrossRef]
5. Liu, F.; Weir, B.S.; Damm, U.; Crous, P.W.; Wang, Y.; Liu, B.; Wang, M.; Zhang, M.; Cai, L. Unravelling *Colletotrichum* species associated with *Camellia*: Employing ApMat and GS loci to resolve species in the *C. gloeosporioides* complex. *Persoonia* **2015**, *35*, 63–86. [CrossRef]
6. Wang, L.; Wang, Y.C.; Cao, H.L.; Hao, X.Y.; Zeng, J.M.; Yang, Y.J.; Wang, X.C. Transcriptome analysis of an anthracnose-resistant tea plant cultivar reveals genes associated with resistance to *Colletotrichum camelliae*. *PLoS ONE* **2016**, *11*, e0148535. [CrossRef]
7. Guo, M.; Pan, Y.M.; Dai, Y.L.; Gao, Z.M. First report of brown blight disease caused by *Colletotrichum* gloeosporioides on *Camellia sinensis* in Anhui Province, China. *Plant Dis.* **2014**, *98*, 284. [CrossRef] [PubMed]
8. Yoshida, K.; Takeda, Y. Evaluation of anthracnose resistance among tea genetic resources by wound-inoculation assay. *JARQ-Jpn. Agr. Res. Quart.* **2006**, *40*, 379–386. [CrossRef]
9. Yamada, K.; Sonoda, R. A fluorescence microscopic study of the infection process of *Discula theae-sinensis* in Tea. *JARQ-Jpn. Agr. Res. Quart.* **2014**, *48*, 399–402. [CrossRef]
10. Yoshida, K.; Ogino, A.; Yamada, K.; Sonoda, R. Induction of disease resistance in tea (*Camellia sinensis* L.) by plant activators. *JARQ-Jpn. Agr. Res. Quart.* **2010**, *44*, 391–398. [CrossRef]
11. Kim, G.H.; Lim, M.T.; Hur, J.S.; Yum, K.J.; Koh, Y.J. Biological control of tea anthracnose using an antagonistic bacterium of *Bacillus subtilis* isolated from tea leaves. *Plant Pathol. J.* **2009**, *25*, 99–102. [CrossRef]
12. Wang, Y.C.; Hao, X.Y.; Lu, Q.H.; Wang, L.; Qian, W.J.; Li, N.N.; Ding, C.Q.; Wang, X.C.; Yang, Y.J. Transcriptional analysis and histochemistry reveal that hypersensitive cell death and H_2O_2 have crucial roles in the resistance of tea plant (*Camellia sinensis* (L.) O. Kuntze) to anthracnose. *Horticul. Res.* **2018**, *5*, 18. [CrossRef] [PubMed]
13. Wang, Y.C.; Qian, W.J.; Li, N.N.; Hao, X.Y.; Wang, L.; Xiao, B.; Wang, X.C.; Yang, Y.J. Metabolic changes of caffeine in tea plant (*Camellia sinensis* (L.) O. Kuntze) as defense response to *Colletotrichum fructicola*. *J. Agric. Food Chem.* **2016**, *64*, 6685–6693. [CrossRef]
14. Lu, Q.; Wang, Y.; Li, N.; Ni, D.; Yang, Y.; Wang, X. Differences in the characteristics and pathogenicity of *Colletotrichum camelliae* and *C. fructicola* isolated from the tea plant [*Camellia sinensis* (L.) O. Kuntze]. *Front. Microbiol.* **2018**, *9*, 3060. [CrossRef] [PubMed]
15. Chen, Y.J.; Qiao, W.J.; Zeng, L.; Shen, D.H.; Liu, Z.; Wang, X.S.; Tong, H.R. Characterization, pathogenicity, and phylogenetic analyses of *Colletotrichum* species associated with brown blight disease on *Camellia sinensis* in China. *Plant Dis.* **2017**, *101*, 1022–1028. [CrossRef] [PubMed]
16. Vlot, A.C.; Dempsey, D.A.; Klessig, D.F. Salicylic acid, a multifaceted hormone to combat disease. *Annu. Rev. Phytopathol.* **2009**, *47*, 177–206. [CrossRef]
17. An, C.F.; Mou, Z.L. Salicylic acid and its function in plant immunity. *J. Integr. Plant Biol.* **2011**, *53*, 412–428. [CrossRef]
18. Pal, M.; Kovacs, V.; Szalai, G.; Soos, V.; Ma, X.; Liu, H.; Mei, H.; Janda, T. Salicylic acid and abiotic stress responses in rice. *J. Agron. Crop. Sci.* **2014**, *200*, 1–11. [CrossRef]
19. Kovacs, V.; Gondor, O.K.; Szalai, G.; Darko, E.; Majlath, I.; Janda, T.; Pal, M. Synthesis and role of salicylic acid in wheat varieties with different levels of cadmium tolerance. *J. Hazard. Mater.* **2014**, *280*, 12–19. [CrossRef]
20. Yalpani, N.; Enyedi, A.J.; Leon, J.; Raskin, I. Ultraviolet-Light And Ozone Stimulate accumulation of salicylic-acid, pathogenesis-related proteins and virus-resistance in tobacco. *Planta* **1994**, *193*, 372–376. [CrossRef]
21. Zhou, N.; Tootle, T.L.; Tsui, F.; Klessig, D.F.; Glazebrook, J. PAD4 functions upstream from salicylic acid to control defense responses in *Arabidopsis*. *Plant Cell* **1998**, *10*, 1021–1030. [CrossRef] [PubMed]

22. Nawrath, C.; Metraux, J.P. Salicylic acid induction-deficient mutants of *Arabidopsis* express PR-2 and PR-5 and accumulate high levels of camalexin after pathogen inoculation. *Plant Cell* **1999**, *11*, 1393–1404. [CrossRef] [PubMed]
23. Shirasu, K.; Nakajima, H.; Rajasekhar, V.K.; Dixon, R.A.; Lamb, C. Salicylic acid potentiates an agonist-dependent gain control that amplifies pathogen signals in the activation of defense mechanisms. *Plant Cell* **1997**, *9*, 261–270. [CrossRef]
24. Loake, G.; Grant, M. Salicylic acid in plant defence-the players and protagonists. *Curr. Opin. Plant Biol.* **2007**, *10*, 466–472. [CrossRef]
25. Gaffney, T.; Friedrich, L.; Vernooij, B.; Negrotto, D.; Nye, G.; Uknes, S.; Ward, E.; Kessmann, H.; Ryals, J. Requirement of salicylic-acid for the induction of systemic acquired-resistance. *Science* **1993**, *261*, 754–756. [CrossRef]
26. Hammerschmidt, R. Systemic acquired resistance. *Adv. Bot. Res.* **2009**, *51*, 173–222. [CrossRef]
27. Westermann, A.J.; Gorski, S.A.; Vogel, J. Dual RNA-seq of pathogen and host. *Nat. Rev. Microbiol.* **2012**, *10*, 618–630. [CrossRef]
28. Marguerat, S.; Bahler, J. RNA-seq: From technology to biology. *Cell. Mol. Life Sci.* **2010**, *67*, 569–579. [CrossRef]
29. Torres, T.T.; Metta, M.; Ottenwalder, B.; Schlotterer, C. Gene expression profiling by massively parallel sequencing. *Genome Res.* **2008**, *18*, 172–177. [CrossRef]
30. Zhang, J.A.; Liang, S.; Duan, J.L.; Wang, J.; Chen, S.L.; Cheng, Z.S.; Zhang, Q.; Liang, X.Q.; Li, Y.R. De novo assembly and characterisation of the transcriptome during seed development, and generation of genic-SSR markers in peanut (*Arachis hypogaea* L.). *BMC Genom.* **2012**, *13*, 90. [CrossRef]
31. Grabherr, M.G.; Haas, B.J.; Yassour, M.; Levin, J.Z.; Thompson, D.A.; Amit, I.; Adiconis, X.; Fan, L.; Raychowdhury, R.; Zeng, Q.D.; et al. Full-length transcriptome assembly from RNA-Seq data without a reference genome. *Nat. Biotechnol.* **2011**, *29*, 644. [CrossRef] [PubMed]
32. Wei, C.L.; Yang, H.; Wang, S.B.; Zhao, J.; Liu, C.; Gao, L.P.; Xia, E.H.; Lu, Y.; Tai, Y.L.; She, G.B.; et al. Draft genome sequence of *Camellia sinensis* var. *sinensis* provides insights into the evolution of the tea genome and tea quality. *Proc. Natl. Acad. Sci. USA* **2018**, *115*, E4151–E4158. [CrossRef] [PubMed]
33. Yang, Y.J.; Liang, Y.R. *Clonal Tea Cultivars Records in China*; Yang, Y.J., Liang, Y.R., Eds.; Shanghai Scientific and Technical Publishers: Shanghai, China, 2014; pp. 27–98. (In Chinese)
34. Mur, L.A.J.; Kenton, P.; Atzorn, R.; Miersch, O.; Wasternack, C. The outcomes of concentration-specific interactions between salicylate and jasmonate signaling include synergy, antagonism, and oxidative stress leading to cell death. *Plant Physiol.* **2006**, *140*, 249–262. [CrossRef]
35. Verberne, M.C.; Brouwer, N.; Delbianco, F.; Linthorst, H.J.M.; Bol, J.F.; Verpoorte, R. Method for the extraction of the volatile compound salicylic acid from tobacco leaf material. *Phytochem. Anal.* **2002**, *13*, 45–50. [CrossRef]
36. Jones, J.D.G.; Dangl, J.L. The plant immune system. *Nature* **2006**, *444*, 323–329. [CrossRef]
37. Rastogi, G.; Coaker, G.L.; Leveau, J.H.J. New insights into the structure and function of phyllosphere microbiota through high-throughput molecular approaches. *FEMS Microbiol. Lett.* **2013**, *348*, 1–10. [CrossRef]
38. Liu, W.; Yuan, D.; Yin, P.; Wang, Z.H.; Guo, G.Y. Research progress on anthracnose of tea plant. *Chin. J. Trop. Agric.* **2016**, *36*, 20–26. (In Chinese) [CrossRef]
39. Bolwell, G.P. Role of active oxygen species and NO in plant defence responses. *Curr. Opin. Plant Biol.* **1999**, *2*, 287–294. [CrossRef]
40. Banerjee, A.K.; Mandal, A.; Chanda, D.; Chakraborti, S. Oxidant, antioxidant and physical exercise. *Mol. Cell. Biochem.* **2003**, *253*, 307–312. [CrossRef]
41. Doke, N. Generation of superoxide anion by potato-tuber protoplasts during the hypersensitive response to hyphal wall components of phytophthora-infestans and specific-inhibition of the reaction by suppressors of hypersensitivity. *Physiol. Plant Pathol.* **1983**, *23*, 359–367. [CrossRef]
42. Bolwell, P.P.; Page, A.; Pislewska, M.; Wojtaszek, P. Pathogenic infection and the oxidative defences in plant apoplast. *Protoplasma* **2001**, *217*, 20–32. [CrossRef]
43. Lamb, C.; Dixon, R.A. The oxidative burst in plant disease resistance. *Annu. Rev. Plant Physiol. Plant Mol. Biol.* **1997**, *48*, 251–275. [CrossRef]

44. Grant, J.J.; Loake, G.J. Role of reactive oxygen intermediates and cognate redox signaling in disease resistance. *Plant Physiol.* **2000**, *124*, 21–29. [CrossRef]
45. Ellinger, D.; Naumann, M.; Falter, C.; Zwikowics, C.; Jamrow, T.; Manisseri, C.; Somerville, S.C.; Voigt, C.A. Elevated early callose deposition results in complete penetration resistance to powdery mildew in *Arabidopsis*. *Plant Physiol.* **2013**, *161*, 1433–1444. [CrossRef]
46. ThordalChristensen, H.; Zhang, Z.G.; Wei, Y.D.; Collinge, D.B. Subcellular localization of H_2O_2 in plants. H_2O_2 accumulation in papillae and hypersensitive response during the barley-powdery mildew interaction. *Plant J.* **1997**, *11*, 1187–1194. [CrossRef]
47. Denness, L.; McKenna, J.F.; Segonzac, C.; Wormit, A.; Madhou, P.; Bennett, M.; Mansfield, J.; Zipfel, C.; Hamann, T. Cell wall damage-induced lignin biosynthesis is regulated by a reactive oxygen species- and jasmonic acid-dependent process in *Arabidopsis*. *Plant Physiol.* **2011**, *156*, 1364–1374. [CrossRef]
48. Bellincampi, D.; Cervone, F.; Lionetti, V. Plant cell wall dynamics and wall-related susceptibility in plant-pathogen interactions. *Front. Plant Sci.* **2014**, *5*, 228. [CrossRef]
49. Torres, M.A.; Dangl, J.L.; Jones, J.D.G. *Arabidopsis* gp91(phox) homologues AtrbohD and AtrbohF are required for accumulation of reactive oxygen intermediates in the plant defense response. *Proc. Natl. Acad. Sci. USA* **2002**, *99*, 517–522. [CrossRef]
50. Marino, D.; Dunand, C.; Puppo, A.; Pauly, N. A burst of plant NADPH oxidases. *Trends Plant Sci.* **2012**, *17*, 9–15. [CrossRef]
51. Kadota, Y.; Sklenar, J.; Derbyshire, P.; Stransfeld, L.; Asai, S.; Ntoukakis, V.; Jones, J.D.G.; Shirasu, K.; Menke, F.; Jones, A.; et al. Direct regulation of the NADPH oxidase RBOHD by the PRR-associated kinase BIK1 during plant immunity. *Mol. Cell* **2014**, *54*, 43–55. [CrossRef]
52. Kobayashi, M.; Ohura, I.; Kawakita, K.; Yokota, N.; Fujiwara, M.; Shimamoto, K.; Doke, N.; Yoshioka, H. Calcium-dependent protein kinases regulate the production of reactive oxygen species by potato NADPH oxidase. *Plant Cell* **2007**, *19*, 1065–1080. [CrossRef]
53. Kim, M.C.; Panstruga, R.; Elliott, C.; Muller, J.; Devoto, A.; Yoon, H.W.; Park, H.C.; Cho, M.J.; Schulze-Lefert, P. Calmodulin interacts with MLO protein to regulate defence against mildew in barley. *Nature* **2002**, *416*, 447–450. [CrossRef]
54. Sun, D.Y.; Bian, Y.Q.; Zhao, B.H.; Zhao, L.Y.; Yu, X.M.; Shengjun, D. The effects of extracellular calmodulin on cell-wall regeneration of protoplasts and cell-division. *Plant Cell Physiol.* **1995**, *36*, 133–138.
55. Lecourieux, D.; Raneva, R.; Pugin, A. Calcium in plant defence-signalling pathways. *New Phytol.* **2006**, *171*, 249–269. [CrossRef]
56. Moffett, P. Mechanisms of recognition in dominant r gene mediated resistance. *Adv. Virus Res.* **2009**, *75*, 1–299. [CrossRef]
57. Hubert, D.A.; Tornero, P.; Belkhadir, Y.; Krishna, P.; Takahashi, A.; Shirasu, K.; Dangl, J.L. Cytosolic HSP90 associates with and modulates the *Arabidopsis* RPM1 disease resistance protein. *EMBO J.* **2003**, *22*, 5679–5689. [CrossRef]
58. Arofatullah, N.A.; Hasegawa, M.; Tanabata, S.; Ogiwara, I.; Sato, T. Heat shock-induced resistance against pseudomonas syringae pv. tomato (Okabe) Young et al. via Heat shock transcription factors in tomato. *Agronomy* **2019**, *9*, 2. [CrossRef]
59. Glazebrook, J. Contrasting mechanisms of defense against biotrophic and necrotrophic pathogens. *Annu. Rev. Phytopathol.* **2005**, *43*, 205–227. [CrossRef]
60. Tsuda, K.; Mine, A.; Bethke, G.; Igarashi, D.; Botanga, C.J.; Tsuda, Y.; Glazebrook, J.; Sato, M.; Katagiri, F. Dual regulation of gene expression mediated by extended MAPK activation and salicylic acid contributes to robust innate immunity in *Arabidopsis thaliana*. *PLoS Genet.* **2013**, *9*, e1004015. [CrossRef]
61. Abad, L.R.; DUrzo, M.P.; Liu, D.; Narasimhan, M.L.; Reuveni, M.; Zhu, J.K.; Niu, X.M.; Singh, N.K.; Hasegawa, P.M.; Bressan, R.A. Antifungal activity of tobacco osmotin has specificity and involves plasma membrane permeabilization. *Plant Sci.* **1996**, *118*, 11–23. [CrossRef]
62. Song, J.T.; Lu, H.; McDowell, J.M.; Greenberg, J.T. A key role for ALD1 in activation of local and systemic defenses in *Arabidopsis*. *Plant J.* **2004**, *40*, 200–212. [CrossRef]
63. Mou, Z.; Fan, W.H.; Dong, X.N. Inducers of plant systemic acquired resistance regulate NPR1 function through redox changes. *Cell* **2003**, *113*, 935–944. [CrossRef]
64. Wu, Y.; Zhang, D.; Chu, J.Y.; Boyle, P.; Wang, Y.; Brindle, I.D.; De Luca, V.; Despres, C. The *Arabidopsis* NPR1 protein is a receptor for the plant defense hormone salicylic acid. *Cell Rep.* **2012**, *1*, 639–647. [CrossRef]

65. Zhou, J.M.; Trifa, Y.; Silva, H.; Pontier, D.; Lam, E.; Shah, J.; Klessig, D.F. NPR1 differentially interacts with members of the TGA/OBF family of transcription factors that bind an element of the PR-1 gene required for induction by salicylic acid. *Mol. Plant-Microbe Interact.* **2000**, *13*, 191–202. [CrossRef]
66. Klessig, D.F.; Durner, J.; Noad, R.; Navarre, D.A.; Wendehenne, D.; Kumar, D.; Zhou, J.M.; Shah, J.; Zhang, S.Q.; Kachroo, P.; et al. Nitric oxide and salicylic acid signaling in plant defense. *Proc. Natl. Acad. Sci. USA* **2000**, *97*, 8849–8855. [CrossRef] [PubMed]
67. Wang, D.; Amornsiripanitch, N.; Dong, X.N. A genomic approach to identify regulatory nodes in the transcriptional network of systemic acquired resistance in plants. *PLoS Path.* **2006**, *2*, 1042–1050. [CrossRef]
68. Rushton, P.J.; Somssich, I.E.; Ringler, P.; Shen, Q.X.J. WRKY transcription factors. *Trends Plant Sci.* **2010**, *15*, 247–258. [CrossRef] [PubMed]
69. Thomma, B.P.H.J.; Eggermont, K.; Penninckx, I.A.M.A.; Mauch-Mani, B.; Vogelsang, R.; Cammue, B.P.A.; Broekaert, W.F. Separate jasmonate-dependent and salicylate-dependent defense-response pathways in *Arabidopsis* are essential for resistance to distinct microbial pathogens. *Proc. Natl. Acad. Sci. USA* **1998**, *95*, 15107–15111. [CrossRef]
70. Hartmann, M.; Zeier, T.; Bernsdorff, F.; Reichel-Deland, V.; Kim, D.; Hohmann, M.; Scholten, N.; Schuck, S.; Brautigam, A.; Holzel, T.; et al. Flavin monooxygenase-generated n-hydroxypipecolic acid is a critical element of plant systemic immunity. *Cell* **2018**, *173*, 456–469. [CrossRef] [PubMed]
71. Brown, J.; Pirrung, M.; McCue, L.A. FQC Dashboard: Integrates FastQC results into a web-based, interactive, and extensible FASTQ quality control tool. *Bioinformatics* **2017**, *33*, 3137–3139. [CrossRef]
72. Haas, B.J.; Papanicolaou, A.; Yassour, M.; Grabherr, M.; Blood, P.D.; Bowden, J.; Couger, M.B.; Eccles, D.; Li, B.; Lieber, M.; et al. De novo transcript sequence reconstruction from RNA-seq using the Trinity platform for reference generation and analysis. *Nat. Protoc.* **2013**, *8*, 1494–1512. [CrossRef] [PubMed]
73. Li, Q.S.; Li, X.M.; Qiao, R.Y.; Shen, E.H.; Lin, X.M.; Lu, J.L.; Ye, J.H.; Liang, Y.R.; Zheng, X.Q. De novo transcriptome assembly of fluorine accumulator tea plant *Camellia sinensis* with fluoride treatments. *Sci. Data* **2018**, *5*, 180194. [CrossRef] [PubMed]
74. Camacho, C.; Coulouris, G.; Avagyan, V.; Ma, N.; Papadopoulos, J.; Bealer, K.; Madden, T.L. BLAST plus: Architecture and applications. *BMC Bioinform.* **2009**, *10*, 421. [CrossRef] [PubMed]
75. Pruitt, K.D.; Tatusova, T.; Maglott, D.R. NCBI reference sequences (RefSeq): A curated non-redundant sequence database of genomes, transcripts and proteins. *Nucleic Acids Res.* **2007**, *35*, D61–D65. [CrossRef]
76. Ashburner, M.; Ball, C.A.; Blake, J.A.; Botstein, D.; Butler, H.; Cherry, J.M.; Davis, A.P.; Dolinski, K.; Dwight, S.S.; Eppig, J.T.; et al. Gene ontology: Tool for the unification of biology. *Nat. Genet.* **2000**, *25*, 25–29. [CrossRef] [PubMed]
77. Kanehisa, M.; Goto, S.; Kawashima, S.; Okuno, Y.; Hattori, M. The KEGG resource for deciphering the genome. *Nucleic Acids Res.* **2004**, *32*, D277–D280. [CrossRef] [PubMed]
78. Boeckmann, B.; Bairoch, A.; Apweiler, R.; Blatter, M.C.; Estreicher, A.; Gasteiger, E.; Martin, M.J.; Michoud, K.; O'Donovan, C.; Phan, I.; et al. The SWISS-PROT protein knowledgebase and its supplement TrEMBL in 2003. *Nucleic Acids Res.* **2003**, *31*, 365–370. [CrossRef] [PubMed]
79. Powell, S.; Forslund, K.; Szklarczyk, D.; Trachana, K.; Roth, A.; Huerta-Cepas, J.; Gabaldon, T.; Rattei, T.; Creevey, C.; Kuhn, M.; et al. eggNOG v4.0: Nested orthology inference across 3686 organisms. *Nucleic Acids Res.* **2014**, *42*, D231–D239. [CrossRef] [PubMed]
80. Gotz, S.; Garcia-Gomez, J.M.; Terol, J.; Williams, T.D.; Nagaraj, S.H.; Nueda, M.J.; Robles, M.; Talon, M.; Dopazo, J.; Conesa, A. High-throughput functional annotation and data mining with the Blast2GO suite. *Nucleic Acids Res.* **2008**, *36*, 3420–3435. [CrossRef] [PubMed]
81. Moriya, Y.; Itoh, M.; Okuda, S.; Yoshizawa, A.C.; Kanehisa, M. KAAS: An automatic genome annotation and pathway reconstruction server. *Nucleic Acids Res.* **2007**, *35*, W182–W185. [CrossRef]
82. Li, B.; Dewey, C.N. RSEM: Accurate transcript quantification from RNA-Seq data with or without a reference genome. *BMC Bioinform.* **2011**, *12*, 323. [CrossRef] [PubMed]
83. Anders, S.; Huber, W. Differential expression analysis for sequence count data. *Genome Biol.* **2010**, *11*, R106. [CrossRef] [PubMed]
84. Li, Q.S.; Lin, X.M.; Qiao, R.Y.; Zheng, X.Q.; Lu, J.L.; Ye, J.H.; Liang, Y.R. Effect of fluoride treatment on gene expression in tea plant (*Camellia sinensis*). *Sci. Rep.* **2017**, *7*, 9847. [CrossRef] [PubMed]

85. Livak, K.J.; Schmittgen, T.D. Analysis of relative gene expression data using real-time quantitative PCR and the 2(T)(-Delta Delta C) method. *Methods* **2001**, *25*, 402–408. [CrossRef] [PubMed]
86. Zawoznik, M.S.; Groppa, M.D.; Tomaro, M.L.; Benavides, M.P. Endogenous salicylic acid potentiates cadmium-induced oxidative stress in *Arabidopsis thaliana*. *Plant Sci.* **2007**, *173*, 190–197. [CrossRef]

International Journal of *Molecular Sciences*

MDPI

Article

"Salicylic Acid Mutant Collection" as a Tool to Explore the Role of Salicylic Acid in Regulation of Plant Growth under a Changing Environment

Kamila Pluhařová [1,2,†], Hana Leontovyčová [1,2,3,†], Věra Stoudková [1,2], Romana Pospíchalová [1], Petr Maršík [4], Pavel Klouček [4], Anastasiia Starodubtseva [1,2], Oksana Iakovenko [1,2,5], Zuzana Krčková [1], Olga Valentová [2], Lenka Burketová [1], Martin Janda [1,2,6,*] and Tetiana Kalachova [1,*]

1 Institute of Experimental Botany of the Czech Academy of Sciences, 16502 Prague, Czech Republic; Kamila.Pluharova@vscht.cz (K.P); leontovycova@ueb.cas.cz (H.L.); stoudkovavera@seznam.cz (V.S.); pospichalova@ueb.cas.cz (R.P.); starodubtseva@ueb.cas.cz (A.S.); iakovenko@ueb.cas.cz (O.I.); krckova@ueb.cas.cz (Z.K.); burketova@ueb.cas.cz (L.B.)

2 Department of Biochemistry and Microbiology, University of Chemistry and Technology, 16628 Prague, Czech Republic; Olga.Valentova@vscht.cz

3 Department of Biochemistry, Faculty of Science, Charles University in Prague, 11000 Prague, Czech Republic

4 Department of Food Science, The Faculty of Agrobiology, Food and Natural Resources, The Czech University of Life Sciences Prague, 16500 Prague, Czech Republic; marsik@ueb.cas.cz (P.M.); kloucek@af.czu.cz (P.K.)

5 V.P. Kukhar Institute of Bioorganic Chemistry and Petrochemistry, National Academy of Sciences of Ukraine, 02094 Kyiv, Ukraine

6 Genetics, Faculty of Biology, Ludwig-Maximilians-University of Munich (LMU), D-82152 Martinsried, Germany

* Correspondence: martin.janda@vscht.cz (M.J.); kalachova@ueb.cas.cz (T.K.); Tel.: +420-603-579-167 (M.J.); Tel.: +420-225-106-816 (T.K.)

† These authors contributed equally to this work.

Received: 31 October 2019; Accepted: 13 December 2019; Published: 17 December 2019

Abstract: The phytohormone salicylic acid (SA) has a crucial role in plant physiology. Its role is best described in the context of plant response to pathogen attack. During infection, SA is rapidly accumulated throughout the green tissues and is important for both local and systemic defences. However, some genetic/metabolic variations can also result in SA overaccumulation in plants, even in basal conditions. To date, more than forty *Arabidopsis thaliana* mutants have been described as having enhanced endogenous SA levels or constitutively activated SA signalling pathways. In this study, we established a collection of mutants containing different SA levels due to diverse genetic modifications and distinct gene functions. We chose prototypic SA-overaccumulators (SA-OAs), such as *bon1-1*, but also "non-typical" ones such as *exo70b1-1*; the selection of OA is accompanied by their crosses with SA-deficient lines. Here, we extensively studied the plant development and SA level/signalling under various growth conditions in soil and in vitro, and showed a strong negative correlation between rosette size, SA content and *PR1*/*ICS1* transcript signature. SA-OAs (namely *cpr5*, *acd6*, *bon1-1*, *fah1*/*fah2* and *pi4kβ1β2*) had bigger rosettes under high light conditions, whereas WT plants did not. Our data provide new insights clarifying a link between SA and plant behaviour under environmental stresses. The presented SA mutant collection is thus a suitable tool to shed light on the mechanisms underlying trade-offs between growth and defence in plants.

Keywords: Salicylic acid; Arabidopsis mutants; light; growth; gene transcription

1. Introduction

Salicylic acid (SA; 2-hydroxybenzoic acid) is a plant hormone (phytohormone) which plays a role in numerous plant physiological processes. It influences plant development, seed germination [1], cell cycle [2], flowering and responses to stresses [3]. In particular, the importance of SA has been reported in various abiotic stresses: response to high and low temperatures, humidity and drought, salt and osmotic stress [4] or responses to UV light [5]. However, since the 1990s, SA research has mainly focused on its role in immunity [6,7].

The role of SA in plant responses to stresses is generally studied using two approaches: a pharmacological approach using SA treatment on plants and a genetic approach that relies on mutants with modulated endogenous SA concentrations or SA-related signalling. These mutants can be either deficient in SA accumulation, or accumulating high levels of SA (in basal conditions and/or upon stimulation). The widely used SA-deficient lines are *sid2*, carrying a T-DNA insertion in *ISOCHORISMATE SYNTHASE 1* and thus showing lower production of SA upon pathogen attack [8]; or *NahG*, expressing bacterial SA hydroxylase that degrades SA to inactive catechol [9,10]. These mutants are generally more susceptible to pathogen attack, especially by biotrophs [11]. On the other hand, mutants synthesising higher basal levels of SA (SA-overaccumulating mutants; SA-OAs) tend to be more resistant to pathogens. The boom of using SA-OA mutants emerged with forward genetics approach, where EMS mutants were screened for spontaneous lesions and small sizes, which often coincided with high levels of SA and enhanced resistance to pathogens [12,13]. However, such resistance often correlates with general dwarfism [14]. This remains a critical disadvantage for potential use of SA pathway modification in agriculture, where the high yield is needed. Hence, the modulation of SA in terms of possible use in agriculture has to be carefully prepared. However, until now, the molecular mechanism of the trade-off is not fully understood yet.

The increased SA level in mutants could be triggered by distinct events (mutations). The SA-OA phenotype can be caused by gain-of-function mutation (typically activation of immune receptor) or by loss-of-function mutation (typically negative regulation of SA pathway). For example, *bon1-1* shows enhanced immunity and SA levels due to the mutation in the negative regulator of SNC1. This phenotype is thus reversed by introducing an *snc1-11* point mutation [15]. Particular interest has been devoted to mutants with altered phospholipid turnover/signalling and vesicular trafficking that were reported to have pleiotropic effects, often connected with SA accumulation. In particular, *pi4kβ1β2* mutation in phosphatidylinositol-4-kinases β1 and β2 was recently reported as an SA overaccumulator [16,17], or *exo70b1-1* [18]. *fah1/fah2* is deficient in fatty acid hydroxylase genes and also showed enhanced immune responses and a modulated sphingolipid profile [19]. Further characterisation of those lines will thus help in mechanistic understanding of the connections between phospholipid metabolism, vesicular trafficking and immunity in plants.

Here, we present a collection of *Arabidopsis thaliana* (hereinafter Arabidopsis) mutants with SA content altered in various ways: affected immunity-related signalling, modified vesicular trafficking and a directly altered SA biosynthesis/accumulation. As controls, we included crosses of SA-OAs with SA-deficient lines. We propose this collection as a tool to investigate the role of SA in plant growth regulation and stress tolerance.

2. Results

2.1. Cultivation Conditions Influence the Phenotype of the SA Collection Mutants

Our aim was to establish a collection from available sources of Arabidopsis mutants with alterations in the SA pathway with special attention to creating a group of highly diverse SA-OA mutants, and not only prototypic SA-OA ones. Thus, we selected two SA-deficient mutants (*NahG*, *sid2*), eight known or putative SA-OA mutants (*cpr5-1*, *acd6-1*, *pi4kβ1β2*, *fah1fah2*, *bon1-1*, *exo70B1-2*, *pmr4-1*, *edr2-6*), and four "reverted" mutants: SA-OA mutants crossed with the SA-deficient ones (*sid2pi4kβ1β2*, *NahGpi4kβ1β2*, *NahGedr2-6*, *bon1-1snc1-11*) (see Table 1).

Table 1. Selected Arabidopsis mutants with potentially affected salicylic acid (SA) signatures.

Mutant Name		Targeted Gene	Targeted Process	Reference
cpr5	*CPR5*	Constitutive Expression of Pathogenesis-related genes 5	Constitutive expression of pathogenesis-related genes 5	Yoshida et el. 2002 [20]
bon1-1	*BON1*	BONZAI 1	Negative regulator of cell death, defence responses and several R genes	Li et al. 2007 [15]
acd6	*ACD6*	Accelerated Cell Death 6	Dose-dependent activation of defence signalling, accelerated cell death observed	Rate et al. 1999 [21]
pi4kβ1β2	*PI4Kβ1, PI4Kβ2*	Phosphatidylinositol-4-kinase β1 and β2	Second messenger, phosphatidyl inositol-4-phosphate production	Preuss et al. 2006 [22]
fah1fah2	*FAH1, FAH2*	Fatty acid5-hydroxylase 1 and 2	Fatty acid hydroxylation	Konig et al. 2012 [19]
edr2-6	*EDR2*	Enhanced Disease Resistance 2	Negative regulation of cell death	Vorwerk et al. 2008 [23]
exo70B1-1	*EXO70B1*	Exocyst Complex Component EXO70B1	Endomembrane trafficking	Kulich et al. 2013 [18]
pmr4-1	*CALS12*	Callose Synthase 12	Pathogen-induced callose synthesis	Nishimura et al. 2003 [24]
sid2 pi4kβ1β2	*ICS1, PI4Kβ1, PI4Kβ2*	Isochorismate synthase 1, phosphatidylinositol-4-kinase β1 and β2	SA biosynthesis, second messenger inositol-1,4,5-trisphosphate production	Sasek et al. 2014 [25]
NahG pi4kβ1β2	*NahG, PI4Kβ1, PI4Kβ2*	SA hydroxylase, phosphatidylinositol-4-kinase β1 and β2	SA degradation, second messenger inositol-1,4,5-trisphosphate production	Sasek et al. 2014 [25]
NahG edr2-6	*NahG, EDR2*	SA hydroxylase, enhanced disease resistance 2	SA degradation, negative regulation of cell death	Vorwerk et al. 2008 [23]
bon1-1 snc1-11	*BON1, SNC1*	BONZAI 1, Suppresssor npr1-1, constitutive 1	*bon1-1* crossed to the *snc1-11*, loss-of-function point mutation of the *SNC1*	Li et al. 2007 [15]
sid2	*ICS1*	Isochorismate synthase 1	SA biosynthesis	Wildermuth et al. 2001 [8]
NahG	*NahG*	SA hydroxylase	SA degradation	Nawrath and Metraux 1999 [26]

First we analysed the growth of selected mutants under long day (LD) and short day (SD) conditions. We initially focused on the 4 week old plants and analysed their rosette size and SA content (Figure 1; Supplementary Figure S1, Supplementary Table S2). Except for SA-deficient lines (*NahG* and *sid2*) and *exo70B1*, all others responded to LD condition with retarded growth. Due to their distinctive dwarf phenotypes (with an 85–50% reduction of rosette area compared to WT), we could clearly identify several SA-OA mutants: *cpr5*, *pi4kβ1β2*, *acd6* and *bon1-1* (Figure 1A,B). In SD conditions, the differences between mutants in growth were comparable to those under LD, although less important by absolute values (Figure S1A,B). In contrast, the differences in SA content were more pronounced at SD conditions. This could be due to a higher basal level of SA in the LD condition connected with a developmental stage; in LD conditions the plants started bolting at 3–3.5 weeks. In the "reverted mutants", the SA level was decreased to the level of WT, which correlated with the WT-like rosette size (Figure 1 and Figure S1).

For better description of the effect of the growth conditions on dwarf phenotypes of SA-OAs, we focused on a subset of four mutants: WT, *sid2*, *pi4kβ1β2* and *sid2pi4kβ1β2*, previously used for studies of SA-related effects [17]. We compared the growth dynamics of this subset in several cultivation conditions often used in stress-related studies: SD, LD and greenhouse (Supplementary Figure S2).

The *pi4kβ1β2* mutant appeared dwarfed under all conditions. Rosettes of *pi4kβ1β2* were smaller than in WT during the early developmental stages (2 week old seedlings), and the difference increased with time. Notably, the smaller rosettes did not result in a delay in development, since all plants started flowering simultaneously. Unexpectedly, *sid2pi4kβ1β2* grew bigger than *pi4kβ1β2* under all conditions but never reached the size of WT plants. This finding was surprising considering the previously published full reversal of growth in *sid2pi4kβ1β2* [25]. This finding indicates a high sensitivity of SA-related phenotypes to cultivation conditions.

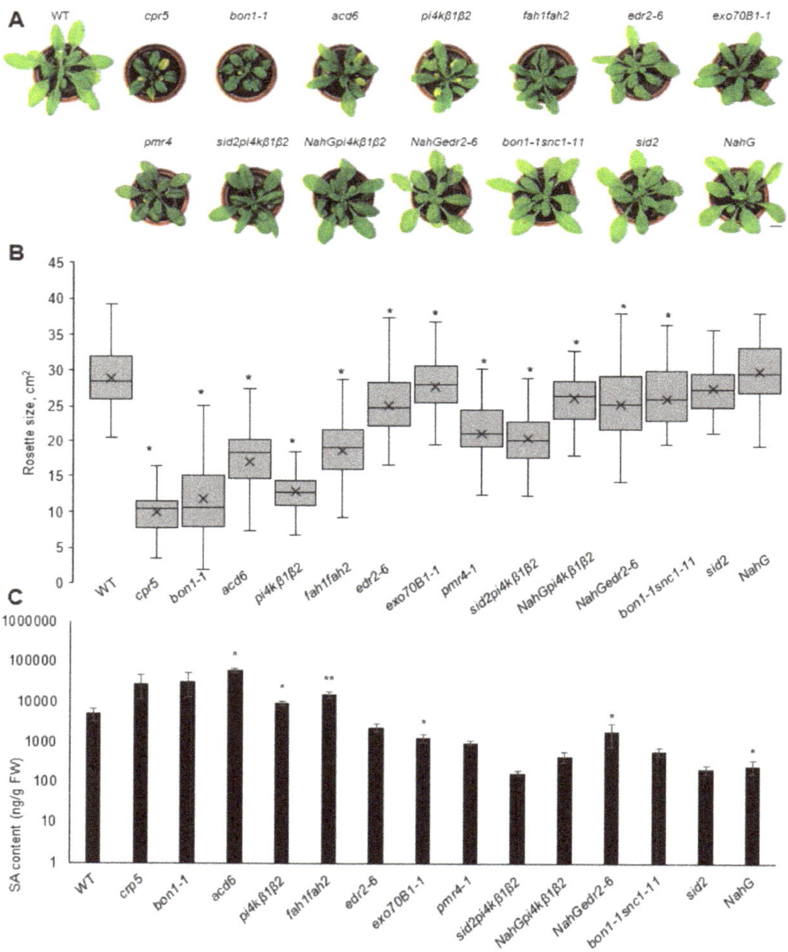

Figure 1. Rosette size and SA content of plants cultivated under long-day conditions. (**A**) Representative images of 4 week old plants cultivated at 22 °C, 16 h light/ 8 h dark. (**B**) Rosette size (area). Data are from three biological replicates, $n \geq 70$. Central line of the boxplot represents the median occupancy, cross represents the mean, bottom and top edges of the box are 25 and 75% of distribution and the ends of whiskers are set at 1.5 times the interquartile range. (**C**) SA content in the leaves, $n = 4$. Data represent means + SEM, asterisks indicate variants different from WT, one-way ANOVA with Tukey's HSD post hoc test, * $p < 0.05$, ** $p < 0.01$.

Particular attention was directed to the *pmr4* mutant, deficient in callose synthase CALS12 and first described as POWDERY MILDEW RESISTANT 4 [24]. Under both SD and LD, the plants appeared slightly smaller than WT; however, no increase of SA was detected (Figure 1, Supplementary Figure S1, Supplementary Table S2). To the best of our knowledge, this is the first report on the (comparative) determination of the SA content in *pmr4* mutants, as the previous studies claimed a constitutively activated SA pathway based on SA-related gene transcription and resistance to pathogens [24].

2.2. SA-Related Gene Transcription Varies in Different Growth Conditions

We analysed the SA marker genes' (*PR1* and *ICS1*) transcription in soil-grown plants under SD and LD conditions. In most cases, the expression of the *PR1* gene coincided with small rosettes and a higher level of SA (Figure 2, Supplementary Figure S3). Generally, our results confirmed those from the studies wherein the mutants were first described. Under SD conditions, the differences between mutants were more pronounced both in *PR1*/*ICS1* transcription and in SA content. As gene transcription data were normalized to WT in both conditions, and WT at LD had almost 5 times higher SA content than in SD, that might have strongly affected basal *PR1* level.

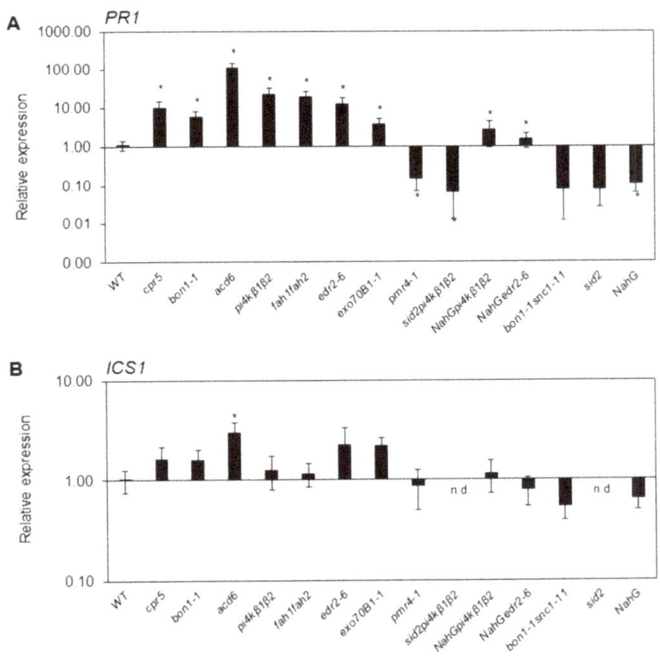

Figure 2. Transcription of *ICS1* and *PR1* in soil-grown plants cultivated under LD conditions. Samples were collected from four 4 week old plants. Values were normalized to WT at the respective conditions. *TIP41* was used as a reference gene. Data represent means + SEM, asterisks indicate values different from WT, *t*-test, * $p < 0.05$, $n = 4$.

To link SA-related signatures to the physiological state of plants, we measured photosynthetic activity. As reliable parameters of photosynthetic state, we chose quantum yield of PSII photochemistry (QY) and non-photochemical quenching (NPQ) [27]. Increase of NPQ can occur as a result either of the processes that protect the leaf from damage or of the damage itself [28]. The changes in NPQ were observed in several studied mutants (Supplementary Figure S4). NPQ at steady state (NPQ_Lss) was decreased in *snc1-1* compared to WT, but increased in several other mutants from the collection. These mutants were mainly "immune-related" mutants *cpr5* and *bon1-1* and "trafficking" mutants *pi4kβ1β2* and *exo70B1-1*. Notably, the values were similar to those in "reverted" mutants *sid2pi4kβ1β2*, *NahGpi4kβ1β2* and *bon1-1snc1-11*, suggesting SA-independent origin of NPQ_Lss increase. The SA-deficient mutants also showed slight (by 10%) increase in NPQ_Lss, indicating SA independency. Maximum quantum efficiency of PSII photochemistry (QY_max) was stable among all studied mutants. Generally, the analysis of photosynthetic parameters did not reveal any drastic differences between the selected mutants in basal conditions.

2.3. Overaccumulation of SA Increases High Light Sensitivity in In-Vitro-Grown Seedlings

To investigate the behaviour of the SA mutant collection under in vitro conditions, we switched to the in vitro setup often used for the study of developmental defects. Continuous illumination of the roots, though widely used in research, can cause diverse effects on the phenotype, including spontaneous production of reactive oxygen species [29]. To study the reliability of our collection for root phenotyping, we studied seedling growth in vitro upon different light conditions. Seedlings were grown in vertically placed Petri dishes under LD light regime and at two light intensities, 450 $\mu E.m^{-2}.s^{-1}$ and 170 $\mu E.m^{-2}.s^{-1}$. Rosette weight and primary root length were evaluated at 11 dpg (Figure 3). While the rosette weight of WT plants did not change in response to light intensity, the mutants exhibiting high changes in SA content showed more intensive growth under 450 $\mu E.m^{-2}.s^{-1}$ light. Interestingly, such an effect was not evident within the majority of the reverted group, except for *bon1-1snc1-11*. SA-deficient genotypes and *pmr4-1* grew similarly at both light intensities, thus highlighting the role of SA in this effect (Figure 3A).

The growth of the primary roots was moderately inhibited by high light intensity in WT and also in both genotypes carrying the *NahG* transgene; however, the roots of other mutants were almost insensitive to different light conditions (Figure 3B). On the other hand, some mutants with pronounced dwarf rosette phenotypes had roots of the same size as WT plants (*bon1-1, acd6*). In contrast, in the *pi4kβ1β2* mutants, which had both small rosettes and short roots in all studied setups, the mechanisms regulating root and rosette growth inhibition seemed to be distinct. Indeed, while the rosette sizes were particularly restored by preventing SA accumulation (*sid2pi4kβ1β2*, *NahGpi4kβ1β2*), the roots remained small, indicating the SA-independent character of the phenotype (Figure 3B). To quantify this in time-course and to further investigate the effect of light regime on root growth, we focused on the phosphatidylinositol-4-kinase- related subset (WT, *sid2*, *pi4kβ1β2* and *sid2pi4kβ1β2*). First, we measured root elongation in kinetics at the light intensity corresponding to 170 $\mu E.m^{-2}.s^{-1}$ (Supplementary Figure S5A). The root defects caused by *pi4kβ1β2* mutations appeared at the very early germination stages and this effect was enhanced in time. To confirm the light independency of the phenotype, we also analysed root length in a semi-dark setup, modified from Silva-Navas et al., 2015 [30]. The Petri dishes were placed in dark plastic covers to shadow roots (Supplementary Figure S5B). In both setups, the growth kinetics were comparable: in the dark setup, *pi4kβ1β2* roots were about 3 times shorter than WT at 4 dpg, and about 5 times shorter at 8 dpg. Furthermore, *sid2pi4kβ1β2* roots were about 20% longer than *pi4kβ1β2* at 4 dpg and 50 % longer at 8 dpg. However, while the difference between *pi4kβ1β2* and WT remained stable in the light setup (up to 6 times at 8dpg), the difference between *pi4kβ1β2* and *sid2pi4kβ1β2* was more pronounced, up to 30% at 4 dpg and up to 200% at 8 dpg. This confirmed the SA-dependent sensitivity to light in in vitro growth conditions, and it also means that the light regime should be seriously considered in various types of experiments, especially those connected with SA.

With the SA collection, we were able to show that the regulation of the rosettes and root size is independent: the SA content mostly influenced the aboveground plant part, while the root length corresponded to SA-independent phenotype. Indeed, while *bon1-1* and *pi4kβ1β2* mutants were similar in terms of rosette growth, the roots of *bon1-1* were twice longer than that of *pi4kβ1β2* at both light intensities (Figures 1B and 3A,B).

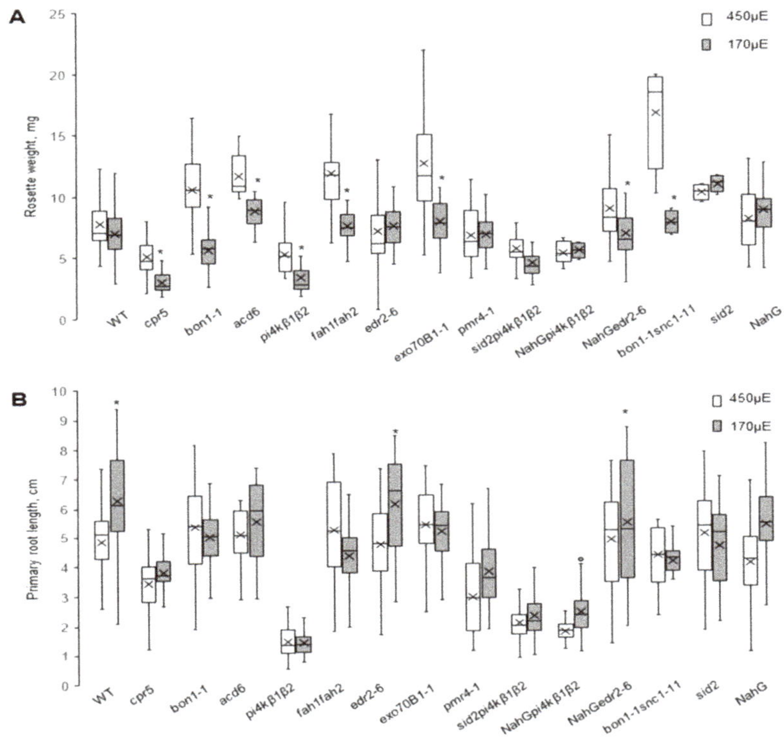

Figure 3. In vitro growth of SA collection mutants under different light intensities. Two week old seedlings were cultivated on $\frac{1}{2}$ MS medium under 450 µE.m^{-2}.s^{-1} or 170 µE.m^{-2}.s^{-1} under 12 h light /12 h dark photoperiod. (**A**) Rosette weight. (**B**) Primary root length. Data represent four biological repetitions; at least 10 seedlings were measured for each variant in each biological repetition. Central line of the boxplot represents the median occupancy, cross represents the mean, bottom and top edges of the box are 25 and 75% of distribution and the ends of whiskers are set at 1.5 times the interquartile range, asterisks indicates variants different from those for the 450 µE.m^{-2}.s^{-1} intensity the same genotype, * $p < 0.01$, *t*-test.

2.4. Salicylic Acid's Effect on the Root Growth and Shoot Growth is Distinct

To evaluate the behaviour of the presented SA collection in various growth setups, we performed a correlation analysis among all studied parameters: rosette size, SA content and expression of *ICS1* and *PR1* genes in soil-grown plants under two light regimes; and rosette weight and primary root length of seedlings cultivated in vitro under two light intensities. Putting together data of three biological repeats of all 15 genotypes in the collection, we quantified Pearson correlations (Figure 4).

The correlation table provided several clear outcomes: the rosette size of plants grown in soil negatively correlated with SA content accompanied with *PR1/ICS1* upregulation, which has been abundantly shown in previous studies [31]. Rosette growth correlated positively in all conditions. In contrast, the root growth in in vitro conditions was SA-independent (Figure 4). Generally, the rosette growth correlated with root growth, despite the above-mentioned difference between *bon1-1* and *pi4kβ1β2*. Interestingly, only seedlings grown under the 170 µE.m^{-2}.s^{-1} intensity strongly correlated with rosette growth of soil-grown plants, suggesting that particular attention needs to be paid to light intensity while comparing data obtained in different growth conditions.

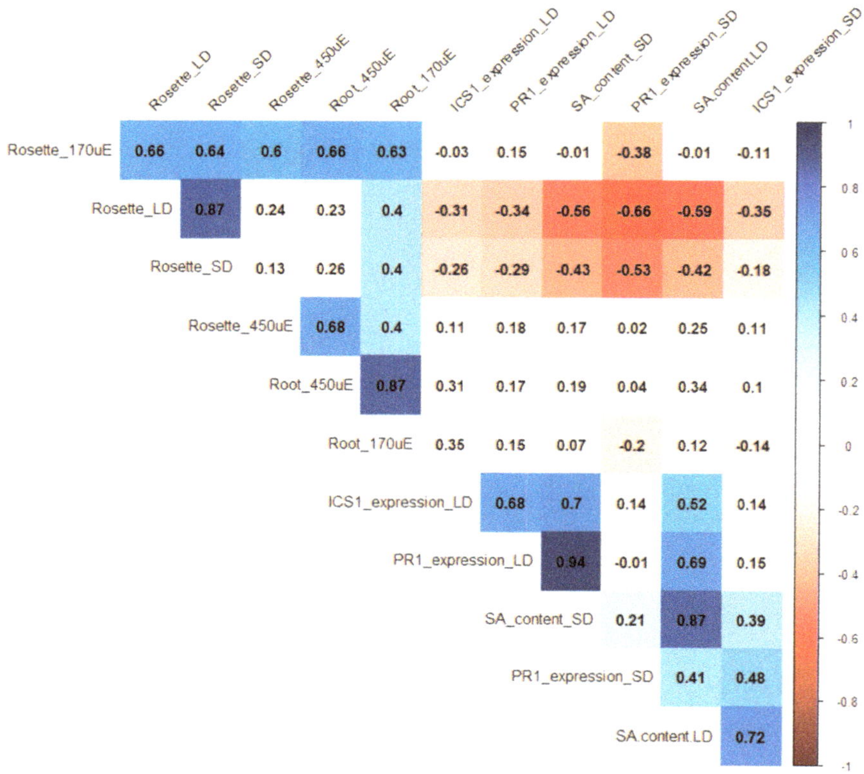

Figure 4. Correlation table of SA effects on growth. The matrix was built using the Pearson correlation for 12 parameters (rosette size, SA content and SA-related gene expression (*ICS1* and *PR1*) for soil-grown plants under short-day (SD) and long-day (LD) conditions; and rosette weight and primary root length for in vitro grown plants grown under an LD photoperiod at 450 uE or 170 uE light intensity). Measurements were taken for 15 genotypes (listed in Table 1). Data are from three biological repetitions for each variant. Positive correlations are displayed in blue and negative correlations in red. Correlation coefficients are indicated. Only results that passed the 0.05 threshold for significance are displayed in colour.

3. Discussion

SA plays a role in many fundamental processes in plants. Nowadays, it is probably the best characterised phytohormone in connection with plant immunity. A great tool which have provided insight into SA signalling pathways and their roles, especially in Arabidopsis, is SA-OA mutants. Interestingly, some of the initially described immunity-related mutants later appeared to have altered SA metabolism/signalling [31]. Changes in SA levels have a very strong impact on plant growth, and the majority of known SA-OA mutants are dwarfs. Because of their clearly distinguishable growth phenotypes, SA-OA mutants have been successfully used to find new components of plant immunity in forward genetic screening [32], in evolutionary studies [33,34] and in studies of ubiquitination cascades [35]. Growth inhibition of SA-OA has been used as a marker of an activated immune state in heat stress experiments: SA-OA mutants exhibit dwarf phenotypes under 22 °C, but have WT-like phenotype under 28 °C [36]. Although they have been studied for more than 30 years, SA-OA mutants still display many features that lack mechanistic explanation. One of them is the impact of cultivation conditions on SA-regulated growth, which has never been extensively studied.

To gain a complex understanding of the connection between growth and SA, we created a collection of 14 SA-modulated Arabidopsis mutants in a Col-0 background. We collected mutants from already published studies, including prototypic SA-deficient mutants *sid2* and *NahG* and prototypic SA-OA mutants *bon1-1*, *cpr5-1* and *acd6-1*. Additionally, we included recently described SA-OAs connected with lipid signalling, *pi4kβ1β2* and *fah1fah2*, and mutants associated with SA signalling based on gene expression analysis and pathogen assays, *edr2-6*, *pmr4-1*, *exo70b1* (Table 1). To complete the picture, we included three SA-OA "reverted lines", in which SA-OA mutants were prevented from accumulating a high SA level by affected biosynthesis (*sid2pi4kβ1β2*) or accumulation (*NahGpi4kβ1β2* and *NahGedr2-6*). All the selected mutants have been reported as having altered resistance to pathogens [31].

While analysing mutant phenotypes under various conditions, it is often difficult to distinguish between "typical" immune response and "just pleotropic" effects caused by mutation. A good example is the *pi4kβ1β2* mutant with impaired vesicle trafficking, which is a ubiquitous process that affects almost everything in plant cells [22]. We studied the SA-(in) dependent effects in *pi4kβ1β2*, showing that resistance to adapted pathogens is strictly SA-dependent, but callose production is SA-independent [17]. By creating this type of collection, we wanted to be able to compare more mutants with modulated SA patterns under exactly the same experimental conditions. For this purpose, we started with characterisation of the plant growth under short-day (SD) and long-day (LD) conditions. In general, our data confirmed previously published data that SA-OA mutants exhibit dwarf phenotypes (Figure 1A,B and Figure S1A,B). The SA content negatively correlated with rosette size (Figure 1 and Supplementary Figure S1,S4) which has been previously shown in literature [31]. However, the SA measurement also revealed that *edr2* and *exo70b1* are SA-OAs only under short-day conditions and *pmr4* is not SA-OA at all. This is particularly important as *edr2* and *pmr4* mutants have previously been described to accumulate high SA under biotic stress conditions, which thus suggests enhanced SA pathways at basal conditions as well. As expected, higher SA content in basal conditions was shown for *edr2* [23], but, surprisingly, we were unable to find in the literature any SA measurement for the *pmr4* mutant, although it is generally referred to as the one with constitutively induced SA pathways [24]. Again, SA marker genes are highly enhanced in *pmr4* under biotic stress conditions [37]. This statement is based on *PR1* gene expression, but not on SA level itself. In our setup, *PR1* expression was not highly induced even in basal conditions.

First, we characterised the SA collection's growth, *PR1/ICS1* transcription and SA content in plants cultivated in soil under LD and SD conditions (Figure 1 and Figure S1). Interestingly, in WT plants PR1 transcription was 5 times higher under LD than SD. That coincides with the fact that under LD conditions, plants tended to bolt at the age of 4 weeks. The induction of flowering is also associated with an increase in SA content and vice versa—SA treatment can trigger flowering [38]. No drastic changes in photosynthesis efficiency were detected (Supplementary Figure S3). For the full set of analysed mutants, we observed a negative correlation between rosette size and SA content under both LD and SD conditions. However, our data suggest that growth phenotype related to SA content would be better investigated under LD conditions. On the other hand, differences in SA content and gene transcription of SA marker genes were more pronounced under SD conditions. In comparison with the literature, our data showed that the mutants with modulated SA pathways were very sensitive to growth conditions. In terms of growth size, this could be clearly seen in the WT, *sid2*, *pi4kβ1β2* and *sid2pi4kβ1β2* subsets. This has been previously used to distinguish between SA-dependent and SA-independent effects of *pi4kβ1β2* deficiency [17,25]. Interestingly, in Šašek et al. (2014) [25], we showed that crossing of *pi4kβ1β2* with *sid2* led to a fully reverted phenotype when plants were grown in soil for 4 weeks. In our current cultivation conditions, we were not able to fully revert the growth (Figure 1A). We studied this in more detail under three distinct growth conditions. Two were in climate chambers with 8 h/16 h (light/dark) (short day) or 16 h/8 h (light/dark) (long day) and one was in greenhouse conditions. In all setups, *sid2pi4kβ1β2* was smaller than WT. In SD, *sid2pi4kβ1β2* had a size comparable even to *pi4kβ1β2* (Figure S2). The data of *ICS1* expression showed that the *ICS1* mutation was functional in both *sid2* and *sid2pi4kβ1β2* lines (Figure 2 and Figure S2).

Such behaviour can affect data interpretation and highlights the importance of checking SA levels while studying pleiotropic phenotypes, especially in a newly discovered mutant lines.

Early studies of SA mutants were mostly done on the rosettes (leaves) of soil-grown plants, while in recent years, the usage of in-vitro-grown seedlings as a model system has been rapidly increasing. The induction of the SA pathway has been shown during infection with root pathogen *Trichoderma* [39]. The sensitivity of Arabidopsis roots to SA treatment was recently demonstrated by a proteomics and metabolomics approach using SA-altered mutants [40]. Furthermore, the role of SA in root morphogenesis was recently shown by Pasternak et al. 2019 [41]. These authors reported SA treatment to modulate root meristem patterning by affecting auxin signalling in a concentration-dependent manner. However, no mutants with modulated SA were used in the study and the usage of only a pharmacological approach often raises questions about appropriate controls. We believe that the SA collection could be a helpful tool to continue studies of hormonal cross-talk in Arabidopsis roots. Here, we showed that root growth in the SA mutant collection is highly variable (Figure 3), and is not correlated with SA levels or SA marker gene expression in the rosettes of soil-grown plants (Figure 4). A clear example is the comparison of *bon1-1* phenotype (small rosette and almost WT-size roots) to *pi4kβ1β2*, which also had small rosettes but impaired root growth (Figures 1 and 3). Our data confirmed the critical role of PI4Kβ1β2 for root growth (Figure 3) [16,25]. To precisely analyse the SA role in seedlings' sensitivity to light, we used the subset of WT, *sid2*, *pi4kβ1β2* and *sid2pi4kβ1β2*. We grew plants in a light growth setup (roots were exposed to light) and dark growth setup (roots were shadowed by placing in dark chambers). The SA-deficient mutant *sid2* grew similarly to WT under dark conditions, but slower in the light setup, and both *pi4kβ1β2* and *sid2pi4kβ1β2* roots grow slower, while the difference between them was more pronounced in a light setup (Figure S4).

As mentioned above, SA-OA mutants are indispensable in studies of SA-related immunity. Additionally, SA's role in biotic stress was also shown via SA-OA involvement in response to abiotic stresses. In particular, the role of SA in cold stress was shown using *acd6* [21], *cpr1* and *pi4kβ1β2* [42]; in potassium stress by using *cpr5* [43], in response to drought and ABA treatment by using *cpr5* and *acd6* [44,45], and in sugar sensing by using *acd6* and *cpr1* [46]. Here, we tested the behaviour of these mutants under distinct growth conditions and under moderate abiotic stress in vitro (distinct light intensities). Surprisingly, the SA-OA mutants *cpr5*, *acd6*, *bon1-1*, *fah1/fah2* and *pi4kβ1β2* tended to form bigger rosettes under higher light intensities, while size of the WT rosettes was not affected (Figure 3). This indicates that SA makes plant more sensitive to high light conditions. In contrast, WT root growth was inhibited by high light but the roots of the above-mentioned SA-OA mutants were not affected (Figure 3B). These findings suggest an opposite effect of light on rosette and root growth.

The trade-off between immunity and growth has been widely discussed [14,47–49]. Our data present a robust quantitative background for this. We have shown strong negative correlation between SA levels and *PR1/ICS1* transcript signature with rosette size, but no correlation with root growth. This is important to take into account while planning phenotyping of mutants on different scales, and also confirms the suitability of putative SA-OA mutants for studies of root growth without impact of SA itself.

4. Materials and Methods

4.1. Plant Material and Growth Conditions

Arabidopsis thaliana Col-0 was used as a wild type (WT), and the collection consisted of following mutants (see Table 1): *crp5 (SALK_071947)*, *bon1-1 (SALK_123132)*, *acd6 (SALK_059132)*, *pi4kβ1β2* (SALK_040479/SALK_09069), *fah1fah2* (SALK_094443, SALK_033090), *exo70B1-1 (CS410875)*, *edr2-6 (CS66944)*; *NahG* [9], *sid2-3* (*SALK_042603*); *bon1-1snc1-11* (SALK_047058, SALK_123132), *NahGpi4kβ1β2*, *sid2pi4kβ1β2* [25] and *NahGedr2-6 (CS66944)*. Prior to experiments, all seeds were propagated for one generation under the same conditions and genotyped as described in the literature (see Table 1).

Plants were grown in two main setups: in a cultivation substrate (soil) (a), and in vitro (b). For both setups, seeds were sterilized in 1.6% sodium hypochloride (30% of SAVO®, Unilever) solution with 0.02%TWEEN20 (Sigma Aldrich, St. Louis, Missouri, USA). Stratification for 2 days at 4 °C in dark conditionswas applied to break dormancy. (a) In soil: seeds were transferred to pots with substrate tablets (Jiffy, Kristiansand, Norway and grown in cultivation chambers (Snijders, Drogenbos, Belgium at 22 °C day temperature, 65–70% humidity and 16 h light/ 8 h dark (LD) or 12 h light/ 12 h dark (SD). After one week, the seedlings were replanted to one plant per pot. Four week old plants were used for analysis. (b) In vitro: seeds were germinated for 3 days in Petri dishes containing a half-strength Mirashige–Skoog medium ($\frac{1}{2}$ Murashige–Skoog basal salts (Duchefa), pH = 5.7) supplemented with 1% sucrose and 0.8% plant agar (Duchefa, Haarlem, Netherlands. At 4 days, seedlings were aseptically transferred to new plates and cultivated in a vertical position in cultivation chambers (Snijders) at 22 °C under long-day light conditions. After one week (11 days after germination), the Petri dishes were scanned (Epson Perfection V700 Photo, Suwa, Japanc), the root length was measured and the rosettes were cut and weighted. Root length was measured by FiJi software [50].

To investigate the effect of light on root development, the seedlings were grown under continuous exposure to light at different intensities, 450 $\mu E.m^{-2}.s^{-1}$ or 170 $\mu E.m^{-2}.s^{-1}$, or in the dark (plates were put in black chambers to shadow roots, Supplementary Figure S5B). To investigate the kinetics of root growth, the primary root length was monitored daily from 4 dpg to 8 dpg and measured using FiJi [50]. At least 10 roots were analysed for each condition.

4.2. Plant Phenotyping

Rosette size of soil-grown plants and primary root length of seedlings were measured by FiJi (area tool) [50]. Rosette weight of 11 day old seedlings was determined using analytical scales.

4.3. SA Concentration Measurements

Leaf tissue was collected from three plants (approximately 100 mg, three 6 mm discs from three leaves) in Eppendorf tubes with 1 g ceramic beads and frozen in liquid nitrogen. Hormone extraction procedure and salicylic acid content measurement were done as in [51]. Briefly, frozen samples were homogenized in tubes with silica beads using a FastPrep-24 instrument (MP Biomedicals, CA, United States) with extraction reagent methanol/water/formic acid (15:4:1, *v*/*v*/*v*) supplemented with stable-isotope-labelled 13C-SA internal standards. Extracts were subjected to solid phase extraction using Oasis MCX cartridges (Waters Co., Milford, MA, United States) and eluted with methanol. The eluate was evaporated to dryness and dissolved in 15% acetonitrile/water (*v*/*v*) immediately before the analysis. Quantification was performed on an Ultimate 3000 high-performance liquid chromatograph (UHPLC, Dionex; Thermo Fisher Scientific, Waltham, MA, United States) coupled to a IMPACT II Q-TOF ultra-high resolution and high-mass-accuracy mass spectrometer (HRAM-MS; Bruker Daltonik, Bremen, Germany). Separation was carried out using an Acclaim RSLC 120 C18 column (2.2 m, 2.1 × 100 mm; Thermo Fisher Scientific, Waltham, MA, United States) mobile phase consisting of 0.1% formic acid (A) and methanol (B) by gradient elution. The full-scan data were recorded in negative electrospray ionization (ESI) mode.

4.4. Gene Transcription Analysis

Total RNA was extracted as in [51]. Briefly, plant tissue was frozen in liquid nitrogen. The tissue was homogenized in plastic Eppendorf tubes with silica beads using a FastPrep-24 instrument (MP Biomedicals, USA). Total RNA was isolated using Spectrum Plant Total RNA kit (Sigma-Aldrich, St Louis, Missouri, USA) and treated with a DNA-free kit (Ambion, Austin, Texas, USA). Subsequently, 1 µg of RNA was converted into cDNA with M-MLV RNase H– Point Mutant reverse transcriptase (Promega Corp., Madison, Wisconsin, USA) and an anchored oligo dT21 primer (Metabion, Planegg, Germany). Transcription of *PR-1* and *ICS1* genes was determined using real-time qPCR. Gene transcription values were normalized to *TIP41*. The primers used are listed in the Supplementary Table S1.

4.5. Photosynthetic Parameter Analysis

Plants were put in the dark for 15 min, and then the photosynthetic parameters were measured using FluorCam Handy FC 1000-H (PSI, Drasov, Czech Republic). Images of whole plants were taken. Chlorophyll fluorescence images were analysed using FluorCam 7.0 (PSI) software. Non-photochemical quenching (NPQ) was calculated as (Fm − Fm')/Fm' and maximum quantum efficiency of PS II photochemistry (QY) was calculated as Fv/Fm. Fm and Fm´ are the maximal fluorescence level from the dark-adapted and light-adapted leaf, respectively, and Fv is variable fluorescence from the dark-adapted leaf [52].

4.6. Statistical Analysis

All experiments were done in three biological repetitions. For soil-grown plants, $n = 24$; for in vitro grown plants, $n \geq 10$ for each genotype. Graphs display analysis of all values together, unless stated otherwise. Student's *t*-test and one-way ANOVA with Tukey's post hoc test were applied for the comparisons, $p < 0.05$. Correlation analysis was done using R software, *Corrplot* package [53]. Pearson coefficients were quantified, and only the values that passed the $p < 0.05$ threshold are displayed.

5. Conclusions

In this study, we introduced a new tool for studying the role of SA role in plants, the Arabidopsis "SA collection". It provides a robust tool benefitting from the distinct origin of the modulated SA pathway in *Arabidopsis thaliana*. The effective usage of the SA collection was demonstrated by phenotyping under different growing conditions, in soil and in vitro, using several light regimes. First, our data confirmed the correlation of SA content and expression of SA-related genes in different cultivation setups. Second, we clearly showed that SA is responsible for the regulation of rosettes, but not growth. Additionally, the SA collection revealed that a high basal SA content makes rosettes more sensitive to light. Surprisingly, we reassessed that *pmr4* mutant is not SA-OA under basal growing conditions. The presented SA collection is a starting point for future research trying to determine the roles of SA in response to environmental changes and to shed light on the complexity of SA-triggered signalling.

Supplementary Materials: Supplementary materials can be found at http://www.mdpi.com/1422-0067/20/24/6365/s1.

Author Contributions: Conceptualization, M.J. and T.K.; methodology, M.J., T.K., Z.K., P.M., H.L.; formal analysis, K.P., V.S., H.L., A.S., O.I., T.K.; resources, L.B., O.V.; data curation, K.P., V.S., R.P., H.L., P.M., P.K., O.I., A.S., Z.K.; writing—original draft preparation, M.J., T.K., K.P.; writing—review and editing, M.J., T.K.; project administration, L.B.; funding acquisition, O.V., L.B. All the authors commented and approved the text.

Funding: This work was supported by Czech Science Foundation [GAČR grant no. 17-05151S]. Salicylic acid measurements were performed in the frame of METROFOOD-CZ research infrastructure project (MEYS Grant No: LM2018100) including access to its facilities. TK benefited from the Program of Postdoctoral Fellowships of the Czech Academy of Sciences [PPPLZ grant no. TK 919220]. OI received the Visegrad scholarship (2018-2019) [grant no. 51810647]. HL benefited from Charles University student grants [GAUK no. 992416 and SVV260427/2019].. The work was also supported by European Regional Development Fund, Project "Centre for Experimental Plant Biology" [grant no. CZ.02.1.01/0.0/0.0/16_019/0000738]. MJ was also supported by the European Structural and Investment Funds, OP RDE-funded project 'CHEMFELLS4UCTP' (No. CZ.02.2.69/0.0/0.0/17_050/0008485).

Acknowledgments: We would like to thank all colleagues who kindly provided us the seeds necessary for establishing "SA-collection". In particular, we thank Jitka Ortmannová (*exo70b1-1*), Xinnian Dong (*cpr5*), Jian Hua (*bon1; bon1/snc1-11*), Ivo Feussner (*fah1/fah2*) and Hua Lu (*acd6*). We also would like to thank Andrea Kung Wai for help with English editing.

Conflicts of Interest: The authors declare that there is no conflict of interest.

Abbreviations

SA	Salicylic acid
SA-OA	Salicylic acid overaccumulating mutants
ICS1	Isochorismate synthase 1
PR1	Pathogenesis related protein 1
SD	Short day conditions
LD	Long day conditions

References

1. Rajjou, L.; Belghazi, M.; Huguet, R.; Robin, C.; Moreau, A.; Job, C.; Job, D. Proteomic Investigation of the Effect of Salicylic Acid on Arabidopsis Seed Germination and Establishment of Early Defense Mechanisms. *Plant Physiol.* **2006**, *141*, 910–923. [CrossRef] [PubMed]
2. Vanacker, H.; Lu, H.; Rate, D.N.; Greenberg, J.T. A role for salicylic acid and NPR1 in regulating cell growth in Arabidopsis. *Plant J. Cell Mol. Biol.* **2001**, *28*, 209–216. [CrossRef] [PubMed]
3. Martínez, C.; Pons, E.; Prats, G.; León, J. Salicylic acid regulates flowering time and links defence responses and reproductive development. *Plant J. Cell Mol. Biol.* **2004**, *37*, 209–217. [CrossRef]
4. Jayakannan, M.; Bose, J.; Babourina, O.; Rengel, Z.; Shabala, S. Salicylic acid improves salinity tolerance in Arabidopsis by restoring membrane potential and preventing salt-induced K+ loss via a GORK channel. *J. Exp. Bot.* **2013**, *64*, 2255–2268. [CrossRef] [PubMed]
5. Bravo, R.E.; Chen, G.; Grosser, K.; Dam, N.M.V.; Leiss, K.A.; Klinkhamer, P.G.L. Ultraviolet radiation enhances salicylic acid-mediated defense signaling and resistance to Pseudomonas syringae DC3000 in a jasmonic acid-deficient tomato mutant. *Plant Signal. Behav.* **2019**, *14*, e1581560. [CrossRef] [PubMed]
6. Klessig, D.F.; Choi, H.W.; Dempsey, D.A. Systemic Acquired Resistance and Salicylic Acid: Past, Present, and Future. *Mol. Plant Microbe Interact. MPMI* **2018**, *31*, 871–888. [CrossRef] [PubMed]
7. Zhang, Y.; Li, X. Salicylic acid: Biosynthesis, perception, and contributions to plant immunity. *Curr. Opin. Plant Biol.* **2019**, *50*, 29–36. [CrossRef]
8. Wildermuth, M.C.; Dewdney, J.; Wu, G.; Ausubel, F.M. Isochorismate synthase is required to synthesize salicylic acid for plant defence. *Nature* **2001**, *414*, 562–565. [CrossRef]
9. Delaney, T.P.; Uknes, S.; Vernooij, B.; Friedrich, L.B.; Weymann, K.; Negrotto, D.V.; Gaffney, T.; Gut-Rella, M.; Kessmann, H.; Ward, E.; et al. A central role of salicylic Acid in plant disease resistance. *Science* **1994**, *266*, 1247–1250. [CrossRef]
10. Fragnière, C.; Serrano, M.; Abou-Mansour, E.; Métraux, J.P.; L'Haridon, F. Salicylic acid and its location in response to biotic and abiotic stress. *FEBS Lett.* **2011**, *585*, 1847–1852. [CrossRef]
11. Glazebrook, J. Contrasting mechanisms of defense against biotrophic and necrotrophic pathogens. *Annu. Rev. Phytopathol.* **2005**, *43*, 205–227. [CrossRef] [PubMed]
12. Canet, J.V.; Dobón, A.; Roig, A.; Tornero, P. Structure-function analysis of npr1 alleles in Arabidopsis reveals a role for its paralogs in the perception of salicylic acid. *Plant Cell Environ.* **2010**, *33*, 1911–1922. [CrossRef] [PubMed]
13. Ding, Y.; Shaholli, D.; Mou, Z. A large-scale genetic screen for mutants with altered salicylic acid accumulation in Arabidopsis. *Front. Plant Sci.* **2015**, *5*, 763. [CrossRef] [PubMed]
14. Rivas-San Vicente, M.; Plasencia, J. Salicylic acid beyond defence: Its role in plant growth and development. *J. Exp. Bot.* **2011**, *62*, 3321–3338. [CrossRef] [PubMed]
15. Li, Y.; Yang, S.; Yang, H.; Hua, J. The TIR-NB-LRR Gene *SNC1* Is Regulated at the Transcript Level by Multiple Factors. *Mol. Plant. Microbe Interact.* **2007**, *20*, 1449–1456. [CrossRef] [PubMed]
16. Lin, F.; Krishnamoorthy, P.; Schubert, V.; Hause, G.; Heilmann, M.; Heilmann, I. A dual role for cell plate-associated PI4Kβ in endocytosis and phragmoplast dynamics during plant somatic cytokinesis. *EMBO J.* **2019**, *38*, e100303. [CrossRef] [PubMed]
17. Kalachova, T.; Janda, M.; Šašek, V.; Ortmannová, J.; Nováková, P.; Dobrev, I.P.; Kravets, V.; Guivarc'h, A.; Moura, D.; Burketová, L.; et al. Identification of salicylic acid-independent responses in an Arabidopsis phosphatidylinositol 4-kinase beta double mutant. *Ann. Bot.* **2019**. [CrossRef]

18. Kulich, I.; Pečenková, T.; Sekereš, J.; Smetana, O.; Fendrych, M.; Foissner, I.; Höftberger, M.; Žárský, V. Arabidopsis Exocyst Subcomplex Containing Subunit EXO70B1 Is Involved in Autophagy-Related Transport to the Vacuole. *Traffic* **2013**, *14*, 1155–1165. [CrossRef]
19. König, S.; Feussner, K.; Schwarz, M.; Kaever, A.; Iven, T.; Landesfeind, M.; Ternes, P.; Karlovsky, P.; Lipka, V.; Feussner, I. Arabidopsis mutants of sphingolipid fatty acid α-hydroxylases accumulate ceramides and salicylates. *New Phytol.* **2012**, *196*, 1086–1097. [CrossRef]
20. Yoshida, S.; Ito, M.; Nishida, I.; Watanabe, A. Identification of a novel gene HYS1/CPR5 that has a repressive role in the induction of leaf senescence and pathogen-defence responses in Arabidopsis thaliana. *Plant J.* **2002**, *29*, 427–437. [CrossRef]
21. Rate, D.N.; Cuenca, J.V.; Bowman, G.R.; Guttman, D.S.; Greenberg, J.T. The Gain-of-Function Arabidopsis acd6 Mutant Reveals Novel Regulation and Function of the Salicylic Acid Signaling Pathway in Controlling Cell Death, Defenses, and Cell Growth. *Plant Cell* **1999**, *11*, 1695–1708. [CrossRef] [PubMed]
22. Preuss, M.L.; Schmitz, A.J.; Thole, J.M.; Bonner, H.K.S.; Otegui, M.S.; Nielsen, E. A role for the RabA4b effector protein PI-4Kbeta1 in polarized expansion of root hair cells in Arabidopsis thaliana. *J. Cell Biol.* **2006**, *172*, 991–998. [CrossRef] [PubMed]
23. Vorwerk, S.; Schiff, C.; Santamaria, M.; Koh, S.; Nishimura, M.; Vogel, J.; Somerville, C.; Somerville, S. EDR2 negatively regulates salicylic acid-based defenses and cell death during powdery mildew infections of Arabidopsis thaliana. *BMC Plant Biol.* **2007**, *7*, 35. [CrossRef] [PubMed]
24. Nishimura, M.T.; Stein, M.; Hou, B.H.; Vogel, J.P.; Edwards, H.; Somerville, S.C. Loss of a callose synthase results in salicylic acid-dependent disease resistance. *Science* **2003**, *301*, 969–972. [CrossRef] [PubMed]
25. Šašek, V.; Janda, M.; Delage, E.; Puyaubert, J.; Guivarc'h, A.; Maseda, E.L.; Dobrev, P.I.; Caius, J.; Bóka, K.; Valentová, O.; et al. Constitutive salicylic acid accumulation in pi4kIIIβ1β2 Arabidopsis plants stunts rosette but not root growth. *New Phytol.* **2014**, *203*, 805–816. [CrossRef] [PubMed]
26. Nawrath, C.; Métraux, J.P. Salicylic Acid Induction—Deficient Mutants of Arabidopsis Express PR-2 and PR-5 and Accumulate High Levels of Camalexin after Pathogen Inoculation. *Plant Cell* **1999**, *11*, 1393–1404.
27. Maxwell, K.; Johnson, G.N. Chlorophyll fluorescence—A practical guide. *J. Exp. Bot.* **2000**, *51*, 659–668. [CrossRef]
28. Hendrickson, L.; Furbank, R.T.; Chow, W.S. A Simple Alternative Approach to Assessing the Fate of Absorbed Light Energy Using Chlorophyll Fluorescence. *Photosynth. Res.* **2004**, *82*, 73. [CrossRef]
29. Yokawa, K.; Kagenishi, T.; Kawano, T.; Mancuso, S.; Baluška, F. Illumination of *Arabidopsis* roots induces immediate burst of ROS production. *Plant Signal. Behav.* **2011**, *6*, 1460–1464. [CrossRef]
30. Silva-Navas, J.; Moreno-Risueno, M.A.; Manzano, C.; Pallero-Baena, M.; Navarro-Neila, S.; Téllez-Robledo, B.; Garcia-Mina, J.M.; Baigorri, R.; Gallego, F.J.; Pozo, J.C. D-Root: A system for cultivating plants with the roots in darkness or under different light conditions. *Plant J.* **2015**, *84*, 244–255. [CrossRef]
31. Janda, M.; Ruelland, E. Magical mystery tour: Salicylic acid signalling. *Environ. Exp. Bot.* **2015**, *114*, 117–128. [CrossRef]
32. Nie, H.; Wu, Y.; Yao, C.; Tang, D. Suppression of edr2-mediated powdery mildew resistance, cell death and ethylene-induced senescence by mutations in ALD1 in Arabidopsis. *J. Genet. Genom. Yi Chuan Xue Bao* **2011**, *38*, 137–148. [CrossRef] [PubMed]
33. Smith, L.M.; Bomblies, K.; Weigel, D. Complex evolutionary events at a tandem cluster of Arabidopsis thaliana genes resulting in a single-locus genetic incompatibility. *PLoS Genet.* **2011**, *7*, e1002164. [CrossRef] [PubMed]
34. Zhu, W.; Zaidem, M.; Van de Weyer, A.L.; Gutaker, R.M.; Chae, E.; Kim, S.T.; Bemm, F.; Li, L.; Todesco, M.; Schwab, R.; et al. Modulation of ACD6 dependent hyperimmunity by natural alleles of an Arabidopsis thaliana NLR resistance gene. *PLOS Genet.* **2018**, *14*, e1007628. [CrossRef] [PubMed]
35. Miura, K.; Lee, J.; Miura, T.; Hasegawa, P.M. SIZ1 controls cell growth and plant development in Arabidopsis through salicylic acid. *Plant Cell Physiol.* **2010**, *51*, 103–113. [CrossRef]
36. Todesco, M.; Kim, S.T.; Chae, E.; Bomblies, K.; Zaidem, M.; Smith, L.M.; Weigel, D.; Laitinen, R.A.E. Activation of the Arabidopsis thaliana Immune System by Combinations of Common ACD6 Alleles. *PLoS Genet.* **2014**, *10*, e1004459. [CrossRef] [PubMed]
37. Ellinger, D.; Naumann, M.; Falter, C.; Zwikowics, C.; Jamrow, T.; Manisseri, C.; Somerville, S.C.; Voigt, C.A. Elevated Early Callose Deposition Results in Complete Penetration Resistance to Powdery Mildew in Arabidopsis. *Plant Physiol.* **2013**, *161*, 1433–1444. [CrossRef]

38. Khurana, J.P.; Cleland, C.F. Role of Salicylic Acid and Benzoic Acid in Flowering of a Photoperiod-Insensitive Strain, Lemna paucicostata LP6 1. *Plant Physiol.* **1992**, *100*, 1541–1546. [CrossRef]
39. Martínez-Medina, A.; Appels, F.V.W.; van Wees, S.C.M. Impact of salicylic acid- and jasmonic acid-regulated defences on root colonization by Trichoderma harzianum T-78. *Plant Signal. Behav.* **2017**, *12*, e1345404. [CrossRef]
40. Badri, D.V.; Loyola-Vargas, V.M.; Du, J.; Stermitz, F.R.; Broeckling, C.D.; Iglesias-Andreu, L.; Vivanco, J.M. Transcriptome analysis of Arabidopsis roots treated with signaling compounds: A focus on signal transduction, metabolic regulation and secretion. *New Phytol.* **2008**, *179*, 209–223. [CrossRef]
41. Pasternak, T.; Groot, E.P.; Kazantsev, F.V.; Teale, W.; Omelyanchuk, N.; Kovrizhnykh, V.; Palme, K.; Mironova, V.V. Salicylic Acid Affects Root Meristem Patterning via Auxin Distribution in a Concentration-Dependent Manner. *Plant Physiol.* **2019**, *180*, 1725–1739. [CrossRef]
42. Liu, P.; Xu, Z.S.; Pan-Pan, L.; Hu, D.; Chen, M.; Li, L.C.; Ma, Y.Z. A wheat PI4K gene whose product possesses threonine autophophorylation activity confers tolerance to drought and salt in Arabidopsis. *J. Exp. Bot.* **2013**, *64*, 2915–2927. [CrossRef] [PubMed]
43. Borghi, M.; Rus, A.; Salt, D.E. Loss-of-Function of Constitutive Expresser of Pathogenesis Related Genes5 Affects Potassium Homeostasis in Arabidopsis thaliana. *PLoS ONE* **2011**, *6*, e26360. [CrossRef] [PubMed]
44. Gao, G.; Zhang, S.; Wang, C.; Yang, X.; Wang, Y.; Su, X.; Du, J.; Yang, C. Arabidopsis CPR5 independently regulates seed germination and postgermination arrest of development through LOX pathway and ABA signaling. *PLoS ONE* **2011**, *6*, e19406. [CrossRef] [PubMed]
45. Okuma, E.; Nozawa, R.; Murata, Y.; Miura, K. Accumulation of endogenous salicylic acid confers drought tolerance to Arabidopsis. *Plant Signal. Behav.* **2014**, *9*. [CrossRef] [PubMed]
46. Aki, T.; Konishi, M.; Kikuchi, T.; Fujimori, T.; Yoneyama, T.; Yanagisawa, S. Distinct modulations of the hexokinase1-mediated glucose response and hexokinase1-independent processes by HYS1/CPR5 in Arabidopsis. *J. Exp. Bot.* **2007**, *58*, 3239–3248. [CrossRef] [PubMed]
47. Miller, R.N.G.; Costa Alves, G.S.; Van Sluys, M.A. Plant immunity: Unravelling the complexity of plant responses to biotic stresses. *Ann. Bot.* **2017**, *119*, 681–687. [CrossRef] [PubMed]
48. Wang, W.; Wang, Z.Y. At the intersection of plant growth and immunity. *Cell Host Microbe* **2014**, *15*, 400–402. [CrossRef]
49. Karasov, T.L.; Chae, E.; Herman, J.J.; Bergelson, J. Mechanisms to Mitigate the Trade-Off between Growth and Defense. *Plant Cell* **2017**, *29*, 666–680. [CrossRef]
50. Schindelin, J.; Arganda-Carreras, I.; Frise, E.; Kaynig, V.; Longair, M.; Pietzsch, T.; Preibisch, S.; Rueden, C.; Saalfeld, S.; Schmid, B.; et al. Fiji: An open-source platform for biological-image analysis. *Nat. Methods* **2012**, *9*, 676–682. [CrossRef]
51. Kalachova, T.; Leontovyčová, H.; Iakovenko, O.; Pospíchalová, R.; Maršík, P.; Klouček, P.; Janda, M.; Valentová, O.; Kocourková, D.; Martinec, J.; et al. Interplay between phosphoinositides and actin cytoskeleton in the regulation of immunity related responses in Arabidopsis thaliana seedlings. *Environ. Exp. Bot.* **2019**, *167*, 103867.
52. Huang, J.; Gu, M.; Lai, Z.; Fan, B.; Shi, K.; Zhou, Y.H.; Yu, J.Q.; Chen, Z. Functional Analysis of the Arabidopsis PAL Gene Family in Plant Growth, Development, and Response to Environmental Stress. *Plant Physiol.* **2010**, *153*, 1526–1538. [CrossRef] [PubMed]
53. Simko, W. R Package "Corrplot": Visualization of a Correlation Matrix (Version 0.84). Available online: https://github.com/taiyun/corrplot (accessed on 2 August 2018).

International Journal of *Molecular Sciences*

MDPI

Article

Polyamine-Induced Hormonal Changes in *eds5* and *sid2* Mutant *Arabidopsis* Plants

Judit Tajti [1], Kamirán Áron Hamow [2], Imre Majláth [1], Krisztián Gierczik [3], Edit Németh [1], Tibor Janda [1] and Magda Pál [1,*]

1 Plant Physiology Department, Agricultural Institute, Centre for Agricultural Research, H-2462 Martonvásár, Hungary; tajti.judit@agrar.mta.hu (J.T.); majlath.imre@agrar.mta.hu (I.M.); nedokka@gmail.com (E.N.); janda.tibor@agrar.mta.hu (T.J.)

2 Plant Protection Institute, Centre for Agricultural Research, H-2462 Martonvásár, Hungary; hamow.kamiran@agrar.mta.hu

3 Department of Plant Molecular Biology, Agricultural Institute, Centre for Agricultural Research, H-2462 Mrtonvásár, Hungary; gierczik@gmail.com

* Correspondence: pal.magda@agrar.mta.hu; Tel.: +36-22-569-502; Fax: +36-22-569-576

Received: 22 October 2019; Accepted: 14 November 2019; Published: 15 November 2019

Abstract: Polyamines are multifaceted compounds which play a role in regulating plant growth and stress tolerance in interactions with plant hormones. The aim of the present study was to reveal how exogenous polyamines influence the synthesis of salicylic acid, with a special emphasis on the effect of salicylic acid deficiency on the polyamine metabolism and polyamine-induced changes in other plant hormone contents. Our hypothesis was that the individual polyamines induced different changes in the polyamine and salicylic acid metabolism of the wild type and salicylic acid-deficient *Arabidopsis* mutants, which in turn influenced other hormones. To our knowledge, such a side-by-side comparison of the influence of *eds5-1* and *sid2-2* mutations on polyamines has not been reported yet. To achieve our goals, wild and mutant genotypes were tested after putrescine, spermidine or spermine treatments. Polyamine and plant hormone metabolism was investigated at metabolite and gene expression levels. Individual polyamines induced different changes in the *Arabidopsis* plants, and the responses were also genotype-dependent. Polyamines upregulated the polyamine synthesis and catabolism, and remarkable changes in hormone synthesis were found especially after spermidine or spermine treatments. The *sid2-2* mutant showed pronounced differences compared to Col-0. Interactions between plant hormones may also be responsible for the observed differences.

Keywords: polyamine; salicylic acid; plant hormone; *Arabidopsis*; *eds5-1*; *sid2-2*

1. Introduction

Polyamines (PAs) are low molecular weight aliphatic amines containing two or more amino groups found in all living cells [1]. In higher plants, PAs are mainly present in their free form, and have several potent biological activities. Thus, PAs are considered to be a new group of growth regulators in plants [2]. The total and individual PAs levels vary depending on the plant species, the organ and on the developmental stage; furthermore, their metabolism is dynamic, due to the polyamine cycle [3]. Experiments on treatments with both PAs and PA synthesis inhibitors revealed the role of PAs in e.g., flower induction [4], embryogenesis [5], regulation of nucleic acid synthesis and protein translation, development of organelles [6], and senescence [7], and numerous studies have been published on the relationship between the enhanced synthesis of PAs and the level of a stress tolerance [8–10]. A positive role of the early stress-induced activation or the overexpressing of certain PA-biosynthesis genes has also been proved (arginine decarboxylase: *ADC*, spermidine synthase: *SPDS* and spermine synthase: *SPMS*) [11–16]. In addition, microarray studies have shown that increased endogenous PA content can alter the gene expression level of genes involved in the synthesis and signalling of several plant hormones, such as abscisic acid (ABA), auxin, ethylene, gibberellins (GAs), jasmonic acid (JA) or salicylic acid (SA) in transgenic *Arabidopsis* plants [17]. It is thus becoming more evident that PAs are also switching points in plant signalling pathways, and the induced plant responses are interconnected at many levels with other metabolic routes and hormonal cross-talk and activate gene expression, which has a predominant role in the PA-signalling processes compared only to PA accumulation [18].

Most studies have focused on the protective effects of exogenous PAs, but their role is more complicated. It is difficult to establish a direct relationship between PAs, especially the individual levels of the most abundant Pas—namely putrescine (PUT), spermidine (SPD) and spermine (SPM)—and the observed beneficial effects. In addition, an excess of PAs can be harmful to the plants [15]. According to these facts, changes in the PA levels are important for fine-tuning PA signalling, which influences the hormonal balance required for them to exert their positive role in regulating plant growth and stress tolerance [15].

Earlier it was demonstrated that SA treatment affects the synthesis and/or catabolism of PAs [19–22]. Vice versa, it was shown that SPD treatment increased SA content in the leaves of wheat, but PUT or SPD treatments decreased it in the roots [23], whereas SPD did not affect endogenous SA content in cucumber [24]. In our previous study, it was found that there is a close, positive relationship between PAs and SA accumulation after SPD and SPM treatments in wheat and maize plants [15]. SA synthesis starts from chorismate as a precursor, after which the synthesis pathway branches into two separate pathways. SA can be synthesized through the phenylalanine ammonia-lyase (PAL) pathway from phenylalanine via benzoic acid (BA), while on the isochorismate pathway the enzyme isochorismate synthsase (ICS) catalyzes isochorismate formation from chorismate. In *Arabidopsis*, the primary route for SA production is the isochorismate pathway, and there are two genes *ICS1* and *ICS2* encoding ICS, but it was demonstrated that ICS1 is responsible for the main source of isochorismate [25]. Excessive SA accumulation can be detrimental to plants under stress conditions. For example, a negative relationship was observed between the SA content and growth parameters of *cpr1 Arabidopsis* mutant plants, which showed higher levels of free and bound SA and increased oxidative damage under low temperature stress conditions [26]. Increased SA accumulation was also responsible for the negative effect of SPD and SPM treatments, especially at higher concentration, manifested in growth inhibition in wheat and maize under control conditions [15], and also for the accelerating effect of SPD treatment on cadmium-induced oxidative stress in wheat [16].

However, during the investigation on the relationship between SA and PAs, it should be also taken into consideration that SA may also influence other plant hormones, e.g SA has an antagonistic relation with ABA and JA [27,28], thus the PA-induced hormonal changes may also be affected.

A positive feedback loop has also been found between ABA and PAs, as ABA activates the PA synthesis genes and also that of polyamine oxidase (*PAO*); moreover PA treatment induces ABA synthesis at the gene expression level of 9-*cis*-epoxycarotenoid dioxygenase (*NCED*) [29–31]. In a recent study, PA treatments differently influenced the plant growth parameters of GA-insensitive dwarf (carrying the severe dwarfing allele *Rht-B1c*, responsible for dwarf phenotype) and semi-dwarf Rht (carrying *Rht-B1b* allele) wheat lines compared to the wild, tall line, which effects were in relation with different changes in ABA and SA contents of the three genotypes [32]. These results provide new insights into the role of PAs in plant growth regulation and confirmed their participation in the hormonal balance, however, still little is known about the interplay between PAs and GAs. Methyl jasmonate treatment in barley induced the expression of genes involved in PUT synthesis *ADC* and ornithine decarboxylase, which in turn led to increased PA content in the conjugated fraction [33]. *OsPAO6* has been also reported to be induced by JA [34]. Exogenous PAs, especially SPM, induced JA accumulation after only 1 h of treatment [35].

In order to reveal even more deeply the relationship between PAs and SA at the metabolite and gene expression levels, and the contribution of SA synthesis deficiency to the interplay of PA with other plant hormones, such as ABA, GAs and JA, in the present study, *eds5* and *sid2* mutants of *Arabidopsis*—which do not accumulate SA after pathogen infection or abiotic stresses [36,37]—were tested after PUT, SPD or SPM treatments. Our hypothesis was that the different Pas would induce different changes in the PA and SA metabolism of the wild type and SA-deficient *Arabidopsis* mutants, which in turn would influence other plant hormones. To our knowledge, such a side-by-side comparison of the influence of *eds5* and *sid2* mutation on the effect of different PAs (PUT, SPD and SPM) has not been reported yet. The main questions were: (1) Is there any difference in the PA metabolism between the SA-deficient mutants and the wild type? (2) How do exogenous PAs influence SA synthesis? and (3) Does SA-deficiency have any influence on the PA-induced changes in the ABA, JA and GA contents?

2. Results

2.1. Treatments with PA Repressed the Effective Quantum Yield of PSII

In order to get information about the physiological status of the plants, certain Chl-*a* fluorescence parameters were determined in Col-0, *sid2* and *eds5* plants treated with or without different PAs. The Fv/Fm chlorophyll-*a* fluorescence induction parameter representing the maximum quantum efficiency of PSII, showed similar values in the three genotypes, and was not influenced by any of the PA treatments, indicating that exogenous PA application did not induce severe stress conditions in *Arabidopsis* plants (Figure 1A). However, the effective PSII quantum yield (YII) was significantly lower in the two mutants compared to the wild type, and decreased by all the applied PAs in Col-0 and *sid2* genotypes, while it slightly increased in PUT-treated *eds5* mutant (Figure 1B, Figure S1).

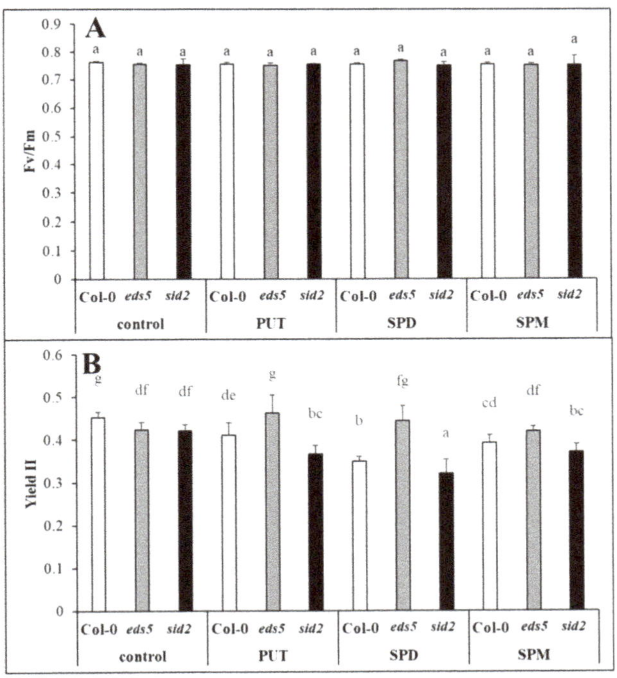

Figure 1. Effect of 0.5 mM 1-day of putrescine (PUT), spermidine (SPD) and spermine (SPM) treatments on the chlorophyll-*a* fluorescence induction parameters ((**A**) Fv/Fm: maximum quantum yield of PSII photochemistry, (**B**) YII: Effective PSII quantum yield) in Col-0, wild type, *eds5-1* (*eds5*) and *sid2-2* (*sid2*) *Arabidopsis* mutants. Data represent mean values ± SD. Different letters indicate significant differences at $p \leq 0.05$ level.

2.2. *Effect of Exogenous PAs on PA Contents and Metabolism*

2.2.1. Exogenous PAs Triggered Endogenous PUT Accumulation by Promoting ADC2 Transcription

Under control conditions, no significant differences were observed between the mutants and the Col-0, wild type regarding on the PUT content or the expression levels of its synthesis genes (*ADC1* and *ADC2*).

After PA treatments, pronounced changes were observed in the PUT contents of *Arabidopsis* leaves, as all the applied treatments increased them, with the highest accumulation being in the case of PUT treatment, followed by SPD and SPM treatments (Figure 2A, Figure S2A). The expression patterns of *ADC1* and *ADC2* showed that PA treatments differentially regulated them (Figure 2B,C, Figure S2B,C). *ADC1* was only slightly induced in PUT- and SPM-treated *sid2* and down-regulated after SPD treatment in all the investigated genotypes, but *ADC2* was significantly induced by all the PA treatments both in the wild type and the mutants. Although the basal gene expression level of *ADC2* did not differ between the three genotypes, interestingly, both after SPD and SPM treatments the highest expression level of *ADC2* was observed in the *sid2* mutant (Figure 2C).

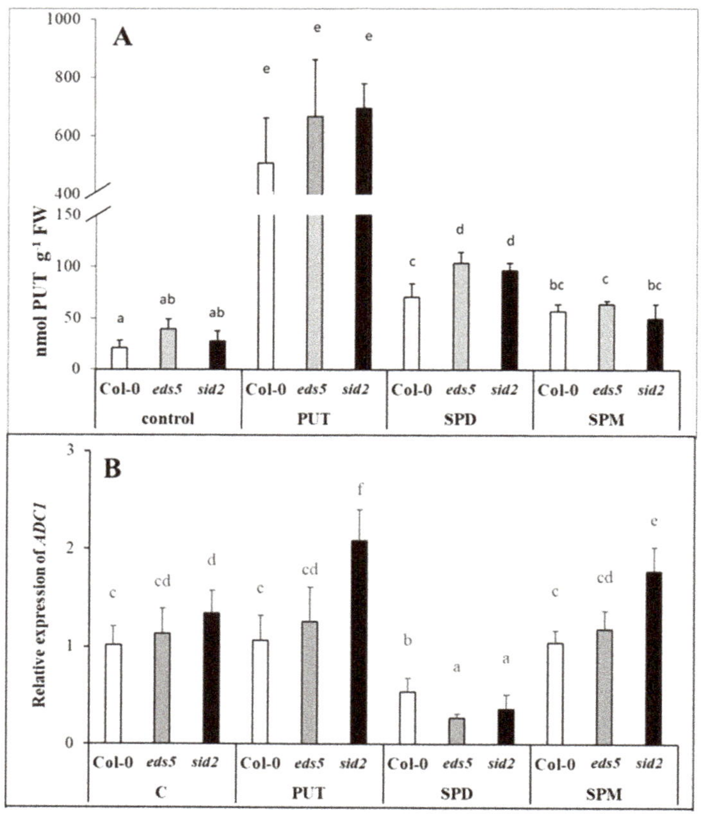

Figure 2. *Cont.*

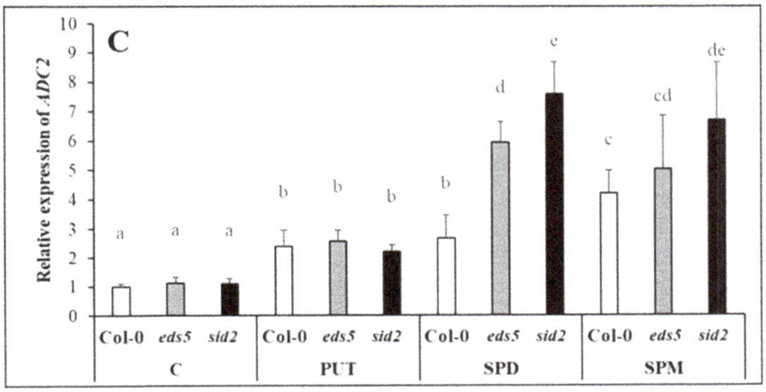

Figure 2. Effect of 0.5 mM 1-day of putrescine (PUT), spermidine (SPD) and spermine (SPM) treatments on the endogenous putrescine content ((**A**) PUT) and on the expression levels of putrescine synthesis genes (**B**,**C**) *ADC1-2: arginine decarboxylase1-2*) in Col-0, wild type, *eds5-1* (*eds5*) and *sid2-2* (*sid2*) *Arabidopsis* mutants. Data represent mean values ± SD. Different letters indicate significant differences at $p \leq 0.05$ level.

2.2.2. Treatment with Higher PAs Down-Regulated the Expression of Genes Involved in SPD Synthesis

The initial SPD content of the Col-0, *eds5* or *sid2* genotypes did not differ (Figure 3A, Figure S3A), and the expression level of *SPDS1* was also similar (Figure 3B, Figure S3B), but a definite difference was found in the level of *SPDS2* transcript between the Col-0/*eds5* and *sid2* genotypes under control conditions (Figure 3C, Figure S3C), as was almost three-fold higher in the latter one. Although the SPD content was not influenced pronouncedly, the gene expression pattern significantly changed after exogenous PA treatments. While PUT treatment decreased the expression of *SPDS1* in either of the genotypes, rather increased that of *SPDS2* in the case of Col-0 and *eds5*. After SPD or SPM treatments the *SPDS1* expression was down-regulated in both the wild and mutant plants, while the initial differences in *SPDS2* expression observed under control conditions were disappeared, as decreased in the *sid2* mutant to a similar level as it was observed for Col-0 (Figure 3B,C). The decrease in *SPDS2* transcript after SPD or SPM treatments was more observable compared to the PUT-treated ones.

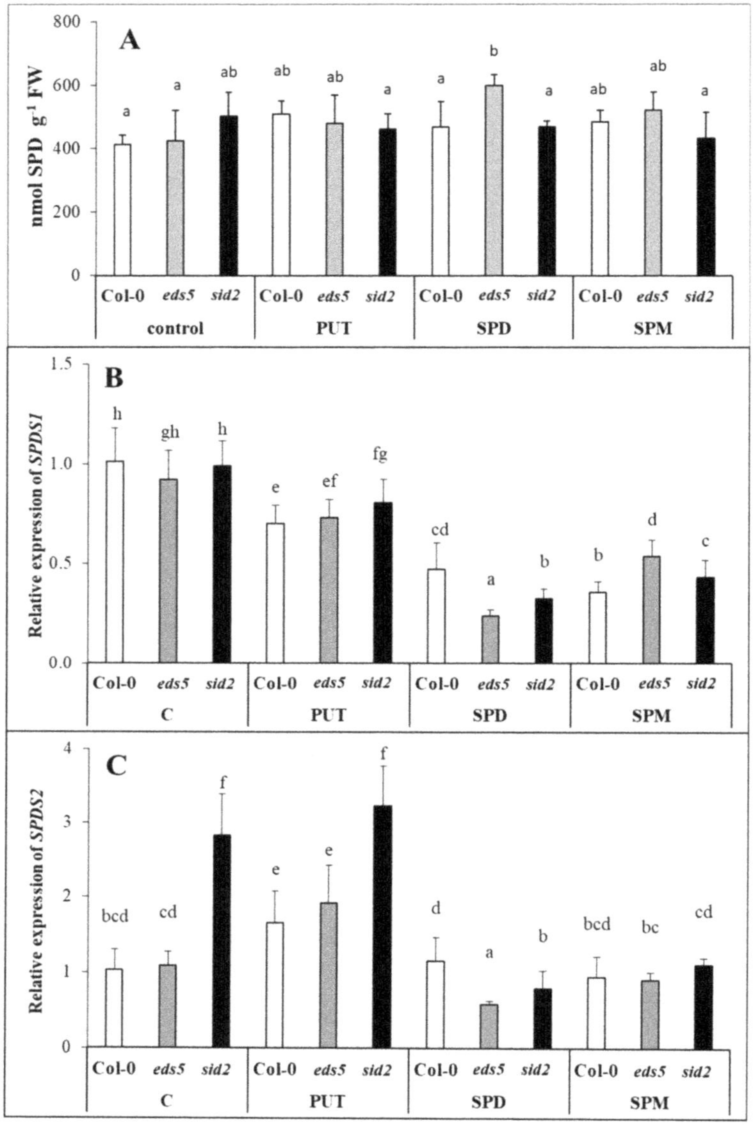

Figure 3. Effect of 0.5 mM 1-day of putrescine (PUT), spermidine (SPD) and spermine (SPM) treatments on spermidine content ((**A**) SPD) and on the expression levels of spermidine synthesis genes (**B**,**C**) *SPDS1-2*: spermidine synthase1-2) in Col-0, wild type, *eds5-1* (*eds5*) and *sid2-2* (*sid2*) *Arabidopsis* mutants. Data represent mean values ± SD. Different letters indicate significant differences at $p \leq 0.05$ level.

2.2.3. Exogenous PAs Up-Regulated the Expression of SPDS

The SPM content in the control plants were in the same range. Significant changes were not detected after PA treatments, except for *sid2*, where SPM treatment increased the SPM content, if it was compared only to the PUT- or SPD-treated ones (Figure 4A, Figure S4A). Despite to these, exogenous application of PUT, SPD or SPM significantly induced the SPMS expression in either of the genotypes, regardless of the type of the mutation, with highest increment in case of the SPM treatment (Figure 4B, Figure S4B).

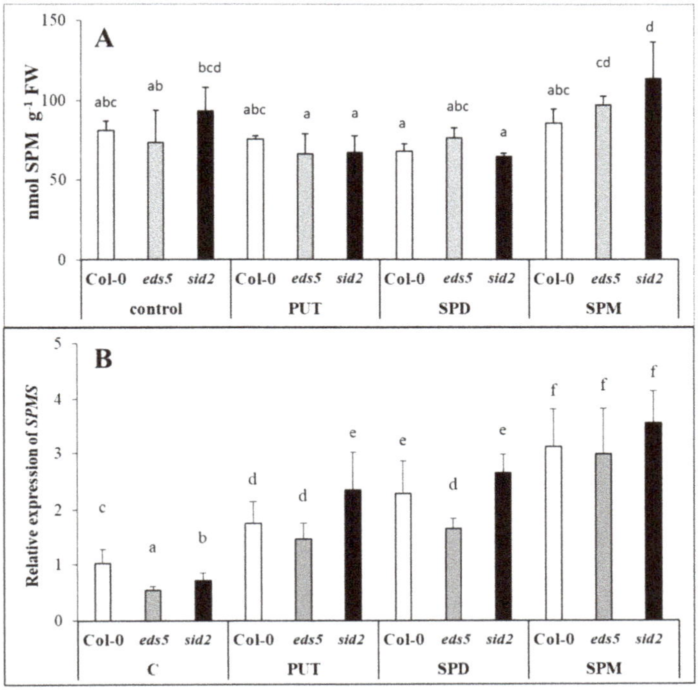

Figure 4. Effect of 0.5 mM 1-day of putrescine (PUT), spermidine (SPD) and spermine (SPM) treatments on spermine content ((**A**) SPM) and on the expression levels of spermine synthase gene ((**B**) *SPMS*) in Col-0, wild type, *eds5-1* (*eds5*) and *sid2-2* (*sid2*) *Arabidopsis* mutants. Data represent mean values ± SD. Different letters indicate significant differences at $p \leq 0.05$ level.

2.2.4. Exogenous PAs Induced the Terminal Catabolism and Not the Back-Conversion of the PAs

The expression levels of the genes encoding PAOs, which are responsible for the back-conversion of the higher polyamines, namely *PAO2* and *PAO5*, and that of *CuAO1*, encoding the copper amine oxidase responsible for terminal catabolism of PUT and SPD, were similar in the three genotypes. *PAO2* and *PAO5* transcript levels did not show remarkable changes after PUT or SPM treatments, but were slightly decreased by SPD in all the genotypes (Figure 5A,B, Figure S4A,B), while the expression of *CuAO1* was induced after all the PA treatments, especially after SPD application (Figure 5C, Figure S5C). An opposite pattern of the expression of *PAOs* and *CuAO1* was seen after SPD treatment, indicating that it rather induced the catabolism and not the conversion back to PUT. Interestingly, the *sid2* mutant showed the highest value in several cases.

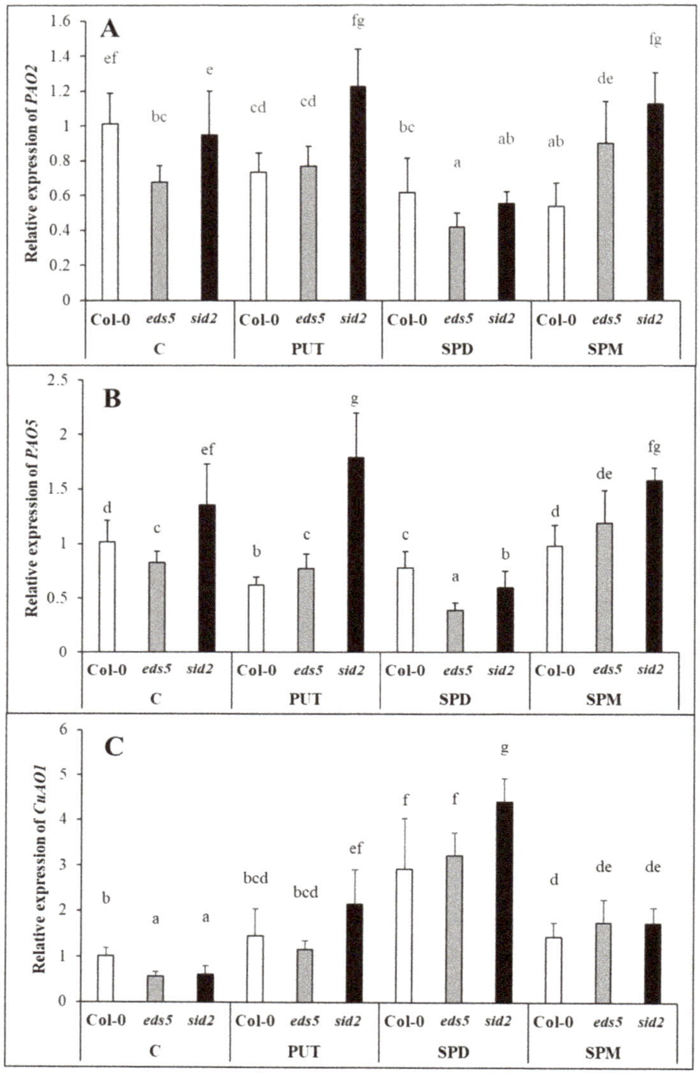

Figure 5. Effect of 0.5 mM 1-day of putrescine (PUT), spermidine (SPD) and spermine (SPM) treatments on the expression levels of polyamine metabolism genes (**A**,**B**) *PAO2-5*: polyamine oxidase2-5, (**C**) *CuAO1*: cooper amine-oxidase1) in Col-0, wild type, *eds5-1* (*eds5*) and *sid2-2* (*sid2*) *Arabidopsis* mutants. Data represent mean values ± SD. Different letters indicate significant differences at $p \leq 0.05$ level.

2.3. Exogenous PAs Induced the PAL Synthesis Pathway

The BA contents of the Col-0 and *eds5* genotypes were almost similar under control conditions and PA treatments did not substantially influenced it (Figure 6A, Figure S6A). Although, as it was expected, pronounced differences were observed in the SA content of the control plants, as its level was lower in the two mutant genotypes, and these differences still remained after the PA treatments (Figure 6B, Figure S6B), exogenous PAs could hardly induce significant changes in the SA contents.

The initial differences in the expression levels of genes involved in SA synthesis were not substantial, except for that *ICS1* was not expressed in the *sid2* mutant. The gene expression level of *CS* was slightly down-regulated in PUT-treated Col-0 and *sid2* plants, but compared to this, it increased in SPM-treated *sid2* plants (Figure 7A, Figure S7A). While the expression of *ICS1* was induced in PUT-treated *eds5*, it decreased after SPD and SPM treatments (Figure 7B, Figure S7B). Interestingly, in the wild type *Arabidopsis* plants the level of SA showed similar pattern to the transcript level of *ICS1*. Compared to these, the *PAL1* expression, likely as a compensation of the mutation of the *ICS1*, showed higher basal level in *sid2* plants (Figure 7C, Figure S7C). In addition, all the PA treatments induced it, with the highest levels in *sid2* mutant, and in case of the SPM treatment.

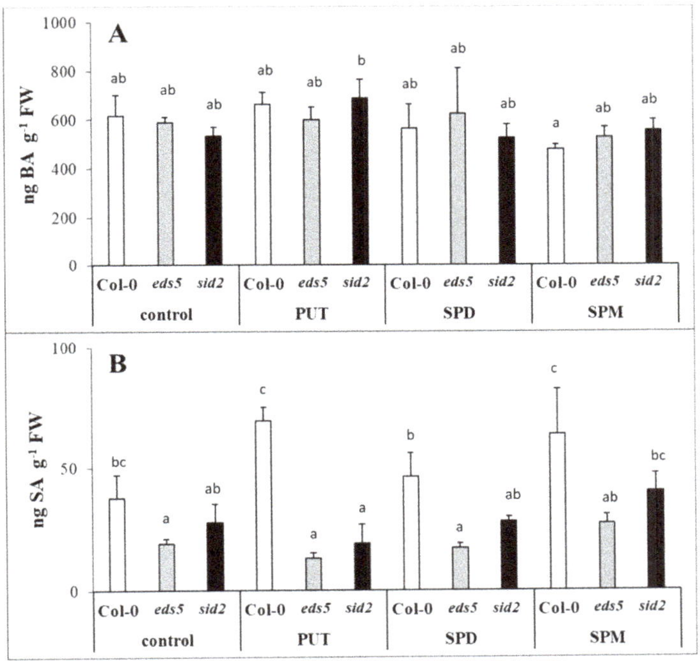

Figure 6. Effect of 0.5 mM 1-day of putrescine (PUT), spermidine (SPD) and spermine (SPM) treatments on the contents of benzoic acid ((**A**) BA) and salicylic acid ((**B**) SA) in Col-0, wild type, *eds5-1* (*eds5*) and *sid2-2* (*sid2*) *Arabidopsis* mutants. Data represent mean values ± SD. Different letters indicate significant differences at $p \leq 0.05$ level.

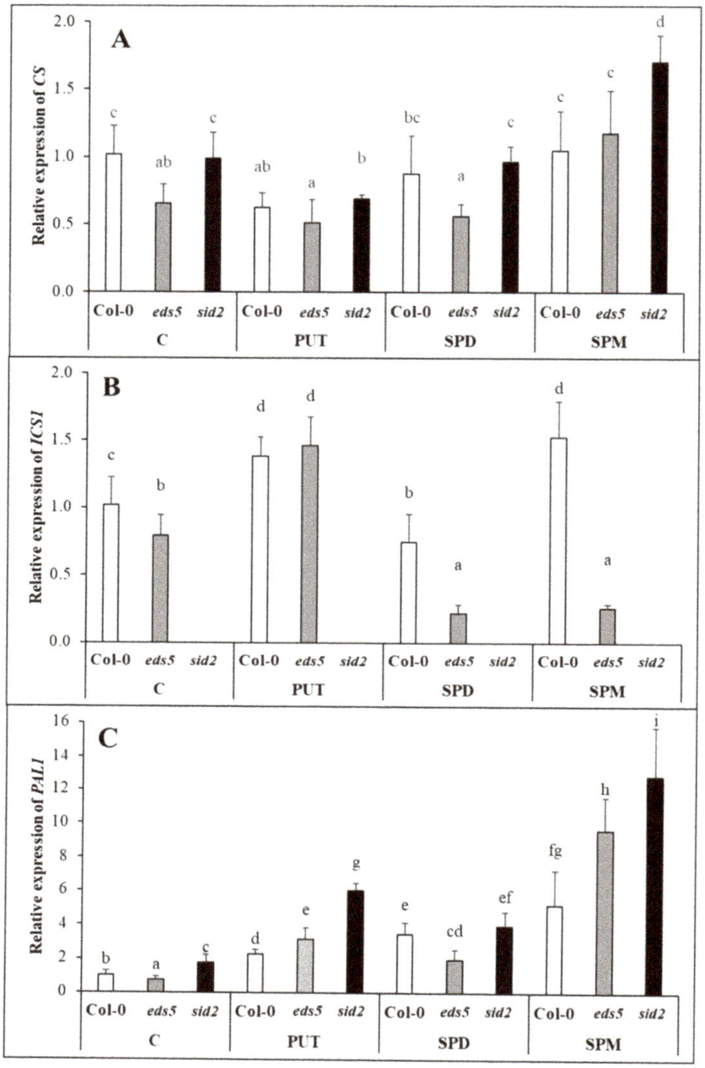

Figure 7. Effect of 0.5 mM 1-day of putrescine (PUT), spermidine (SPD) and spermine (SPM) treatments on the expression levels of genes involved in salicylic acid synthesis ((**A**) *CS*: chorismate synthase, (**B**) *ICS1*: isochorismate synhtase1, (**C**) *PAL1*: phenylalanine ammonia-lyase1) in Col-0, wild type, *eds5-1* (*eds5*) and *sid2-2* (*sid2*) *Arabidopsis* mutants. Data represent mean values ± SD. Different letters indicate significant differences at $p \leq 0.05$ level.

2.4. Exogenous PAs Induced ABA Synthesis

The three genotypes had a similar initial ABA content. All the PA treatments increased the ABA level, interestingly with the highest accumulation in SPD-treated *sid2* mutant (Figure 8A, Figure S8A), where a pronounced difference was observed between the wild types and the *sid2* mutant. The final plastid-localized steps in ABA synthesis is catalysed by NCED. The expression level of *NCED* (Figure 8B, Figure S8B) was in accordance with the changes in ABA content, as the PA treatments up-regulated it, with the highest level in SPD-treated *sid2* mutant.

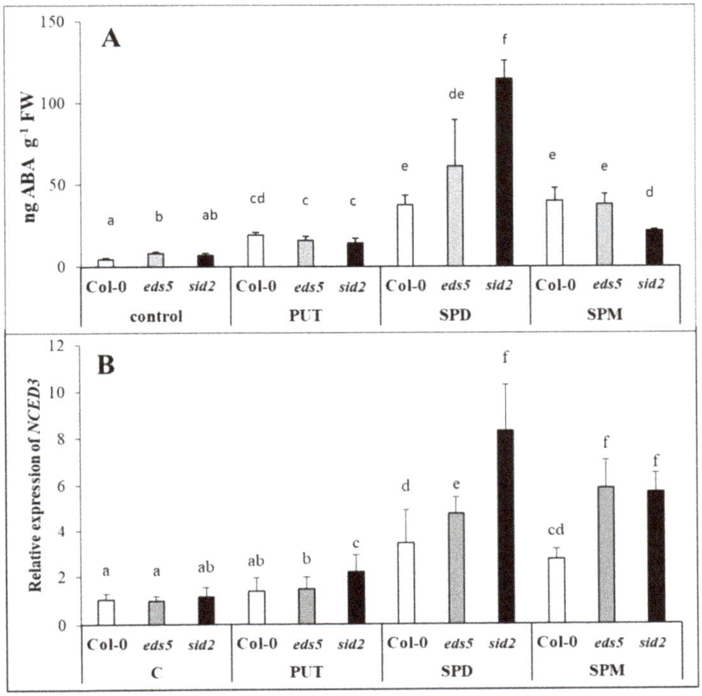

Figure 8. Effect of 0.5 mM 1-day of putrescine (PUT), spermidine (SPD) and spermine (SPM) treatments on the content of abscisic acid ((**A**) ABA) and the expression level of gene involved in abscisic acid synthesis ((**B**) *NCED3:* 9-cis-epoxycarotenoid dioxygenase3) in Col-0, wild type, *eds5-1* (*eds5*) and *sid2-2* (*sid2*) *Arabidopsis* mutants. Data represent mean values ± SD. Different letters indicate significant differences at $p \leq 0.05$ level.

2.5. Exogenous PAs Differently Influenced JA Content and AOS Expression

The JA content did not show any differences between the three genotypes under control conditions or after PA treatments (Figure 9A, Figure S9A). However, a decreasing trend was observed after all the PA treatments, with the highest degree in the case of the SPM treatment. The expression level of the gene encoding allene oxide synthase (AOS), which is one of the synthesis enzymes of JA biosynthesis (Figure 9B, Figure S9B), showed an opposite trend to the changes observed in JA content, as a remarkable induction of it was found in the SPM-treated plants, where the lowest JA accumulation was detected. In addition, remarkable difference was observed in the *AOS* expression of the SPM-treated wild type and *sid2* mutant.

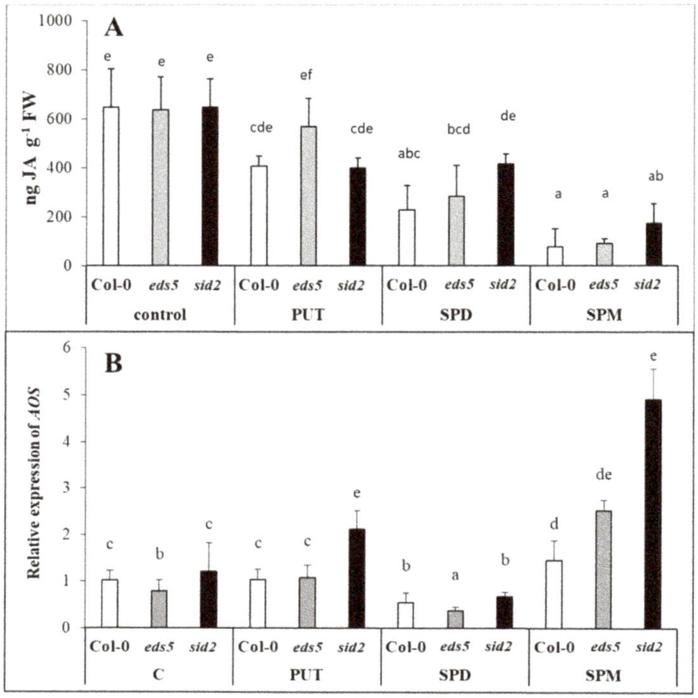

Figure 9. Effect of 0.5 mM 1-day of putrescine (PUT), spermidine (SPD) and spermine (SPM) treatments on the content of jasmonic acid ((**A**) JA) and the expression level of gene involved in jasmonic acid synthesis ((**B**) *AOS:* allene oxide synthase) in Col-0, wild type, *eds5-1* (*eds5*) and *sid2-2* (*sid2*) *Arabidopsis* mutants. Data represent mean values ± SD. Different letters indicate significant differences at $p \leq 0.05$ level.

2.6. PUT and SPD Induced, While SPM Down-Regulated the Expression of GA3ox1

The major bioactive GAs in plants are GA_1 and GA_4; and the final step of the synthesis of these bioactive GAs is catalysed by gibberellin 3-oxidase (GA3ox). Among the monitored $GA_{1,4,8,3,20}$, only the GA_1 could be detected. Although under our experimental conditions no significant difference was detected between the three genotypes, and the PA treatments did not induce remarkable changes in the GA_1 content, a trend towards a higher level of GA_1 could be detected in the *sid2* mutant *Arabidopsis* (Figure 10A, Figure S10A). Interestingly, the *GA3ox1* expression level increased after PUT treatment, and slightly increased also after SPD, but decreased after SPM treatment. In PUT- and SPD-treated plants, the highest up-regulation was found again in *sid2* mutants (Figure 10B, Figure S10B).

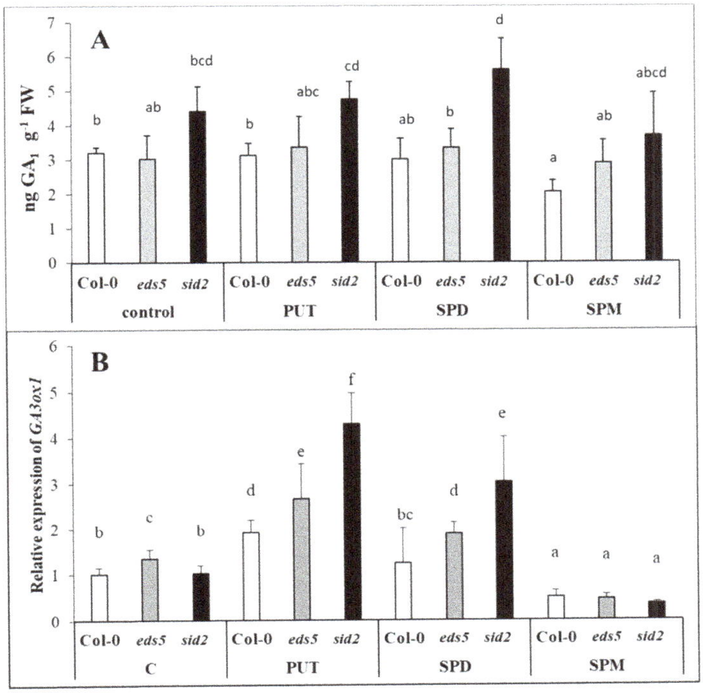

Figure 10. Effect of 0.5 mM 1-day of putrescine (PUT), spermidine (SPD) and spermine (SPM) treatments on the content of gibberellic acid ((**A**) GA1) and the expression level of gene involved in gibberellic acid ((**B**) *GA3ox1:* gibberellin 3-oxidase) in Col-0, wild type, *eds5-1* (*eds5*) and *sid2-2* (*sid2*) *Arabidopsis* mutants. Data represent mean values ± SD. Different letters indicate significant differences at $p \leq 0.05$ level.

3. Discussion

The main aims of the present study were to reveal the possible effects of the *eds5* and *sid2* mutations on the polyamine metabolism in *Arabidopsis* plants, and to evaluate the effects of exogenous PA on SA synthesis, and the influence of SA-deficiency on the PA-induced hormone synthesis.

Besides the visually observed phenotypical differences between the two SA-deficient mutants and wild type, the Y(II) parameter also showed that the two mutants have lower PS II quantum efficiency compared to the Col-0. Similarly, although no differences were observed in the Fv/Fm, but slightly lower effective PSII quantum yields were measured for the SA-deficient *sid2* mutant and *NahG* transgenic line compared to the wild type [38]. Parallel with these, slightly lower stomatal conductance and CO_2 assimilation rate, and lower biomass parameters were measured for *sid2* mutant. These results confirmed that controlled levels of SA are required for optimal photosynthesis. However, SA deficient mutants, namely *eds5* and *sid2* have not been tested under PA treatments yet. Under the present conditions, PA treatments repressed the efficiency of PSII, based on the Y(II) parameter, especially in Col-0 and *sid2* genotypes, while a slight positive effect of PUT treatment has been detected in the case of *eds5* mutant compared the its control. PUT treatment at the same concentration has been reported to have a corroborating effect under control conditions in wheat plants, manifested in higher shoot fresh and dry weight and CO_2 assimilation rate, and the beneficial effect of PUT was accompanied with the lowest SA accumulation [15]. In the present study, PAs induced the lowest decrease in Y(II) in the case of *eds5*, where interestingly the lowest SA values were detected.

Under control conditions, the *eds5* and *sid2* mutations did not influence the PA contents, among the genes involved in PUT, SPD and SPM synthesis, back-conversion or terminal catabolism, only the expression levels of SPMS showed higher transcript level in *sid2* compared to the *eds5* and the wild type. PA treatments induced remarkable increments only in the PUT content, but these were similar in the three genotypes, with the highest values in the case of PUT-treated plants. Besides the PUT uptake and translocation into the leaves, which was predominant in PUT-treated plants, *de novo* PUT synthesis was also occurred, as the expression level of *ADC2* was significantly induced by all the PA treatments, especially after SPD and SPM treatments. Our results regarding the different inducibility of the expression of *ADC1* and *ADC2*, are in accordance with the literature, as it seems that *AtADC1* is constitutively expressed, while *AtADC2* is responsive to abiotic stresses, plant hormones or pathogens [39]. Interestingly, under the present conditions, the initial transcript level of *ADC1* was slightly higher in *sid2*, and this difference became more pronounced after PUT and SPM treatment, while *ADC2* expression level was higher in *sid2* mutants after SPD or SPM treatments compared to the wild type. The *PAO2* and *PAO5* expressions in the present experiment showed exactly the same pattern, and did not show remarkable changes, except for in SPD-treated plants, where they were down-regulated. As *AtPAO2* and *AtPAO5* encoding polyamine oxidases [40,41], which are responsible for the back-conversion of higher PAs to PUT, it is not surprising that their expression was not induced upon increased PUT content and there was no need for back-conversion [42]. Except for SPD treatment, the expression level of the *PAO*s, similarly to the genes of PUT synthesis, was the highest in the *sid2* mutant. At the same time, the expression of *CuAO1*, that encoding an amine oxidase catalyses the terminal catabolism of PUT and SPD, increased after all the PA treatments, especially in case of SPD, the most pronouncedly in the *sid2* genotypes. Although, the activation of terminal catabolism was not sufficient to compensate the increased PUT content, and the differences in its pattern were not manifested in the PUT contents.

Parallel with these, although the level of higher PAs did not change in the leaves of the plants, the *SPDS1* expression level decreased after SPD or SPM treatments, suggesting that higher PAs, uptaken by the roots is probably not transported to the leaves, inducing the down-regulation of the SPD synthesis in the leaves. A decrease in *SPDS1* expression together with *CuAO1* expression in the SPD- and SPM-treated plants may be responsible for the maintenance of optimal SPD content. It has been reported that significantly induced expression of *CuAO1* was observed after ABA treatment in *Arabidopsis* [43]. In the present experiment, the highest *CuAO1* transcript level was detected parallel with the highest ABA content in the SPD-treated plants. Interestingly, under control conditions and after PUT treatment, the *SPDS2* transcript level was higher in the *sid2* mutant than in the wild type, but this difference disappeared in SPD- or SPM-treated plants, as the application of higher amounts of PAs inhibited its expression in this mutant. The up-regulation of *SPMS* was observed after all the PA treatments, which is understandable in PUT- and SPD-treated plants, where in order to decrease the uptaken excess of PUT or SPD, further synthesis was needed to SPM. Despite of these, the SPM content did not show significant increment in either of the SPM-treated plants after 1 day of treatment.

In the present experiment, the SA level was lower in the *eds5* and *sid2* mutants compared to the wild type. As *EDS5* encodes a membrane protein, located at the chloroplast envelope and responsible for SA transport [44,45], in the *eds5* mutant, the accumulated SA is trapped in the chloroplast, which in turn can inhibit *ICS* expression. While in *sid2* mutants, the SA accumulation can be only a fraction of that of the wild type, because of the lack of the ICS enzyme/pathway. Different concentrations of SA have been reported to have different effects on PA metabolism [21,46]. In addition, PA treatments, both as seed soaking or applied hydroponically efficiently enhanced SA content in wheat or maize [15,16,47]. Under the present conditions, PUT and SPM treatments after 1 day could only cause a slight, but statistically not significant increase in SA content of Col-0 genotype. Although, *CS* expression did not change after either of the PA treatments, the *ICS* transcript level in Col-0 was in correlation with the changes in SA content, as it increased after PUT or SPM treatments. Interestingly, PUT also increased the *ICS* expression in the *eds5* mutant, but SPD and SPM treatments decreased it. The latter was probably due to the inhibitory feedback effect of the SA accumulation in the chloroplasts, which was not manifested in statistically significant increase in the total SA content. The *PAL1* expression, maybe in order to compensate the mutation of the *ICS1*, showed slightly higher basal level in *sid2* plants. Although, PA treatments induced *PAL1* expression in all the genotypes and its up-regulation was the highest in SPM-treated plants, in case of all the treatments it was the highest in the *sid2* mutant. The activity of PAL, a crucial enzyme in the synthesis of flavonoids, anthocyanins and simple phenolic acids, increased after PUT treatment in the leaves of maize [15]. In addition, in *Atpao4* plants with increased SPM content, genes involved in flavonoid and/or lignin biosynthesis, such as *PAL1* were induced [48], suggesting that the accumulation of PAs increased the synthesis of phenolic compounds. According to the present results, PA treatments differently induced the SA synthesis pathways. In the Col-0 both pathways activated, especially in PUT- and SPM-treated ones. In the *eds5* mutant, after PUT treatment both pathways induced, and after SPD and SPM treatments the expression level of *PAL* further increased, but that of *ICS1* decreased. In the case of *sid2* mutant, the drastic increment in *PAL1* transcript level will be responsible for SA synthesis.

PUT and ABA are integrated in a positive feedback loop [18]. Modulation of PA metabolism at transcriptional level by ABA has been proved in *Arabidopsis* in case of PA biosynthesis genes, such as *ADC2*, *SPDS1* and *SPMS* [49]. The transcriptional regulation *NCED* in PA-overproducer plants has been demonstrated [17,50,51]. Conversely, the suppression of *ADC* resulted in the reduced expression of *NCED* and the down-regulation of ABA-regulated genes [30]. In the present experiment, PAs induced ABA accumulation and the up-regulation of *NCED*, which was the most pronounced in SPD-treated plants, where it was the highest in *sid2* mutant. This can be resulted from the well-known antagonistic relationship between SA and ABA [27,52,53]. ABA has been also reported to influence the catabolism of PAs, as its exogenous application induced the expression of *CuAO1* in *Arabidopsis* [42,43], and *PAO* in *Medicago sativa* [54] and wheat [29]. Interestingly, under the present conditions, the highest *CuAO1* expression was detected in the SPD-treated plants, where the highest ABA accumulation and *NCED* expression was found. Nevertheless, all these parameters were the highest in the *sid2* mutant plants. Similarly, higher ABA content has been also detected in *sid2 Arabidopsis* mutant compared to the wild type during the transition from pre-reproductive to reproductive stages [55], confirming that SA content may influence, at least to some extent, the endogenous concentrations of ABA.

Similar JA content was detected in *sid2* and Col-0 genotypes despite of the different SA level under control conditions and after infection with *Pythium irregular*, indicating that SA did not inhibit the JA accumulation [55,56]. In the present experiment, different SA contents were also detected parallel with almost the same JA level in the three genotypes. All the PA treatments decreased the JA content, with the highest decrement in SPM-treated plants, but regardless of the genotypes. Exogenous methyl jasmonate has also been reported to increase the production of conjugated PAs in barley [33]. In addition, genes encoding enzymes for synthesis of PAs, such as ADC1 and SPDS1 have been proved to be affected by JA signalling in *Arabidopsis* [57], and exogenous JA induced PAO in rice [34], and regulated the expression of chickpea *CuAO* [58]. However, there are only a few studies about the effects of PAs on the synthesis of JA. PA treatments, especially SPM elicited the biosynthesis of JA in lima bean [35], while constitute overexpression of *SPMS* in *Arabidopsis* increased the levels of expression of genes involved in JA synthesis and signalling [17,59]. Under the present conditions, it was found that all the PAs induced the expression of *AOS*, with the highest level in SPM-treated plants, especially in the SPM-treated *sid2* mutant. The *Arabidopsis AOS* promoter was found to be activated by a variety of signals, including JA, wounding, and even exogenous SA [60], however, SA-induced repression of the JA-signalling pathway is independent of JA biosynthesis, as occurs downstream of JA perception [61]. Despite this, in *eds4* mutant *Arabidopsis* (also with reduced SA content) relieved inhibition of JA-dependent signalling responses were detected, confirming that the SA signalling and JA signalling can be mutually inhibitory [62].

ADC2-overexpressing transgenic plants exhibited a reduction in both the contents of $GA_{1,4,9}$ contents, and in the expression levels of the *GA20ox1*, *GA3ox1* and *GA3ox3* transcripts [63], suggesting that PUT accumulation represses GA synthesis. SPD treatment decreased GA_3 content, and *GA3ox* expression in apple terminal buds during floral induction [4]. However, on the other hand, an increased content of GAs was observed in PA-treated plants under drought conditions in creeping bentgrass [64]. SPD increased GA_3 content in maize [65], and in sweet corn seed embryos [66]. Nevertheless, accumulation or deficiency of PUT, SPD or SPM did not influence the expression of GA signalling gene in transgenic *Arabidopsis* leaves or tomato fruits [67]. These data indicate that the effect of PAs on GA synthesis is highly dependent on plant species and developmental stage. Under the present conditions, PAs treatments did not influence the GA_1 content, except for a slight decrease in SPM-treated wild type. Although, the *GA3ox1* expression increased after PUT treatment, in each of the genotypes, especially in *sid2* mutant, and after SPD treatment in the mutants, with still a higher extent in case of *sid2*, a dramatic decrease was observed after SPM treatment in all the genotypes. According to these, no negative correlation was found between the accumulation of PUT and the level of GA_1, while the inhibitory effect of SPM treatment on *GA3ox1* was pronounced in all genotypes. The existence of crosstalk between GAs and SA signalling in *Arabidopsis* was also suggested, as GA treatment increased the endogenous levels of SA and the expression of the *ICS1* gene in Col-0, and the transcript levels of the *GA3ox1* gene were greatly elevated by SA treatments [68,69]. However, under the present conditions, in each cases GA_1 accumulation was higher in *sid2* mutant than in the wild type, indicating that SA also influences to some extent GA levels in plants. Nevertheless, here, the GA-ABA antagonism should be also taken into consideration, which has been reported during seed development, plant growth and stress responses. Stress-induced increases of ABA level were parallel with decreases in the GA level and the suppression of the GA synthetic enzyme genes in *Arabidopsis* [70,71].

4. Materials and Methods

4.1. Plant Materials, Growth Conditions and Treatments

In our experiment two *Arabidopsis* SA-deficient mutants were investigated, *eds5-1* (*eds5*), which is a SA transport mutant [44,72] and *sid2-2* (*sid2*), which is a SA biosynthesis mutant [73]. *Arabidopsis* Col-0 was used as control, wild type. Seeds of mutants were obtained through the European *Arabidopsis* Stock Centre (NASC, Sutton Bonington Campus, Loughborough, LE125RD, United Kingdom). The plants were self-pollinated for two generations and the presence of mutation was revealed by genotyping. In the case of *eds5-1* the full length coding sequence was amplified (*EDS* FL_F 5′-ATGCTAATCAAATCCCAAAGA-3′ and *EDS* FL_R 5′-TTTAA TCTTCTCCACCGTGTAT-3′) and a deletion of eight bp was proved by sequencing resulting in a frameshift error. Genotyping of *sid2-2* was carried out with *sid2-2* F (5′-acagcaggataattacggatacc-3′) and *sid2-2* R (5′-ccactctgaagatgggtcact-3) primers [74].

Plants were cultivated hydroponically using an Araponics system (Araponics, Liège, Belgium). Cultures were grown in a 25% modified Hoagland-solution [75] in a Conviron GB-48 plant growth chamber (Controlled Environments, Winnipeg, MB, Canada) under control conditions at 22 °C/20 °C with 8/16 h light/dark period and 75% humidity for 28 days. The photosynthetic photon flux density (PPFD) was 100 μmol m^{-2} s^{-2}. The 28-day-old plants were treated with nutrition solution containing 0.5 mM PUT, SPD or SPM. After a one-day exposure of different PA treatments, fully developed leaves and roots were collected. Leaves were frozen immediately in liquid nitrogen, while the roots were washed in distilled water before freezing. Samples were stored at −80 °C until further analysis.

4.2. Chlorophyll-A Fluorescence Induction Measurements

The chlorophyll-*a* fluorescence was measured by using a pulse amplitude modulated fluorometer (Imaging-PAM M-Series fluorometer; Walz, Effeltrich, Germany). The maximum quantum yield of PSII photochemistry Fv/Fm was measured on 20 min dark-adapted leaves. Fv/Fm= (Fm-F0)/Fm, where Fm is the maximum fluorescence induced by a saturating flash (8000 μmol m^{-2} s^{-1} PPFD for 0.8s) in dark adapted leaves, F0 is the minimum chlorophyll fluorescence yield in the dark (PPFD < 1 μmol m^{-2} s^{-1}). The effective PSII quantum yield (YII) was measured at a light intensity of 250 μmol m^{-2} s^{-1} and represents the proportion of absorbed light energy being used in photochemistry. It is calculated as: (Fm'−F)/Fm', Fm' is the maximum fluorescence level induced by a saturating light pulse at the steady state, and F is the steady state chlorophyll fluorescence immediately prior to the flash.

4.3. Polyamine Analysis

Samples preparation and pre-column derivatisation with dansyl chloride and HPLC analyses were performed according to Némethetal [19]. The most abundant polyamines, namely PUT, spermidine (SPD) and spermine (SPM) were analysed by HPLC using a W2690 separation module (Waters, Milford, MA, USA) equipped with a 100 × 2.1 mm Kinetex reverse phase column 5 μm (C18) (Phenomenex, Inc., Torrance, CA, USA) and a W474 scanning fluorescence detector with excitation at 340 nm and emission at 515 nm.

4.4. Benzoic Acid, Jasmonic Acid, Gibberellin 1, Salicylic Acid and Abscisic Acid Extraction and Analytical Procedure

The sample extraction was done using methanol:water (2:1 *v*/*v* %), 100 mg FW/mL final sample ratio. Ultra-performance liquid chromatography with tandem mass spectrometry (Waters Acquity I class UPLC system coupled to a Waters Xevo TQ-XS instrument equipped with a UniSpray (US) ion source operated in timed MRM mode) analyses were carried out according to Vrhovsek et al. [76] with slight modifications as described in detail by Pál et al. [32]. Separation was achieved on a Waters Acquity HSS T3 column (1.8 μm, 100 mm × 2.1 mm), kept at 40 °C. Mobile phases both contained 0.1 *v*/*v* % formic acid while a water and acetonitrile gradient was used. For quantitation the transition exhibiting the highest S/N ratio was utilized (Table S1). Data processing was performed using Waters MassLynx 4.2 and TargetLynx softwares.

4.5. Gene Expression Analysis

Total RNA was extracted from fully developed leaves using TRI Reagent®. The samples were treated with DNase I and cleaned with a Direct-zol™ RNA MiniPrep Kit (Zymo Research, Irvine, CA, USA) according to the manufacturer's instructions. The quality and integrity of RNA was monitored using agarose gel and the samples were quantified with a Nanodrop 2000 spectrophotometer (Thermo Fisher Scientific Inc., Wilmington, MA, USA). Total RNA (1000 ng) was reverse transcribed by using M-MLV Reverse Transcriptase (Promega Corporation, Madison, WI, USA) and oligo(dT)18 (Thermo Fisher Scientific) 1 μL of 2-fold diluted cDNA, gene-specific primers and housekeeping primers (Table S1), PCRBIO SyGreen Mix (PCR Biosystems, London, UK) and CFX96 Touch™ Real-Time PCR Detection System (Bio-Rad, Hercules, CA, USA) were used for quantitative real-time PCR reaction. Melt curve analysis was also performed to confirm the presence of a single PCR product. The relative gene expression values were determined with the $2^{-\Delta\Delta Ct}$ method [77]. Ct values were normalized by the Ct values of housekeeping gene *Atactin8* [78]. All reactions were performed in triplicate using three biological and three technical repetitions.

4.6. Statistical Analysis

Three independent biological experiments were performed, and representative data are presented. The results are the means of at least five replicates for spectrophotometric and chromatographic determinations. The data were statistically evaluated using the standard deviation in Microsoft Excel. Different letters indicate statistically significant differences ($p < 0.05$) between multiple groups (one-way ANOVA with Duncan post hoc test was performed using SPSS 16.0. Box-plot presentation for all the investigated parameters in Supplementary Figures S1–S10. (Boxes represent Q1 and Q3 quartiles and the middle line of the box is the median (Q2). Whiskers show the minimum and maximum values.) Analysis of variance of metabolite accumulation and changes in gene expression was performed by using SPSS 16.0, where mean squares (MS) followed by asterisks (*) are significantly different ($p < 0.05$). Supplementary heat map (Supplementary Figure S11), presenting the metabolite accumulation and changes in gene expression, was evaluated using the membership function value (MFV) using the fuzzy comprehensive evaluation method [79]. The MFV was calculated using the following equation: Xi = (X − Xmin)/(Xmax − Xmin) × 100.

5. Conclusions

The present study indicates that the individual PAs, namely PUT, SPD and SPM, induced different changes in the investigated *Arabidopsis* plants. Table S1 shows the results of variance analysis in order to detect the effect of genotype, treatment and genotype x treatment interaction. Under control conditions, *sid2* mutant showed a remarkable difference in PA metabolism compared to the other two genotypes only regarding *SPDS2* expression, but this higher transcript level was not accompanied by a higher SPD level. The exogenous PA treatments upregulated the PA metabolism, as de novo synthesis of PUT and SPM is induced, while that of SPD is inhibited, and in parallel with these events, the induction of the terminal catabolism instead of back-conversion is responsible for the unchanged level of higher PAs. The SA deficient *sid2* mutant showed pronouncedly different responses to the individual PA treatments compared to the Col-0 wild type. Pronounced differences were observed in the SA content between the two mutant and the wild type. Although the PA treatments could hardly influence the SA levels, the initial differences still remained. However, the *ICS1* and *PAL1* expression showed PA treatment and genotype dependent changes, suggesting that the induction of the PAL pathway is more predominant upon PA treatment, with the highest upregulation in *sid2* mutant. Remarkable changes in hormone synthesis were also found after PA treatments. Interestingly PUT treatment increased the *GA3ox1* expression, SPD treatment has spectacular effect on ABA content and *NCED* expression, while SPM application influenced rather JA content and *AOS* expression. In several cases the most pronounced difference in the hormone biosynthesis after PA treatments were found in *sid2* mutant compared to the Col-0 wild type (Figure S11). However, these differences at transcript levels were not always in accordance with the hormone contents, suggesting, that synergetic or antagonistic interactions between these plant hormones should also be taken into consideration. Thus, understanding the link between PA metabolism/signalling and plant hormone signalling needs further studies from this point of view, using plant hormone synthesis mutants.

Supplementary Materials: Supplementary materials can be found at http://www.mdpi.com/1422-0067/20/22/5746/s1.

Author Contributions: Conceptualization, M.P. and J.T.; investigation, J.T., K.Á.H., I.M. and M.P.; methodology, K.Á.H., I.M., K.G., and E.N.; writing—original draft preparation, M.P.; review, T.J.; visualization, J.T. and M.P.; supervision, M.P.

Funding: This work was financed by grants from the Hungarian National Scientific Research Foundation (NKFIH,KH124472).

Conflicts of Interest: The authors declare no conflict of interest. The funders had no role in the design of the study; in the collection, analyses, or interpretation of data; in the writing of the manuscript, or in the decision to publish the results.

Abbreviations

ABA	abscisic acid
ADC	arginine decarboxylase
AOS	allene oxide synthase
CuAO	cooper amine-oxidase
CS	chorismate synthase
GAs	gibberellins
GA3ox	gibberellin 3-oxidase
ICS	isochorismate synthase
JA	jasmonic acid

ADC	arginine decarboxylase
NCED	9-*cis*-epoxycarotenoid dioxygenase
PAL	phenylalanine ammonia-lyase
PAO	polyamine oxidase
PUT	putrescine
SA	salicylic acid
SPD	spermidine
SPDS	spermidine synthase
SPM	spermine
SPMS	spermine synthase

References

1. Chen, D.; Shao, Q.; Yin, L.; Younis, A.; Zheng, B. Polyamine function in plants: Metabolism, regulation on development, and roles in abiotic stress responses. *Front. Plant Sci.* **2019**, *9*, 1945. [CrossRef]
2. Xu, L.; Xing, S.T.; Sun, X. Effects of polyamines on hormones contents and the relationship with the flower bud differentiation in *Chrysanthemum*. *Plant Physiol. J.* **2014**, *50*, 1195–1202. [CrossRef]
3. Pál, M.; Szalai, G.; Janda, T. Speculation: Polyamines are important in abiotic stress signaling. *Plant Sci.* **2015**, *237*, 16–23. [CrossRef]
4. Qin, L.; Zhang, X.; Yan, J.; Fan, L.; Rong, C.; Mo, C.; Zhang, M. Effect of exogenous spermidine on floral induction, endogenous polyamine and hormone production, and expression of related genes in 'Fuji' apple (*Malus domestica* Borkh.). *Sci. Rep.* **2019**, *9*, 12777. [CrossRef] [PubMed]
5. De Oliveira, L.F.; Elbl, P.; Navarro, B.V.; Al, E. Elucidation of the polyamine biosynthesis pathway during Brazilian pine (*Araucaria angustifolia*) seed development. *Tree Physiol.* **2016**, *37*, 116–130. [CrossRef] [PubMed]
6. Vondráková, Z.; Eliášová, K.; Vágner, M.; Al, E. Exogenous putrescine affects endogenous polyamine levels and the development of *Picea abies* somatic embryos. *Plant Growth Regul.* **2015**, *75*, 405–414. [CrossRef]
7. Duan, G. Effect of speridine on protein contents and protease during senescence of excised wheat leaves. *J. Sichuan Teach. Coll.* **2000**, *21*, 44–47. [CrossRef]
8. Liu, J.H.; Nakajima, I.; Moriguchi, T. Effects of salt and osmotic stresses on free polyamine content and expression of polyamine biosynthetic genes in *Vitis vinifera*. *Biol. Plant.* **2011**, *55*, 340–344. [CrossRef]
9. Fariduddin, Q.; Varshney, P.; Yusuf, M.; Ahmad, A. Polyamines: Potent modulators of plant responses to stress. *J. Plant Interact.* **2013**, *8*, 1–16. [CrossRef]
10. Minocha, R.; Majumdar, R.; Minocha, S.C. Polyamines and abiotic stress in plants: A complex relationship. *Front. Plant Sci.* **2014**, *5*, 175. [CrossRef]
11. Ye, B.; Muller, H.; Zhang, J.; Gressel, J. Constitutively elevated levels of putrescine and putrescine-generating enzymes correlated with oxidant stress resistance in *Conyza bonariensis* and wheat. *Plant Physiol.* **1997**, *115*, 1443–1451. [CrossRef] [PubMed]
12. Kasukabe, Y.; He, L.; Nada, K.; Misawa, S.; Ihara, I.; Tachibana, S. Overexpression of spermidine synthase enhances tolerance to multiple environmental stresses and up-regulates the expression of various stress-regulated genes in transgenic *Arabidopsis thaliana*. *Plant Cell Physiol.* **2004**, *45*, 712–722. [CrossRef] [PubMed]
13. Pillai, M.A.; Akiyama, T. Differential expression of an S-adenosyl-L-methionine decarboxylase gene involved in polyamine biosynthesis under low temperature stress in japonica and indica rice genotypes. *Mol. Gen. Genom.* **2004**, *271*, 141–149. [CrossRef] [PubMed]
14. Pottosin, I.; Shabala, S. Polyamines control of cation transport across plant membranes: Implications for ion homeostasis and abiotic stress signaling. *Front. Plant Sci.* **2014**, *5*, 154. [CrossRef] [PubMed]
15. Szalai, G.; Janda, K.; Darkó, E.; Janda, T.; Peeva, V.; Pál, M. Comparative analysis of polyamine metabolism in wheat and maize plants. *Plant Physiol. Biochem.* **2017**, *112*, 239–250. [CrossRef] [PubMed]
16. Tajti, J.; Janda, T.; Majláth, I.; Szalai, G.; Pál, M. Comparative study on the effects of putrescine and spermidine pre-treatment on cadmium stress in wheat. *Ecotoxicol. Environ. Saf.* **2018**, *148*, 546–554. [CrossRef]
17. Marco, F.; Alcázar, R.; Tiburcio, A.F.; Carrasco, P. Interactions between polyamines and abiotic stress pathway responses unraveled by transcriptome analysis of polyamine overproducers. *OMICS* **2011**, *15*, 775–781. [CrossRef]

18. Alcázar, R.; Altabella, T.; Marco, F.; Bortolotti, C.; Reymond, M.; Koncz, C.; Carrasco, P.; Tiburcio, A.F. Polyamines: Molecules with regulatory functions in plant abiotic stress tolerance. *Planta* **2010**, *231*, 1237–1249. [CrossRef]
19. Németh, M.; Janda, T.; Horváth, E.; Páldi, E.; Szalai, G. Exogenous salicylic acid increases polyamine content but may decrease drought tolerance in maize. *Plant Sci.* **2002**, *162*, 569–574. [CrossRef]
20. Szepesi, Á.; Gémes, K.; Orosz, G.; Pető, A.; Takács, Z.; Vorák, M.; Tari, I. Interaction between salicylic acid and polyamines and their possible roles in tomato hardening processes. *Acta Biol. Szeged.* **2011**, *55*, 165–166.
21. Wang, X.; Zhang, Y. Regulation of salicylic acid on polyamine synthesize under NaCl stress in leaves of the yali pear. *Res. J. Appl. Sci. Eng. Technol.* **2012**, *4*, 3704–3708.
22. Hassannejad, S.; Bernard, F.; Mirzajani, F.; Gholami, M. SA improvement of hyperhydricity reversion in *Thymus daenensis* shoots culture may be associated with polyamines changes. *Plant Physiol. Biochem.* **2011**, *51*, 40–46. [CrossRef] [PubMed]
23. Rahdari, P.; Hoseini, S.M. Roll of polyamines (spermidine and putrescine) on protein, chlorophyll and phenolic compounds in wheat (*Triticum aestivum* L.) under salinity stress. *Sci. Res. Rep.* **2013**, *1*, 19–24.
24. Radhakrishnan, R.; Lee, I.J. Spermine promotes acclimation to osmotic stress by modifying antioxidant, abscisic acid, and jasmonic acid signals in soybean. *J. Plant Growth Regul.* **2013**, *32*, 22–30. [CrossRef]
25. Garcion, C.; Lohmann, A.; Lamodière, E.; Catinot, J.; Buchala, A.; Doermann, P.; Métraux, J.P. Characterization and biological function of the ISOCHORISMATE SYNTHASE2 gene of *Arabidopsis*. *Plant Physiol.* **2008**, *147*, 1279–1287. [CrossRef]
26. Scott, I.M.; Clarke, S.M.; Wood, J.E.; Mur, L.A. Salicylate accumulation inhibits growth at chilling temperature in *Arabidopsis*. *Plant Physiol.* **2004**, *135*, 1040–1049. [CrossRef]
27. Pál, M.; Szalai, G.; Kovács, V.; Gondor, O.K.; Janda, T. Salicylic acid-mediated abiotic stress tolerance. In *Salicylic Acid—Plant Growth and Development*; Hayat, S., Ahmad, A., Alyemeni, M.N., Eds.; Springer: Amsterdam, The Netherlands, 2013; pp. 183–247.
28. Suzuki, N.; Bassil, E.; Hamilton, J.S.; Inupakutika, M.A.; Zandalinas, S.I.; Tripathy, D.; Luo, Y.; Dion, E.; Fukui, G.; Kumazaki, A.; et al. ABA is required for plant acclimation to a combination of salt and heat stress. *PLoS ONE* **2016**, *11*, e0147625. [CrossRef]
29. Pál, M.; Tajti, J.; Szalai, G.; Peeva, V.; Végh, B.; Janda, T. Interaction of polyamines, abscisic acid and proline under osmotic stress in the leaves of wheat plants. *Sci. Rep.* **2018**, *8*, 12839. [CrossRef]
30. Cuevas, J.C.; López-Cobollo, R.; Alcázar, R.; Zarza, X.; Koncz, C.; Altabella, T.; Salinas, J.; Tiburcio, A.F.; Ferrando, A. Putrescine is involved in Arabidopsis freezing tolerance and cold acclimation by regulating abscisic acid levels in response to low temperature. *Plant Physiol.* **2008**, *148*, 1094–1105. [CrossRef]
31. Diao, Q.; Song, Y.; Shi, D.; Q, H. Interaction of polyamines, abscisic acid, nitric oxide, and hydrogen peroxide under chilling stress in tomato (*Lycopersicon esculentum* Mill.) seedlings. *Front. Plant. Sci.* **2017**, *8*, 203. [CrossRef]
32. Pál, M.; Ivanovska, B.; Oláh, T.; Tajti, J.; Hamow, K.A.; Szalai, G.; Khalil, R.; Vanková, R.; Dobrev, P.; Misheva, S.P.; et al. Role of polyamines in plant growth regulation of Rht wheat mutants. *Plant Physiol. Biochem.* **2019**, *137*, 189–202. [CrossRef] [PubMed]
33. Walters, D.; Cowley, T.; Mitchell, A. Methyl jasmonate alters polyamine metabolism and induces systemic protection against powdery mildew infection in barley seedlings. *J. Exp. Bot.* **2002**, *53*, 747–756. [CrossRef] [PubMed]
34. Sagor, G.H.M.; Kusano, T.; Berberich, T. Identification of the actual coding region for polyamine oxidase 6 from rice (*OsPAO6*) and its partial characterization. *Plant Signal. Behav.* **2017**, *12*, e1359456. [CrossRef] [PubMed]
35. Ozawa, R.; Bertea, C.M.; Foti, M.; Narayana, R.; Arimura, G.I.; Muroi, A.; Horiuchi, J.I.; Nishioka, T.; Maffei, M.E.; Takabayashi, J. Exogenous polyamines elicit herbivore-induced volatiles in lima bean leaves: Involvement of calcium, H_2O_2 and jasmonic acid. *Plant Cell Physiol.* **2009**, *50*, 2183–2199. [CrossRef] [PubMed]
36. Nawrath, C.; Métraux, J.P. Salicylic acid induction-deficient mutants of *Arabidopsis* express PR-2 and PR-5 and accumulate high levels of camalexin after pathogen inoculation. *Plant Cell.* **1999**, *11*, 1393–1404. [CrossRef] [PubMed]

37. Jang, I.C.; Choi, W.B.; Lee, K.H.; Song, S.I.; Nahm, B.H.; Kim, J.K. High-level and ubiquitous expression of the rice Cytochrome c gene *OsCc1* and its promoter activity in transgenic plants provides a useful promoter for transgenesis of monocots. *Plant Physiol.* **2002**, *29*, 1473–1481. [CrossRef] [PubMed]
38. Mateo, A.; Funck, D.; Mühlenbock, P.; Kular, B.; Mullineaux, P.M.; Karpinski, S. Controlled levels of salicylic acid are required for optimal photosynthesis and redox homeostasis. *J. Exp. Bot.* **2006**, *57*, 1795–1807. [CrossRef]
39. Takahashi, T.; Tong, W. Regulation and diversity of polyamine biosynthesis in plants. In *Polyamines*; Kusano, T., Suzuki, H., Eds.; Springer: Berlin, Germany, 2015; pp. 27–44.
40. Tavladoraki, P.; Rossi, M.N.; Saccuti, G.; Perez-Amador, M.A.; Polticelli, F.; Angelini, R.; Federico, R. Heterologous expression and biochemical characterization of a polyamine oxidase from Arabidopsis involved in polyamine back conversion. *Plant Physiol.* **2006**, *141*, 1519–1532. [CrossRef]
41. Kim, D.W.; Watanabe, K.; Murayama, C.; Izawa, S.; Niitsu, M.; Michael, A.J.; Berberich, T.; Kusano, T. Polyamine oxidase 5 regulates Arabidopsis growth through thermospermine oxidase activity. *Plant Physiol.* **2014**, *165*, 1575–1590. [CrossRef]
42. Planas-Portell, J.; Gallart, M.; Tiburcio, A.F.; Altabella, T. Copper-containing amine oxidases contribute to terminal polyamine oxidation in peroxisomes and apoplast of *Arabidopsis thaliana*. *BMC Plant Biol.* **2013**, *13*, 109. [CrossRef]
43. Wimalasekera, R.; Villar, C.; Begum, T.; Scherer, G.F. Copper amine oxidase1 (*CuAO1*) of *Arabidopsis thaliana* contributes to abscisic acid- and polyamine-induced nitric oxide biosynthesis and abscisic acid signal transduction. *Mol. Plant* **2011**, *4*, 663–678. [CrossRef] [PubMed]
44. Nawrath, C.; Heck, S.; Parinthawong, N.; Métraux, J.P. *EDS5*, an essential component of salicylic acid-dependent signaling for disease resistance in *Arabidopsis*, is a member of the MATE transporter family. *Plant Cell.* **2002**, *14*, 275–286. [CrossRef] [PubMed]
45. Serrano, M.; Wang, B.; Aryal, B.; Garcion, C.; Abou-Mansour, E.; Heck, S.; Geisler, M.; Mauch, F.; Nawrath, C.; Métraux, J.P. Export of salicylic acid from the chloroplast requires the multidrug and toxin extrusion-like transporter EDS5. *Plant Physiol.* **2013**, *162*, 1815–1821. [CrossRef]
46. Francisco, J.C.; Montilla-Bascón, G.; Rispail, N.; Prats, E. Salicylic acid regulates polyamine biosynthesis during drought responses in oat. *Plant Signal. Behav.* **2019**, *14*, e1651183. [CrossRef]
47. Iqbal, M.; Ashraf, M.; Jamil, A.; Rehman, S. Does seed priming induce changes in the levels of some endogenous plant hormones in hexaploid wheat plants under salt stress? *J. Int. Plant Biol.* **2006**, *48*, 181–189. [CrossRef]
48. Kamada-Nobusada, T.; Hayashi, M.; Fukazawa, M.; Sakakibara, H.; Nishimura, M. A putative peroxisomal polyamine oxidase, *AtPAO4*, is involved in polyamine catabolism in *Arabidopsis thaliana*. *Plant Cell Physiol.* **2008**, *49*, 1272–1282. [CrossRef]
49. Alcázar, R.; Marco, F.; Cuevas, J.C.; Patron, M.; Ferrando, A.; Carrasco, P.; Tiburcio, A.F.; Altabella, T. Involvement of polyamines in plant response to abiotic stress. *Biotechnol. Lett.* **2006**, *28*, 1867. [CrossRef]
50. Alet, A.I.; Sanchez, D.H.; Cuevas, J.C.; Del Valle, S.; Altabella, T.; Tiburcio, A.F.; Marco, F.; Ferrando, A.; Espasandín, F.D.; González, M.E.; et al. Putrescine accumulation in *Arabidopsis thaliana* transgenic lines enhances tolerance to dehydration and freezing stress. *Plant Signal. Behav.* **2011**, *6*, 278–286. [CrossRef]
51. Espasandin, F.D.; Maiale, S.J.; Calzadilla, P.; Ruiz, O.A.; Sansberro, P.A. Transcriptional regulation of 9-cis-epoxycarotenoid dioxygenase (*NCED*) gene by putrescine accumulation positively modulates ABA synthesis and drought tolerance in *Lotus tenuis* plants. *Plant Phys. Biochem.* **2014**, *76*, 29–35. [CrossRef]
52. Yasuda, M.; Ishikawa, A.; Jikumaru, Y.; Seki, M.; Umezawa, T.; Asami, T.; Maruyama-Nakashita, A.; Kudo, T.; Shinozaki, K.; Yoshida, S.; et al. Antagonistic interaction between systemic acquired resistance and the abscisic acid-mediated abiotic stress response in Arabidopsis. *Plant Cell* **2008**, *20*, 1678–1692. [CrossRef]
53. Lee, B.R.; Islam, M.T.; Park, S.H.; Lee, H.; Bae, D.W.; Kim, T.H. Antagonistic shifting from abscisic acid- to salicylic acid-mediated sucrose accumulation contributes to drought tolerance in *Brassica napus*. *Environ. Exp. Bot.* **2019**, *162*, 38–47. [CrossRef]
54. Guo, Z.; Tan, J.; Zhuo, C.; Wang, C.; Xiang, B.; Wang, Z. Abscisic acid, H_2O_2 and nitric oxide interactions mediated cold-induced S-adenosyl methionine synthetase in *Medicago sativa* subsp. falcata that confers cold tolerance through up-regulating polyamine oxidation. *Plant Biotechnol. J.* **2014**, *12*, 601–612. [CrossRef] [PubMed]

55. Abreu, M.E.; Munné-Bosch, S. Salicylic acid deficiency in *NahG* transgenic lines and *sid2* mutants increases seed yield in the annual plant *Arabidopsis thaliana*. *J. Exp. Bot.* **2009**, *60*, 1261–1271. [CrossRef] [PubMed]
56. Adie, B.A.; Pérez-Pérez, J.; Pérez-Pérez, M.M.; Godoy, M.; Sánchez-Serrano, J.J.; Schmelz, E.A.; Solano, R. ABA is an essential signal for plant resistance to pathogens affecting JA biosynthesis and the activation of defenses in Arabidopsis. *Plant Cell.* **2007**, *19*, 1665–1681. [CrossRef]
57. Hu, Y.; Jiang, L.; Wang, F.; Yu, D. Jasmonate regulates the inducer of CBF expression–c-repeat binding factor/DRE binding factor1 cascade and freezing tolerance in Arabidopsis. *Plant Cell.* **2013**, *25*, 2907–2924. [CrossRef] [PubMed]
58. Rea, G.; Metoui, O.; Infantino, A.; Federico, R.; Angelini, R. Copper amine oxidase expression in defense responses to wounding and *Ascochyta rabiei* invasion. *Plant Physiol.* **2002**, *128*, 865–875. [CrossRef] [PubMed]
59. Gonzalez, M.E.; Marco, F.; Minguet, E.G.; Carrasco-Sorli, P.; Blázquez, M.A.; Carbonell, J.; Ruiz, O.A.; Pieckenstain, F.L. Perturbation of spermine synthase gene expression and transcript profiling provide new insights on the role of the tetraamine spermine in *Arabidopsis* defense against *Pseudomonas viridiflava*. *Plant Physiol.* **2011**, *156*, 2266–2277. [CrossRef]
60. Laudert, D.; Weiler, E.W. Allene oxide synthase: A major control point in *Arabidopsis thaliana* octadecanoid signalling. *Plant J.* **1998**, *15*, 675–684. [CrossRef]
61. Leon-Reyes, A.; Du, Y.; Koornneef, A.; Proietti, S.; Körbes, A.P.; Memelink, J.; Pieterse, C.M.J.; Ritsema, T. Ethylene signaling renders the jasmonate response of *Arabidopsis* insensitive to future suppression by salicylic acid. *Mol. Plant Microbe Interact.* **2000**, *23*, 187–197. [CrossRef]
62. Gupta, V.; Willits, M.G.; Glazebrook, J. *Arabidopsis thaliana* EDS4 contributes to salicylic acid (SA)-dependent expression of defense responses: Evidence for inhibition of jasmonic acid signaling by SA. *Mol. Pant Microbe Interact.* **2000**, *13*, 503–511. [CrossRef]
63. Alcázar, R.; Garcia-Martinez, J.L.; Cuevas, J.C.; Tiburcio, A.F.; Altabella, T. Overexpression of *ADC2* in Arabidopsis induces dwarfism and late-flowering through GA deficiency. *Plant J.* **2005**, *43*, 425–436. [CrossRef]
64. Sanalkumar, K.; Merewitz, E.B. Polyamine application effects on gibberellic acid content in creeping bentgrass during drought stress. *J. Am. Soc. Hortic. Sci.* **2017**, *142*, 135–142. [CrossRef]
65. Li, L.; Gu, W.; Li, J.; Li, C.; Xie, T.; Qu, D.; Meng, Y.; Li, C.; Wei, S. Exogenously applied spermidine alleviates photosynthetic inhibition under drought stress in maize (*Zea mays* L.) seedlings associated with changes in endogenous polyamines and phytohormones. *Plant Physiol. Biochem.* **2018**, *129*, 35–55. [CrossRef] [PubMed]
66. Huang, Y.; Lin, C.; He, F.; Li, Z.; Guan, Y.; Hu, Q.; Hu, J. Exogenous spermidine improves seed germination of sweet corn via involvement in phytohormone interactions, H_2O_2 and relevant gene expression. *BMC Plant Biol.* **2017**, *17*, 1. [CrossRef]
67. Anwar, R.; Mattoo, A.K.; Handa, A.K. Polyamine interactions with plant hormones: Crosstalk at several levels. In *Polyamines*; Kusano, T., Suzuki, H., Eds.; Springer: Berlin, Germany, 2015; pp. 267–302. [CrossRef]
68. Alonso-Ramírez, A.; Rodríguez, D.; Reyes, D.; Jiménez, J.A.; Nicolás, G.; López-Climent, M.; Gómez-Cadenas, A.; Nicolás, C. Evidence for a role of gibberellins in salicylic acid-modulated early plant responses to abiotic stress in Arabidopsis seeds. *Plant Physiol.* **2009**, *150*, 1335–1344. [CrossRef]
69. Lee, S.; Park, C.M. Modulation of reactive oxygen species by salicylic acid in Arabidopsis seed germination under high salinity. *Plant Signal. Behav.* **2010**, *5*, 1534–1536. [CrossRef]
70. Toh, S.; Imamura, A.; Watanabe, A.; Nakabayashi, K.; Okamoto, M.; Jikumaru, Y.; Hanada, A.; Aso, Y.; Ishiyama, K.; Tamura, N.; et al. High temperature-induced abscisic acid biosynthesis and its role in the inhibition of gibberellin action in Arabidopsis seeds. *Plant Physiol.* **2008**, *146*, 1368–1385. [CrossRef]
71. Golldack, D.; Li, C.; Mohan, H.; Probst, N. Gibberellins and abscisic acid signal crosstalk: Living and developing under unfavorable conditions. *Plant Cell Rep.* **2013**, *32*, 1007–1016. [CrossRef]
72. Glazebrook, J.; Rogers, E.E.; Ausubel, F.M. Isolation of Arabidopsis mutants with enhanced disease susceptibility by direct screening. *Genetics* **1996**, *143*, 973–982.
73. Wildermuth, M.C.; Dewdney, J.; Wu, G.; Ausubel, F.M. Isochorismate synthase is required to synthesize salicylic acid for plant defence. *Nature* **2001**, *414*, 562–565. [CrossRef]
74. Gallego-Giraldo, L.; Escamilla-Trevino, L.; Jackson, L.A.; Dixon, R.A. Salicylic acid mediates the reduced growth of lignin down-regulated plants. *Proc. Natl. Acad. Sci. USA* **2011**, *108*, 20814–20819. [CrossRef] [PubMed]

75. Pál, M.; Horváth, E.; Janda, T.; Páldi, E.; Szalai, G. Cadmium stimulates the accumulation of salicylic acid and its putative precursors in maize (*Zea mays* L.) plants. *Physiol. Plant* **2005**, *125*, 356–364. [CrossRef]
76. Vrhovsek, U.; Masuero, D.; Gasperotti, M.; Franceschi, P.; Caputi, L.; Viola, R.; Mattivi, F. A versatile targeted metabolomics method for the rapid quantification of multiple classes of phenolics in fruits and beverages. *J. Agric. Food Chem.* **2012**, *12*, 8831–8840. [CrossRef] [PubMed]
77. Livak, K.J.; Schmittgen, T.D. Analysis of relative gene expression data using real-time quantitative PCR and the $2^{-\Delta\Delta CT}$ method. *Methods* **2001**, *25*, 402–408. [CrossRef] [PubMed]
78. Jubault, M.; Hamon, C.; Gravot, A.; Lariagon, C.; Delourme, R.; Bouchereau, A.; Manzanares-Dauleux, M.J. Differential regulation of root arginine catabolism and polyamine metabolism in clubroot-susceptible and partially resistant Arabidopsis genotypes. *Plant Physiol.* **2008**, *146*, 2008–2019. [CrossRef] [PubMed]
79. Chen, X.; Min, D.; Yasir, T.A.; Hu, Y.G. Evaluation of 14 morphological, yield-related and physiological traits as indicators of drought tolerance in Chinese winter bread wheat revealed by analysis of the membership function value of drought tolerance (MFVD). *Field Crops Res.* **2012**, *137*, 195–201. [CrossRef]

International Journal of
Molecular Sciences

MDPI

Article

Involvement of Salicylic Acid and Other Phenolic Compounds in Light-Dependent Cold Acclimation in Maize

Magda Pál, Tibor Janda, Imre Majláth and Gabriella Szalai *

Agricultural Institute, Centre for Agricultural Research, 2462 Martonvásár, Brunszvik 2, Hungary; pal.magda@agrar.mta.hu (M.P.); janda.tibor@agrar.mta.hu (T.J.); majlath.imre@agrar.mta.hu (I.M.)
* Correspondence: szalai.gabriella@agrar.mta.hu

Received: 13 February 2020; Accepted: 10 March 2020; Published: 12 March 2020

Abstract: The exposure of plants to non-lethal low temperatures may increase their tolerance to a subsequent severe chilling stress. To some extent, this is also true for cold-sensitive species, including maize. In the present work, based on our previous microarray experiment, the differentially expressed genes with phenylpropanoid pathways in the focus were further investigated in relation to changes in certain phenolic compounds and other plant growth regulators. Phenylalanine ammonia lyase (PAL) was mainly activated under limited light conditions. However, light-induced anthocyanin accumulation occurred both in the leaves and roots. Chilling stress induced the accumulation of salicylic acid (SA), but this accumulation was moderated in the cold-acclimated plants. Acclimation also reduced the accumulation of jasmonic acid (JA) in the leaves, which was rather induced in the roots. The level of abscisic acid (ABA) is mainly related to the level of the stress, and less indicated the level of the acclimation. The highest glutathione (GSH) amount was observed during the recovery period in the leaves of plants that were cold acclimated at growth light, while their precursors started to accumulate GSH even during the chilling. In conclusion, different light conditions during the cold acclimation period differentially affected certain stress-related mechanisms in young maize plants and changes were also light-dependent in the root, not only in the leaves.

Keywords: acclimation; antioxidants; chilling; phenylpropanoid pathway; salicylic acid; *Zea mays*

1. Introduction

Due to its subtropical origin, chilling is one of the most important factors limiting the spread and production of maize plants. Long-term exposure to temperatures around 10–15 °C may already decrease the capacity for biomass production. Lower chilling temperatures (0–5 °C) may lead to severe irreversible damage and the death of the plants [1,2]. Especially at continental climates, chilling tolerance at early stages of growth is a critical part of resistance to low temperature stress in maize plants.

Exposure of plants to low, but non-lethal, acclimating temperatures may increase their tolerance to a subsequent severe chilling stress [3–5]. To some extent, this is also true for cold-sensitive species, including maize. Better understanding the mechanisms that play a role in cold acclimation processes may help us to develop crop plants with higher levels of cold tolerance. It has been known for a long time that, without enough light during the cold hardening period, winter cereals—even winter cereals with a potentially high level of frost hardiness—are incapable of achieving a high level of freezing tolerance [6,7]. Acclimation to low temperatures also responds to light and temperature signals [8,9]. Light has been shown to mediate the development of freezing tolerance via several biological processes. These include photosynthesis-related processes, the expression level of stress-related genes and the synthesis of various protective compounds [10]. In the case of chilling sensitive plants, light during the cold period is mainly known as an extra stress factor inducing photoinhibition of photosynthesis [11].

As a consequence, photoinhibition may also contribute to the development of the chilling injury. However, we have recently shown that, in spite of the photoinhibitory effects, light during the cold acclimation period could also enhance the effectiveness of acclimation processes in young maize plants [12]. Similarly, moderate photoinhibition could also protect Photosystem I from photodamage under low temperature conditions in tobacco plants [13]. Furthermore, photoinhition of Photosystem I in *Arabidopsis* also protected the chloroplasts from oxidative damage [14].

It seems that light is at least as important factor as the temperature during the cold acclimation period. During the exposure of maize plants to cold acclimating temperatures, light influenced various light-related cold acclimation processes not only directly, but also at the gene expression and metabolomics levels. A microarray study showed that complex regulation mechanisms and interactions between cold and light signaling processes exist during the acclimation period. Numerous significantly differentially expressed genes that are involved in most of the assimilation and metabolic pathways were detected [12]. However, the exact mechanisms regarding how light may regulate the cold acclimation processes are still poorly understood.

Plants often react to biotic or abiotic stresses with an increase in the secondary metabolite levels. In relation to this, the increased activity of phenylalanine ammonia lyase (PAL) and other related enzymes can be observed. Phenolic compounds are naturally occurring substances in plants, and many of them play important roles in defence mechanisms and the scavenging of oxidizing molecules [15]. Salicylic acid (SA) is also a phenolic compound, and it plays an important signalling role in plants in various abiotic and biotic stresses [16,17]. Among other effects, SA may also induce the production of plant defensive metabolites, including other phenolic compounds and antioxidant systems [18,19].

In the present work, based on our previous microarray experiment [12], we further analysed the differentially expressed genes, focusing on the phenylpropanoid pathways. Furthermore, the changes in certain phenolic compounds, and other plant growth regulators, such as plant hormones and thiol compounds were analysed, in order to reveal their possible role in light-regulating signalling during the cold acclimation processes in young maize plants.

2. Results

2.1. Plant Hormones

Growing young maize plants at cold acclimating temperatures (15 °C) did not cause significant changes in the free SA contents in the leaves compared to the control plants (Figure 1A). However, SA increased in a high manner during chilling at 5 °C in the non-acclimated plants, and it remained at this high level during the recovery period. Plants acclimated at low light (LL) had slightly higher free SA levels during recovery than the cold-acclimated one at growth light (GL). The bound SA level elevated only during recovery and only in non-acclimated plants (Figure 1B). While the bound SA was usually higher than the free SA in the leaves, this difference was less pronounced in the roots (Figure 1C,D). However, a substantial increase in the SA level could be detected in the free SA in the roots after a one-day recovery period. After a longer recovery, of 4 days or more, only low SA levels could be detected. The bound SA levels after the acclimation and chilling periods were significantly lower in the LL plants than in the control.

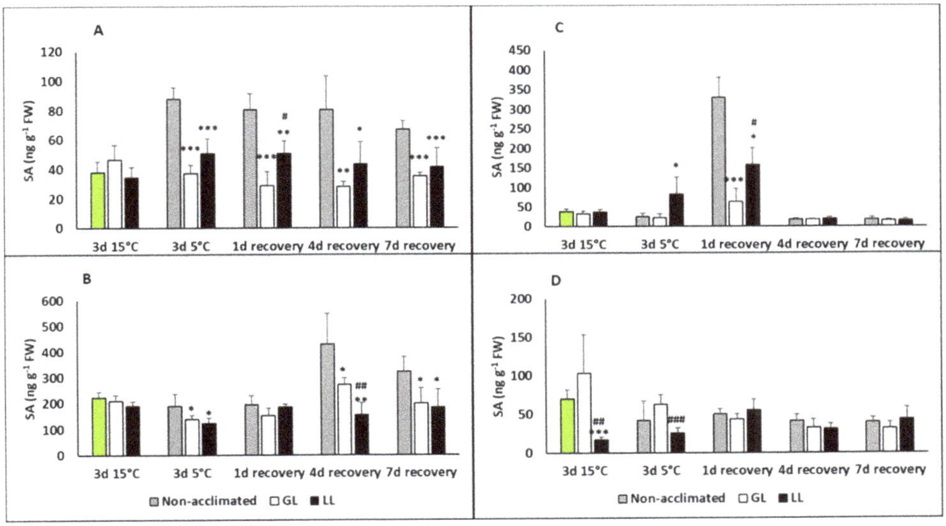

Figure 1. Changes in the salicylic acid (SA) contents during cold acclimation (15/13 °C), chilling (5 °C) and recovery in the leaves and roots of young maize plans: control plants (22/20 °C, 387 µmol m^{-2} s^{-1}). Light intensities during hardening: growth light (GL): 387 µmol m^{-2} s^{-1}; low light (LL): 107 µmol m^{-2} s^{-1}. *, **, *** significant differences compared to the control plants on the same day at the $p < 0.05$, 0.01 and 0.001 levels, respectively. #, ##, ### significant differences compared to the GL plants on the same day at the $p < 0.05$, 0.01 and 0.001 levels, respectively. (**A**: free SA in the leaves; **B**: bound SA in the leaves; **C**: free SA in the roots; **D**: bound SA in the roots).

The level of ortho-hydroxycinnamic acid (oHCA), a putative pre-cursor of SA, also increased during acclimation and chilling in the leaves and dropped back to the initial level during recovery (Figure 2A). A much higher increase was detected in non-acclimated plants at 5 °C and on the first day of recovery, but it also dropped later. No significant changes were detected in the roots, but the oHCA level was twice as high than in the leaves (Figure 2B).

The level of abscisic acid (ABA) did not change during the acclimation in the leaves of GL plants and it remained at the same level during the chilling and recovery period (Figure 3A). A slight increase could be detected in LL plants during the acclimation, but it was much more pronounced after 1d recovery. The ABA level of non-acclimated plants increased only during the recovery, especially on the first day, but it returned to the initial level in all the treatments on the seventh day. No changes were measured during acclimation in the roots (Figure 3B). Although the chilling increased the ABA level, no difference could be observed between the acclimated and non-acclimated plants. A slight increase in ABA level was determined in non-acclimated and GL plants during recovery, which dropped back at the end of recovery.

The jasmonic acid (JA) level did not change in the leaves during acclimation, but chilling significantly increased it in the cold-acclimated GL plants and especially in the non-acclimated ones (Figure 4A). On the first day of recovery, the JA level dropped back to the initial level only in the acclimated plants, but from the fourth day its level was sometimes even less than before the acclimation period. No effect of the light could be observed on the amount of JA during the acclimation, chilling and recovery periods. On the contrary, a slight elevation could be seen in the roots acclimated under GL conditions, but no changes in the LL plants (Figure 4B). The JA level in the roots increased during the chilling in non-acclimated plants, and more pronouncedly in GL, while no changes could be seen in LL plants. Interestingly, some rise could be observed in non-acclimated and LL plants on the first day of recovery, but a decrease in GL roots. On the seventh day of recovery, the JA level was similar to the initial one.

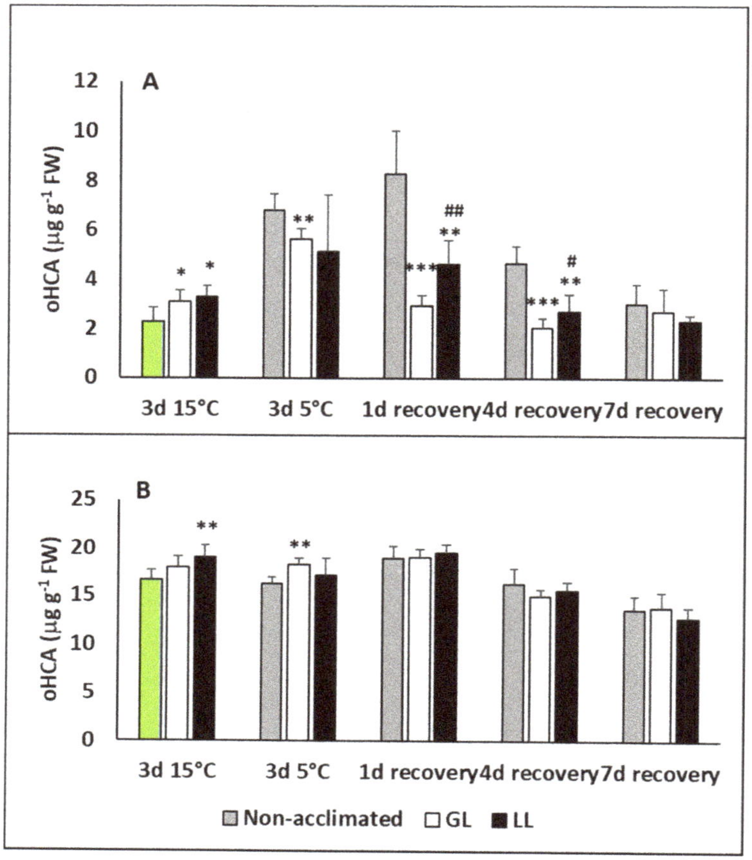

Figure 2. Changes in the ortho-hydroxycinnamic acid (oHCA) contents during cold acclimation (15/13 °C), chilling (5 °C) and recovery in the leaves and roots of young maize plants: control plants (22/20 °C, 387 μmol m^{-2} s^{-1}). Light intensities during hardening: GL: 387 μmol m^{-2} s^{-1}; LL: 107 μmol m^{-2} s^{-1}. *, **, *** significant differences compared to the control plants on the same day at the $p < 0.05$, 0.01 and 0.001 levels, respectively. #, ## significant differences compared to the GL plants on the same day at the $p < 0.05$, and 0.01 levels, respectively. (**A**: free oHCA in the leaves; **B**: free oHCA in the roots).

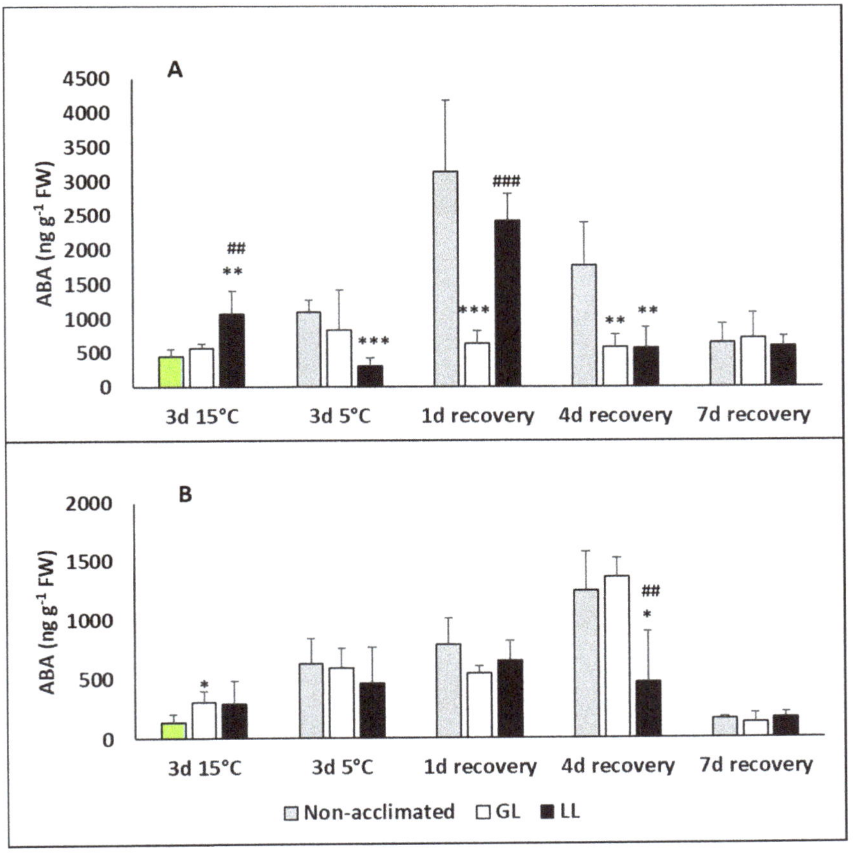

Figure 3. Changes in the abscisic acid (ABA) contents during cold acclimation (15/13 °C), chilling (5 °C) and recovery in the leaves and roots of young maize plans: control plants (22/20 °C, 387 μmol m^{-2} s^{-1}). Light intensities during hardening: GL: 387 μmol m^{-2} s^{-1}; LL: 107 μmol m^{-2} s^{-1}. *, **, *** significant differences compared to the control plants on the same day at the $p < 0.05$, 0.01 and 0.001 levels, respectively. ##, ### significant differences compared to the GL plants on the same day at the $p < 0.05$ and 0.001 levels, respectively. (**A**: free ABA in the leaves; **B**: free ABA in the roots).

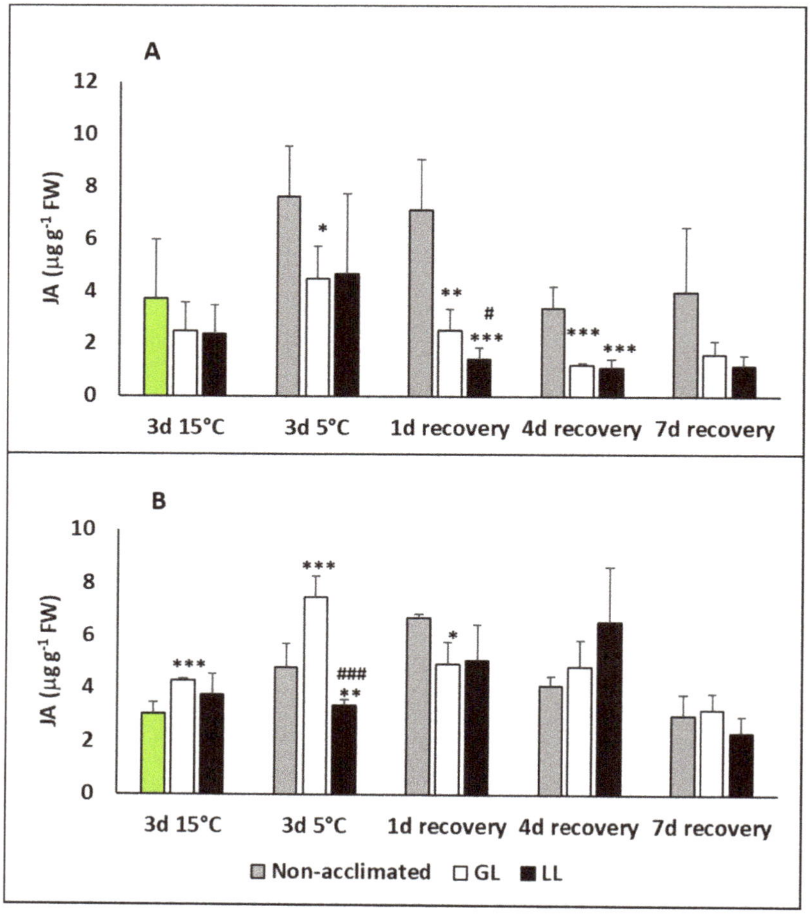

Figure 4. Changes in the jasmonic acid (JA) contents during cold acclimation (15/13 °C), chilling (5 °C) and recovery in the leaves and roots of young maize plans: control plants (22/20 °C, 387 μmol m^{-2} s^{-1}). Light intensities during hardening: GL: 387 μmol m^{-2} s^{-1}; LL: 107 μmol m^{-2} s^{-1}. *, **, *** significant differences compared to the control plants on the same day at the $p < 0.05$, 0.01 and 0.001 levels, respectively. #, ### significant differences compared to the GL plants on the same day at the $p < 0.05$ and 0.001 levels, respectively. (**A**: free JA in the leaves; **B**: free JA in the roots).

2.2. Oxidative Stress and Antioxidants

For detection of the oxidative stress, the malondialdehyde (MDA) level was measured in the leaves and roots of plants during acclimation, chilling and recovery. No changes were detected in the leaves during the acclimation and chilling periods, and only a slight increase was detected in the leaves of hardened plants on the fourth day of recovery (Figure 5A). However, the MDA level increased significantly in the roots of GL plants (Figure 5B).

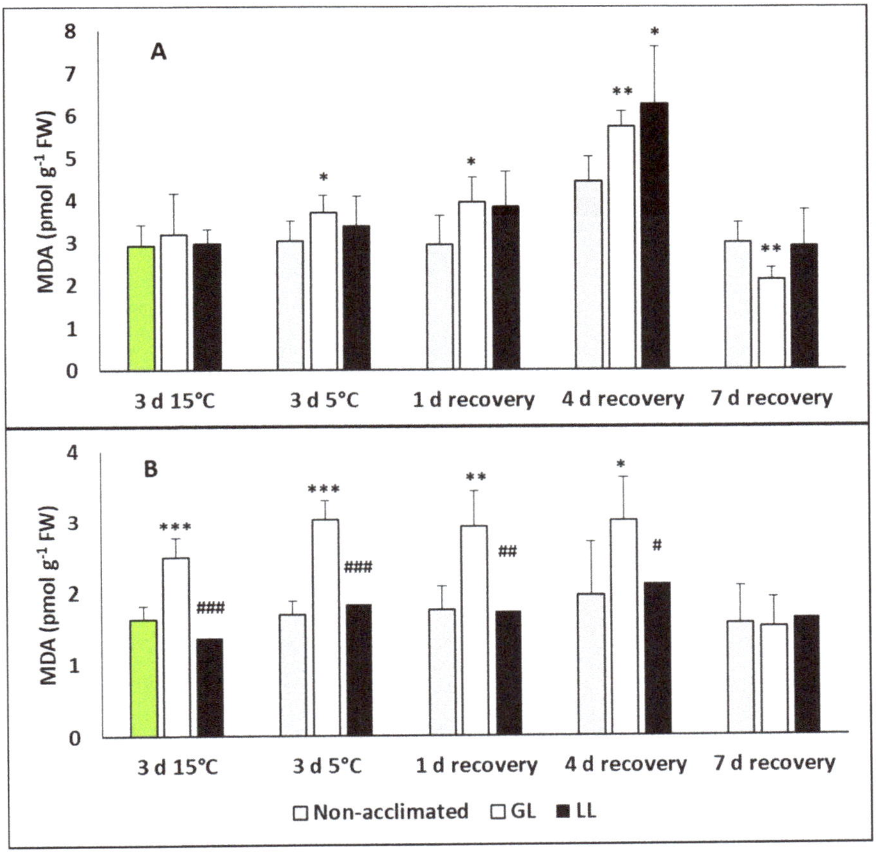

Figure 5. Changes in the malondialdehyde (MDA) contents during cold acclimation (15/13 °C), chilling (5 °C) and recovery in the leaves and roots of young maize plans: control plants (22/20 °C, 387 μmol m^{-2} s^{-1}). Light intensities during hardening: GL: 387 μmol m^{-2} s^{-1}; LL: 107 μmol m^{-2} s^{-1}. *, **, *** significant differences compared to the control plants on the same day at the $p < 0.05$, 0.01 and 0.001 levels, respectively. #, ##, ### significant differences compared to the GL plants on the same day at the $p < 0.05$, 0.01 and 0.001 levels, respectively. (**A**: leaves; **B**: roots).

Thiols, especially glutathione (γ-L-glutamyl-L-cysteinyl–glycine (GSH)), have an important role to play in the defence processes against oxidative stress. Thus, thiol compounds, namely GSH and its precursors, cysteine (Cys) and γ-L-glutamyl-L-cysteine (γEC) were also analysed. Cys levels increased after chilling in the leaves of GL plants and was still higher on the first day of recovery (Figure 6A). Similar tendencies could be seen in the amount of γEC (Figure 6B). Compared to these, the GSH level increased in LL plants after acclimation and decreased during the chilling (Figure 6C), and a dramatic rise was observed on the first day of recovery, both in GL and LL plants. Cysteinyl–glycine (CysGly), a degradation product of GSH, had the highest amount after the chilling in GL plants, but dropped back to the initial level during the recovery (Figure 6D). Cys increased after acclimation in the roots of LL plants, but after chilling it was at the initial level (Figure 6E). An increased level of it was detected in non-acclimated plants after chilling and it was still high on the first day of recovery. The GL plants had elevated levels of root Cys on the first and fourth days of recovery, while root γEC increased only in LL plants after chilling and it was also higher on the fourth and seventh days of recovery (Figure 6F).

Elevated levels could be also seen in the non-acclimated and GL plants during recovery, but this level was still lower than in LL plants. A big enlargement in the GSH amount was measured in LL plants after acclimation, but after the chilling, and during recovery, the initial level was detected in the roots of the plants (Figure 6G). The amount of CysGly increased during acclimation and remained at the same level in LL plants after chilling and during the recovery (Figure 6H). It dropped to the initial level in GL plants after chilling, but it started to increase during the recovery, and the highest level was detected after seven days.

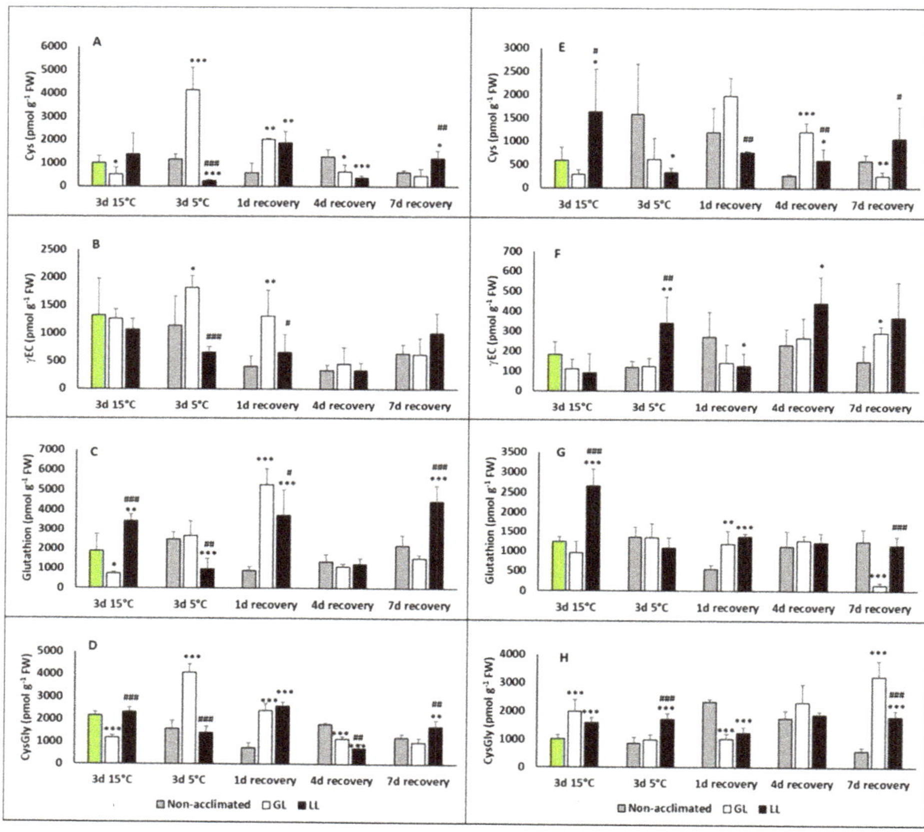

Figure 6. Changes in the thiol contents during cold acclimation (15/13 °C), chilling (5 °C) and recovery in the leaves and roots of young maize plants. Control plants (22/20 °C, 387 µmol m^{-2} s^{-1}). Light intensities during hardening: GL: 387 µmol m^{-2} s^{-1}; LL: 107 µmol m^{-2} s^{-1}. *, **, *** significant differences compared to the control plants on the same day at the $p < 0.05$, 0.01 and 0.001 levels, respectively. #, ##, ### significant differences compared to the GL plants on the same day at the $p < 0.05$, 0.01 and 0.001 levels, respectively. (**A**: cysteine in the leaves; **B**: γ-glutamyl-cysteine in the leaves; **C**: gluthathione in the leaves; **D**: cysteinyl–glycine in the leaves; **E**: cysteine in the roots; **F**: γ-glutamyl-cysteine in the roots; **G**: gluthathione in the roots; **H**: cysteinyl–glycine in the roots).

PAL is a key enzyme of the phenylpropanoid metabolism. It has a role either in the SA or flavonol/anthocyanin biosynthesis. The PAL activity increased in the leaves of LL plants after acclimation, but showed a much higher amount on the first day of recovery (Figure 7A). Its activity was higher in the roots than in the leaves and an increase could be seen after acclimation, chilling and 1 d recovery, mainly in LL plants (Figure 7B).

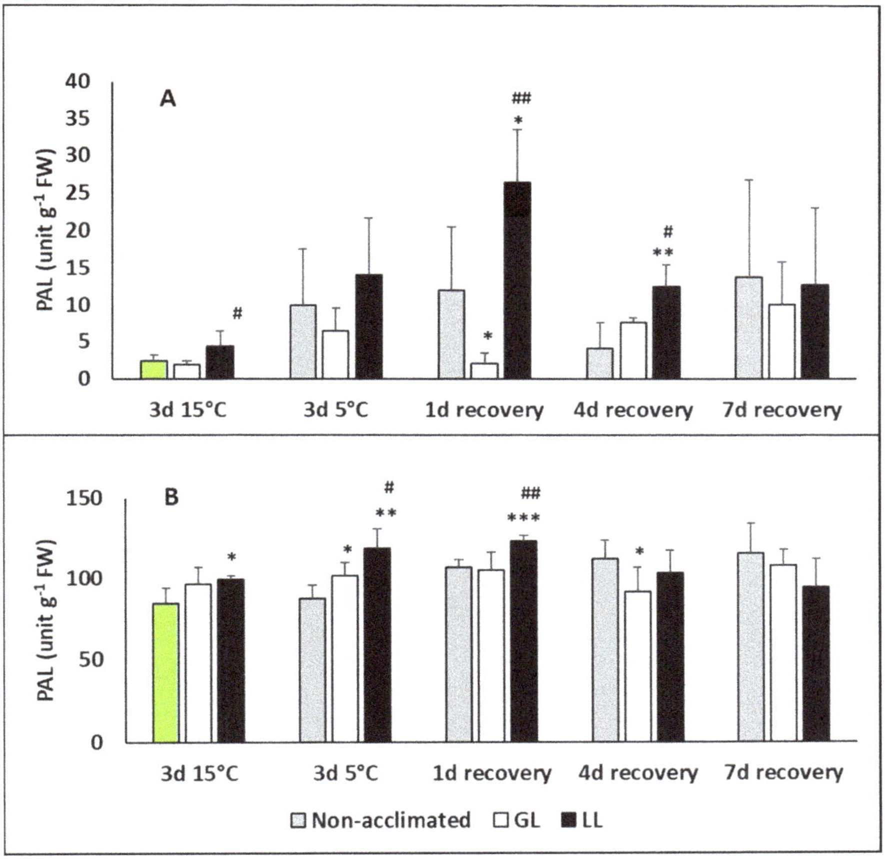

Figure 7. Changes in the phenylalanine ammonia lyase (PAL) contents during cold acclimation (15/13 °C), chilling (5 °C) and recovery in the leaves and roots of young maize plans: control plants (22/20 °C, 387 μmol m^{-2} s^{-1}). Light intensities during hardening: GL: 387 μmol m^{-2} s^{-1}; LL: 107 μmol m^{-2} s^{-1}. *, **, *** significant differences compared to the control plants on the same day at the $p < 0.05$, 0.01 and 0.001 levels, respectively. #, ## significant differences compared to the GL plants on the same day at the $p < 0.05$ and 0.01 levels, respectively. (**A**: leaves; **B**: roots).

Flavonols (kaempferol (K), quercetin (Q), myricetin (M) and rutin (R)) and anthocyanins were also analysed. K, Q and M were in a much lower amount (0.1–2 μg g^{-1} FW) both in the leaves and roots than rutin (100–130 μg g^{-1} FW). Only the R level increased in the leaves, especially in LL plants, but after 4 d recovery it was lower than the initial level (Table 1). There were no substantial changes in the roots, only the M increased at 3 d and at 15 °C in GL and LL plants (Table 1).

Table 1. Changes in the flavonol contents during cold acclimation (15/13 °C), chilling (5 °C) and recovery in the leaves and roots of young maize plants. Control: 22/20 °C, 387 μmol m^{-2} s^{-1}. NA: non-acclimated plants. Light intensities during hardening: GL: 387 μmol m^{-2} s^{-1}; LL: 107 μmol m^{-2} s^{-1}. *, **, *** significant differences compared to the control plants on the same day at the $p < 0.05$, 0.01 and 0.001 levels, respectively. #, ##, ### significant differences compared to the GL plants on the same day at the $p < 0.05$, 0.01 and 0.001 levels, respectively.

Leaf		3 d 15 °C			3 d 5 °C			1 d recovery			4 d recovery			7 d recovery	
	Control	GL	LL	NA	GL	LL	NA	GL	LL	NA	GL	LL	NA	GL	LL
Kaempferol	0.76 ± 0.11	1.04 ± 0.09 **	0.9 ± 0.07 *#	0.92 ± 0.04	0.96 ± 0.16*##	0.725 ± 0.1	1.1 ± 0.46	0.96 ± 0.15	0.92 ± 0.11	2.125 ± 0.44	1.04 ± 0.09 ***	0.96 ± 0.18 ***	1.4 ± 0.34	0.74 ± 0.05 **	0.72 ± 0.08 **
Quercetin	1.43 ± 0.1	1.52 ± 0.38	1.52 ± 0.16	1.76 ± 0.15	1.52 ± 0.23	1.625 ± 0.17	1.66 ± 0.4	1.4 ± 0.12	1.28 ± 0.13	1.42 ± 0.08	1.25 ± 0.06	1.36 ± 0.29	1.12 ± 0.04	1.02 ± 0.11	0.94 ± 0.15 *
Myricetin	0.31 ± 0.07	0.59 ± 0.33	0.71 ± 0.42	0.24 ± 0.05	0.24 ± 0.09	0.19 ± 0.22	nd	0.58 ± 0.21 ***	0.65 ± 0.35 **	nd	nd	nd	nd	0.43 ± 0.12 ***	0.3 ± 0.2 *
Rutin	83.28 ± 10.9	101.38 ± 8.3 **	104.38 ± 7.77 **	87.4 ± 10.57	97.12 ± 18.51	121.83 ± 3.38 ***#	89.78 ± 6.44	82.12 ± 0.82 *	96.68 ± 5.1 ###	41.98 ± 15.44	31.9 ± 8.24	50.28 ± 8.63 #	37.52 ± 12.59	38.54 ± 6.79	37.12 ± 3.07
Root		**3 d 15 °C**			**3 d 5 °C**			**1 d recovery**			**4 d recovery**			**7 d recovery**	
	Control	**GL**	**LL**	**NA**	**GL**	**LL**	**NA**	**GL**	**LL**	**NA**	**GL**	**LL**	**NA**	**GL**	**LL**
Kaempferol	0.04 ± 0.02	0.05 ± 0.01	0.06 ± 0.02	0.13 ± 0.03	0.13 ± 0.01 *##	0.09 ± 0.02	0.19 ± 0.03	0.16 ± 0.03	0.14 ± 0.03	0.21 ± 0.03	0.25 ± 0.05	0.15 ± 0.09	0.06 ± 0.02	0.04 ± 0.02	0.07 ± 0.04
Quercetin	1.18 ± 0.14	1.36 ± 0.06 *	1.38 ± 0.1 *	1.36 ± 0.26	1.44 ± 0.27	1.18 ± 0.16	1.45 ± 0.17	1.02 ± 0.13 **	1.26 ± 0.13#	1.14 ± 0.11	1.2 ± 0.19	0.93 ± 0.05**#	0.8 ± 0.1	0.9 ± 0.19	1.08 ± 0.08 **
Myricetin	8.02 ± 0.92	26.52 ± 8.99 **	29.34 ± 13.61 **	11.28 ± 6.04	7.76 ± 2.38	8.68 ± 6.07	4.5 ± 4.6	2 ± 1.57	2.84 ± 1.8	3.28 ± 1.15	6.58 ± 2.6*	7.72 ± 8.01	7.94 ± 4.25	7.72 ± 1.87	11.04 ± 2.9
Rutin	132.55 ± 18.47	130.86 ± 13.14	118.8 ± 8.76	110.1 ± 6.81	105.16 ± 8.88	97.18 ± 15.11	104.75 ± 2.06	97.68 ± 3.48 *	95.02 ± 6.85	106.56 ± 8.13	100.44 ± 3.23	117.17 ± 20.45	126.12 ± 4.68	119.36 ± 13.55	107.8 ± 12.77 *

Anthocyanins could only be detected during the recovery period. It accumulated mainly in the roots of GL plants but changes in the leaves were also observed (Figure 8). The highest amount was detected on the fourth day of recovery, but it was still high on the seventh day in the acclimated plants.

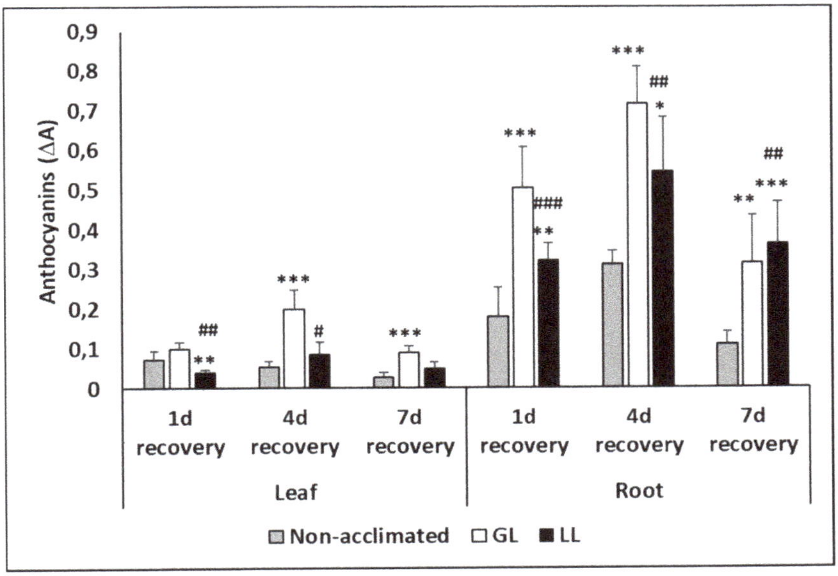

Figure 8. Changes in the anthocyanin contents during cold acclimation (15/13 °C), chilling (5 °C) and recovery in the leaves and roots of young maize plants. Light intensities during hardening: GL: 387 μmol m^{-2} s^{-1}; LL: 107 μmol m^{-2} s^{-1}. *, **, *** significant differences compared to the control plants on the same day at the $p < 0.05$, 0.01 and 0.001 levels, respectively. #, ##, ### significant differences compared to the GL plants on the same day at the $p < 0.05$, 0.01 and 0.001 levels, respectively.

2.3. Gene Expression Studies

Based on a microarray assay, the changes in the gene expression levels during the cold acclimation period under various light conditions have been analysed in our previous study [12]. In the present work—based on the same microarray database—gene expressions related to the above characterised changes in plant hormone metabolisms (Section 2.1.) and antioxidant capacity (Section 2.2.) have been further analysed in the leaves at the end of acclimation period.

A gene encoding a PAL enzyme was downregulated and genes encoding the enzyme related to the SA-MeSA (methyl-SA) conversation (salicylate carboxymethyltransferase), and enzymes related to the synthesis of ABA (zeaxanthin epoxidase (ZEP)) and JA (12-oxo-phytodienoic acid reductase7) and phenypropanoid metabolism (cinnamyl-alcohol dehydrogenase) were upregulated in GL compared to the control plants (Table S1). Chorismate mutase 1 (related to the SA biosynthesis as a part of the shikimic acid pathway) and violaxanthin de-epoxidase were downregulated, while enzymes of ABA (aldehyde oxidase, ZEP) and JA (12-oxo-phytodienoic acid reductase7, allene oxide synthase 1) biosynthesis were upregulated in LL compared to the control plants. In a comparison of GL and LL plants, salicylate carboxymethyltransferase, aldehyde oxidase, allene oxide synthase 1 were upregulated in GL plants, while chorismate mutase 1 and violaxanthin de-epoxidase were downregulated.

Gene expression levels related to the antioxidants were also examined at the end of the acclimation period in the leaves of the plants. Glutathione-S-transferase (GST) was downregulated and ascorbate peroxidases were upregulated in GL plants compared to the control (Table S2). Mainly the enzymes of the ascorbate–glutathione cycle (glutathione reductase, monodehydroascorbate reductase 5, dehydroascorbate reductase) were downregulated and GSTs were upregulated in LL compared to the control. Enzymes related to the flavonol/anthocyanin biosynthesis (isoflavone reductase, dihydroflavonol-4-reductase) were upregulated in GL and LL plants compared to the control (Table S3). In a comparison of GL–LL plants, GST and flavonol/anthocynin biosynthesis enzymes were upregulated and some enzymes of the ascorbate (ASC)–GSH cycle (glutathione reductase, monodehydroascorbate reductase 5) were downregulated in GL compared to the LL plants.

3. Discussion

Growing plants at suboptimal but non-lethal temperatures may help them to prepare for a subsequent, more severe cold stress. In the case of hardy winter cereals, it may also lead to development of frost tolerance, which enables plants to survive low temperatures well below 0 °C. Light during the hardening period is also essential for the development of adequate freezing tolerance. Its role and mode of action have been widely studied [6,20,21]. But the role of light during the cold acclimation period in the case of chilling sensitive plants is hardly documented. When light is accompanied by low growth temperature, usually, photoinhibitory damage can be detected, which may substantially contribute to the chilling injury in cold-sensitive plants. However, our previous study [12] indicated that, in spite of its photoinhibitory effects, light is also essential for the high level of cold acclimation in young maize plants. In order to better understand the molecular mechanisms of the light-related cold acclimation processes, in recent works, certain stress-related plant hormones, focusing mainly on SA and its related compounds, were also followed during the cold acclimation, chilling, and a subsequent recovery periods. SA is also a Janus-faced compound: its exogenous application provided protection against chilling injury in young maize plants [22], but its relatively high endogenous level inhibited the growth of *Arabidopsis* plants at low temperatures [23]. In the present work, the main changes could be observed in the free SA levels in the leaves, and it was lower in acclimated plants than in the controls. Although, the gene expression level of PAL enzyme was downregulated in GL plants under cold conditions, the PAL enzyme activity did not show remarkable differences, which could have been the reason for the lower SA contents in the leaves. However, at the end of the hardening period, the gene expression of the salicylate caboxymethyltransferase, an enzyme responsible for SA-MeSA conversation, increased in the leaves of GL plants both under cold and control conditions, suggesting the dominant effect of light, rather than the cold MeSA, is a volatile form of SA and is part of the SA signalling pathway [24] and the decreased level of free SA could at least be partly a result of this SA-MeSA conversation. Interestingly, parallel to this, the expression of chorismate mutase 1 was also downregulated in GL plants, suggesting that the branch point of the shikimate pathway to channel amino acid precursors to the biosynthesis of phenylalanine and tyrosine and away from that of tryptophan was lower at GL [25].

ABA is a stress hormone, playing a role in the responses to various environmental stimuli [26]. In our experiment, the level of ABA did not correlate with the level of tolerance. Its level mainly increased in the leaves during the post-chilling recovery period. It seems that it is mainly related to the level of the stress, and less related to the level of the acclimation. At the end of acclimation, the synthesis gene of ABA (ZEP) was upregulated both in GL and LL plant compared to the non-acclimated plants, but there was no difference between the GL and LL plants. In addition, the expression of violaxanthin de-epoxidase, which is responsible for the catalysation of the reaction in the opposite direction, was downregulated in LL compared to the control under cold conditions, while it was upregulated in the GL compared to the LL plants. It has been found that the gene expression of ZEP showed mainly circadian rhythms in the leaves rather than changes in the daily periods of stress, and no changes were detected in the amount of the enzyme [27]. On the other hand, light was required for the gene expression of ZEP, because cold treatment only induced the genes of the ABA biosynthesis in light, as was demonstrated in *Arabidopsis* plants [28].

A similar trend could be observed in the JA levels, which mainly increased in the leaves of non-acclimated plants. In addition, at the end of acclimation, the gene of JA biosynthesis (12-oxo-phytodienoic acid reductase7) was upregulated both in GL and LL plants compared to the non-acclimated plants, but there was no difference between GL and LL plants, while another gene encoding allene oxide synthase 1 was shown to exhibit higher upregulation in GL plants. Chilling temperatures may increase the generation of the reactive oxygen species (ROS), especially in light. Plants generally respond to such conditions by activating antioxidant systems. oHCA, a precursor of SA, belongs to the hydroxycinnamic acid family, and it also has antioxidant properties. Although the oHCA level was lower in the acclimated plants, and no pronounced differences were observed between the GL and LL plants, the gene of cinnamyl-alcohol dehydrogenase, related to the cinnamic acid metabolism, was upregulated in GL plants. This latter result is in accordance with earlier findings, where two genes playing roles in the synthesis of hydroxycinnamic acids were also upregulated at a low temperature in light in *Arabidopsis* plants [28]. Parallel to the induction of the gene, a cold-induced increase in the oHCA level was also detected in young maize plants. However, this increase also occurred in the dark. This response was entirely different from that which was earlier found during the cold hardening period in winter wheat plants, where a substantial increase in the oHCA level only occurred in light, but not in the dark [29]. However, the pattern of leaf oHCA content under the recovery period showed similar differences and changes, as was found for the SA. It seems that the exact role of oHCA as a precursor and/or antioxidant compound during the cold acclimation period still requires intensive further research.

Hydroxycinnamic acid biosynthesis, like SA metabolism, is part of the phenylpropanoid pathway which involves the synthesis of various antioxidants, such as flavonols, anthocyanins, etc., and one of the key enzymes of this pathway is PAL. It was found earlier that light had a role not only in the development of chilling injury, but also in the appearance of the post-chilling symptoms in young maize plants [30]. Furthermore, light was also required for the synthesis of anthocyanins when plants were transferred from chilling conditions to optimum growth temperatures [31]. In the present work, although the recovery was in light, and only the acclimation was carried out at different illuminations, the differences in anthocyanin level could be seen both in the leaves and roots. Plants acclimated at low light had much lower anthocyanin content compared to the GL plants. Interestingly, differences in the anthocyanin levels did not follow the changes in the activity of the PAL enzyme. Furthermore, the changes in the various flavonols, such as K, Q, M or R, either in the leaves or in the roots, were more temperature- than light-dependent. This was in accordance with the gene expression changes, as the transcript level of the genes related to flavonoid metabolism was also upregulated in GL and LL plants under cold conditions.

GSH also acts as an antioxidant by quenching ROS. The ASC–GSH cycle plays an important role in this respect. In *Arabidopsis*, cold treatment in light induced the expression of genes of two GSH-dependent peroxidases, while the genes of the ASC–GSH cycle were not induced [28]. In our case similar changes were found in maize plants, the genes of the ASC–GSH cycle enzymes were downregulated, while genes encoding GST were upregulated in GL plants compared to the LL conditions during cold acclimation. However, the accumulation of GSH was also light dependent. The synthesis of GSH is a two-step reaction: the first step is the formation of γEC from Cys catalysed by γEC synthetase then the GSH synthetase catalyses the addition of a glycine. The rate-limiting step is the γEC synthesis [32]. In a recent study, the highest GSH level was found at optimum light conditions in wheat plants. Either the higher or lower illumination caused lower GSH accumulation [33]. In the present work, the highest GSH level was found in the leaves of GL plants on the first day of recovery, but the pre-cursors of GSH (Cys, γEC) started to accumulate during chilling in GL plants. However, this accumulation did not cause an increase in the GSH level, because the enhancement of the degradation processes led to high levels of CysGly. In contrast to the GL plants, under low light conditions, higher GSH accumulation occurred during the acclimation period.

GSH is not only an antioxidant, it can also interact with plant hormones. GSH regulates SA accumulation via the expression of genes of isochorismate synthetase 1 in *Arabidopsis* plants and it can lead to an increase in the intracellular hydrogen peroxide level for the activation of SA signalling [34]. In our case, GSH accumulation was not accompanied by SA accumulation either in the free or bound forms. In another study, the GSH content was higher after SA treatment in maize plants, while it was lower after the application of ABA [35]. In the present work, it was also found that a higher GSH content was coupled with a lower ABA level. However, under the present conditions, the antagonistic relationship between SA and ABA, or between SA and JA, could not be detected.

4. Materials and Methods

4.1. Plant Material and Growth Conditions

Maize plants were grown for 11 days at 22/20 °C at higher photosynthetic photon flux density (PPFD), (growth light, GL = 387 μmol m^{-2} s^{-1}) with 16/8 h light/dark periods. Some of the plants were then cold acclimated at 15/13 °C for 3 days, either at GL or at moderate low light intensity (LL) 107 μmol m^{-2} s^{-1}. Afterwards, all plants were transferred to 5 °C at a continuous growth light (GL) for 3 days, followed by a seven-day recovery period at 22 °C under GL conditions. Samples were collected for biochemical analysis from the control plants, and from treated plants after the acclimation (GL and LL plants), chilling and during recovery periods (non-acclimated, GL and LL plants).

All the chemicals were purchased from Sigma-Aldrich or Merck (Merck Group, Darmstadt, Germany).

4.2. Estimation of Lipid Peroxidation

The lipid peroxidation analysis was based on the measurement of the MDA level according to Gondor et al. [36] using 0.5 g plant material. Leaves and roots were extracted with 1.5 mL 0.1% (*w*/*v*) tricholoracetic acid and, after incubation with 0.5% (*w*/*v*) thiobarbituric acid at 90 °C for 30 min, the MDA equivalent compounds were measured spectrophotometrically at 532 nm, with the subtraction of non-specific absorption at 600 nm. MDA was then quantified using an extinction coefficient of 155 mM^{-1}·cm^{-1} and expressed as nM·g^{-1} fresh weight. Five replicates were measured from each treatment and at least three leaves were used for one replicate.

4.3. Extraction and Analytical Procedure of Salicylic Acid and Flavonols

Flavonoids, JA, ABA, SA and its precursors were measured according to Meuwly and Métraux [37] and Pál et al. [38] using 1 g plant material. The leaves and the roots were ground in liquid nitrogen in a mortar and pestle, in the presence of 0.5 g quartz sand. The tissue powder was transferred to a centrifugation tube and mixed with 2 mL of 70% methanol. The extract was centrifuged at 10,000× *g* for 20 min. The pellet was resuspended in 2 mL of 90% methanol, vortexed and centrifuged as above. The methanol content was evaporated from the mixed supernatants at room temperature under a vacuum. One millilitre of 5% (*w*/*v*) trichloroacetic acid was added to the residual aqueous phase, and the mixture was centrifuged at 15,000× *g* for 10 min. The supernatant was gently partitioned against 3.6 mL of a 1:1 (*v*/*v*) mixture of ethyl acetate/cyclohexane. The upper organic layers contained the free phenolic portion. The aqueous phases containing the methanol-soluble bound phenolics were acid-hydrolysed. 1.3 mL of 8 N HCl was added to the aqueous phase and incubated for 60 min at 80 °C, before partitioning as above. The organic phases were evaporated to dryness under a vacuum and stored at −20 °C. Just prior to the HPLC analysis, the evaporated samples were resuspended in 0.5 mL 15% acetonitrile and filtered through a 0.45 m pore size polytetrafluoroethylene (PTFE) membrane filter (Millipore, Merck Group, Darmstadt, Germany). The detailed description of the HPLC analysis was described by Janda et al. [39].

4.4. Measurement of Thiols

A total of 200 mg plant material was extracted with 1 mL of 0.1 M HCl and centrifuged for 20 min with 10,000 rpm. Total thiol content was determined after the supernatant was reduced (120 µL) with dithiothreitol, and derivatization was carried out with monobromobimane [40]. The samples were analysed after the separation of Cys, γEC, CysGly (a degradation product of GSH) and GSH by reverse-phase HPLC (W996, Waters, Milford, MA, USA) using a scanning fluorescence detector (W2474, Waters, Milford, MA, USA). The detailed parameters of the analysis were described by Janda et al. [39].

4.5. Determination of Phenylalanine Ammonia Lyase (PAL) Activity

PAL activity was measured according to Gao et al. [41] using 1 g leaves and roots homogenised with 4 mL 50 mM TRIS-HCl buffer (pH 8.8 containing 5 mM β–mercaptoethanol and 4% (*v*/*v*) polyvinylpyrrolidone). After centrifugation (10 min 10,000× *g*), 250 µL supernatant was used for enzyme activity determination in a 3 mL total volume of reaction mixture containing 50 mM phosphate buffer (pH 8.8) and 50 mM phenylalanine. The reaction was stopped with 10% trichloroacetic acid after a one-hour incubation at 37 °C and the produced *trans*-cinnamic acid was detected at 290 nm spectrophotometrically and expressed as enzyme units per g fresh weight (U g^{-1} FW).

4.6. Determination of Anthocyanin Content

The method described by Kho et al. [42] was used for the anthocyanin content determination using 0.5 g plant tissue extracting with 2 × 300 µL of methanol containing 1% HCl and was kept at 4 °C for overnight. After removing the chlorophylls with 500 µL chloroform and centrifugation at 10,000× *g* for 30 min, the total anthocyanin content was determined from the supernatant at 530 nm spectrophotometrically.

4.7. Statistical Analysis

The experiments were repeated three times and representative data are shown. The results were the means of five measurements. The data were statistically evaluated using the standard deviation, ANOVA and *t*-test methods.

5. Conclusions

Different light conditions during the cold acclimation period differentially affected certain stress-related mechanisms in young maize plants (Figure 9). Interestingly, changes were also light dependent in the root, not only in the leaves. PAL has a key role in the synthesis of various secondary metabolic routes, including compounds with antioxidant activities, and was mainly activated under limited light conditions. However, light-induced anthocyanin accumulation occurred not only in the leaves, but in the roots, too. Chilling induced the accumulation of SA and one of its putative precursors, oHCA. However, this accumulation was moderate in the cold-acclimated plants. Low temperature reduced the accumulation of JA in the leaves, but induced it in the roots. The highest GSH level was found on the first day of the recovery period in the leaves of GL plants, but its pre-cursors started to accumulate even during the chilling. The level of ABA is mainly related to the level of stress, and less related to the level of the acclimation. According to these results, it was demonstrated that LL condition partly modifies the acclimation processes after only 1 day at the gene expression level, which, in turn, may result in hormonal and metabolic shifts, sometimes in different ways, in the leaves and roots of the plants.

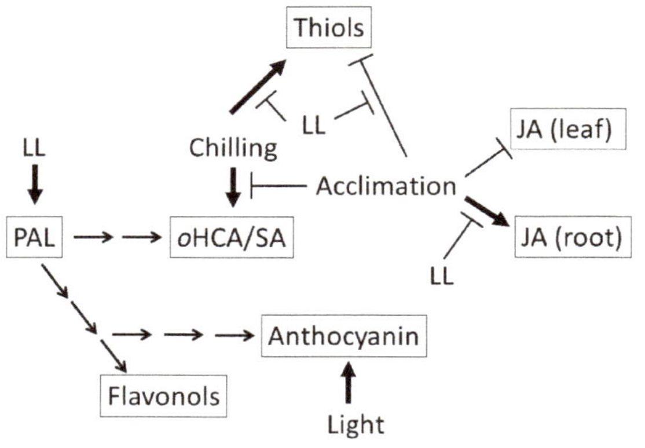

Figure 9. Schematic representation of the effect of the light on phenolic metabolism in young maize plants during cold acclimation. JA: jasmonic acid; LL: low light intensity (107 μmol m^{-2} s^{-1}) during cold acclimation; ortho-hydroxycinnamic acid (oHCA); phenylalanine ammonia lyase (PAL).

Supplementary Materials: Supplementary Materials can be found at http://www.mdpi.com/1422-0067/21/6/1942/s1. Table S1: BLASTX search hits of cold- and light-induced significantly differentially expressed transcripts related to hormone biosynthesis. Table S2: BLASTX search hits of cold- and light-induced significantly differentially expressed transcripts related to thiol metabolism. Table S3: BLASTX search hits of cold- and light-induced significantly differentially expressed transcripts related to flavonoid metabolism.

Author Contributions: Conceptualization, G.S. and T.J.; methodology, M.P.; formal analysis, I.M.; investigation, G.S. and M.P.; resources, G.S.; writing—original draft preparation, G.S. and T.J.; writing—review and editing, M.P. and I.M.; visualization, G.S. and I.M; supervision, G.S.; project administration, G.S.; funding acquisition, G.S. All authors have read and agreed to the published version of the manuscript.

Funding: This research was funded by National Research, Development and Innovation Office, Hungary, grant number K124430 and 2018-2.1.14-TÉT-CN2018-00010.

Acknowledgments: Thanks are due to Timea Oláh for her technical assistance.

Conflicts of Interest: There is no conflict of interest. The funders had no role in the design of the study; in the collection, analyses, or interpretation of data; in the writing of the manuscript, or in the decision to publish the results.

Abbreviations

ABA	abscisic acid
ASC	ascorbate
CysGly	cysteinyl–glycine
GL	growth light, 387 μmol m^{-2} s^{-1}
GSH	gluthathione (γ-L-glutamyl-L-cysteinyl–glycine)
GST	glutathione-S-transferase
JA	jasmonic acid
K	kaempferol
LL	low light intensity, 107 μmol m^{-2} s^{-1}
MDA	malondialdehyde
M	myricetin
oHCA	ortho-hydroxycinnamic acid
PAL	phenylalanine ammonia lyase
Q	quercetin
R	rutin
SA	salicylic acid
ZEP	zeaxanthin epoxidase
γEC	γ-L-glutamyl-L-cysteine

References

1. Bredenkamp, G.J.; Baker, N.R. Temperature-sensitivity of D1 protein metabolism in isolated *Zea mays* chloroplasts. *Plant Cell Environ.* **1994**, *17*, 205–210. [CrossRef]
2. Greaves, J.A. Improving suboptimal temperature tolerance in maize—The search for variation. *J. Exp. Bot.* **1996**, *47*, 307–323. [CrossRef]
3. Anderson, M.D.; Prasad, T.K.; Martin, B.A.; Stewart, C.R. Differential gene expression in chilling-acclimated maize seedlings and evidence for the involvement of abscisic acid in chilling tolerance. *Plant Physiol.* **1994**, *105*, 331–339. [CrossRef] [PubMed]
4. Prasad, T.K. Mechanisms of chilling-induced oxidative stress injury and tolerance in developing maize seedlings: Changes in antioxidant system, oxidation of proteins and lipids, and protease activity. *Plant J.* **1996**, *10*, 1017–1026. [CrossRef]
5. Janda, T.; Szalai, G.; Ducruet, J.-M.; Páldi, E. Changes in photosynthesis in inbred maize lines with different degrees of chilling tolerance grown at optimum and suboptimum temperatures. *Photosynthetica* **1998**, *35*, 205–212. [CrossRef]
6. Gray, G.R.; Chauvin, L.P.; Sarhan, F.; Huner, N. Cold acclimation and freezing tolerance (a complex interaction of light and temperature). *Plant Physiol.* **1997**, *114*, 467–474. [CrossRef]
7. Apostol, S.; Szalai, G.; Sujbert, L.; Popova, L.P.; Janda, T. Noninvasive monitoring of the light-induced cyclic photosynthetic electron flow during cold hardening in wheat leaves. *Z. Nat. C* **2006**, *61*, 734–740.
8. Catalá, R.; Medina, J.; Salinas, J. Integration of low temperature and light signaling during cold acclimation response in *Arabidopsis*. *Proc. Natl. Acad. Sci. USA* **2011**, *108*, 16475–16480. [CrossRef]
9. Lee, C.M.; Thomashow, M.F. Photoperiodic regulation of the C-repeat binding factor (CBF) cold acclimation pathway and freezing tolerance in *Arabidopsis thaliana*. *Proc. Natl. Acad. Sci. USA* **2012**, *109*, 15054–15059. [CrossRef]
10. Janda, T.; Majláth, I.; Szalai, G. Interaction of temperature and light in the development of freezing tolerance in plants. *J. Plant Growth Regul.* **2014**, *33*, 460–469. [CrossRef]
11. Li, L.; Aro, E.M.; Millar, A.H. Mechanisms of Photodamage and Protein Turnover in Photoinhibition. *Trends Plant Sci.* **2018**, *23*, 667–676. [CrossRef] [PubMed]
12. Szalai, G.; Majláth, I.; Pál, M.; Gondor, O.K.; Rudnóy, S.; Oláh, C.; Vanková, R.; Kalapos, B.; Janda, T. Janus-Faced Nature of Light in the Cold Acclimation Processes of *Maize*. *Front. Plant Sci.* **2018**, *9*, 850. [CrossRef] [PubMed]
13. Huang, W.; Yang, Y.J.; Hu, H.; Zhang, S.B. Moderate photoinhibition of photosystem II protects photosystem I from photodamage at chilling stress in tobacco leaves. *Front. Plant Sci.* **2016**, *7*, 182. [CrossRef] [PubMed]
14. Lima-Melo, Y.; Alencar, V.T.C.B.; Lobo, A.K.M.; Sousa, R.H.V.; Tikkanen, M.; Aro, E.-M.; Silveira, J.A.G.; Gollan, P.J. Photoinhibition of Photosystem I provides oxidative protection during imbalanced photosynthetic electron transport in *Arabidopsis thaliana*. *Front. Plant Sci.* **2019**, *10*, 916. [CrossRef] [PubMed]
15. Solecka, D.; Kacperska, A. Phenylpropanoid deficiency affects the course of plant acclimation to cold. *Physiol. Plant.* **2003**, *119*, 253–262. [CrossRef]
16. Seyfferth, C.; Tsuda, K. Salicylic acid signal transduction: The initiation of biosynthesis, perception and transcriptional reprogramming. *Front. Plant Sci.* **2014**, *5*, 697. [CrossRef]
17. Poór, P.; Czékus, Z.; Tari, I.; Ördög, A. The multifaceted roles of plant hormone salicylic acid in endoplasmic reticulum stress and unfolded protein response. *Int. J. Mol. Sci.* **2019**, *20*, 5842. [CrossRef]
18. Cappellari, L.D.R.; Santoro, M.V.; Schmidt, A.; Gershenzon, J.; Banchio, E. Improving phenolic total content and monoterpene in Mentha x piperita by using salicylic acid or methyl jasmonate combined with Rhizobacteria inoculation. *Int. J. Mol. Sci.* **2020**, *21*, 50. [CrossRef]
19. Palmer, I.A.; Chen, H.; Chen, J.; Chang, M.; Li, M.; Liu, F.; Fu, Z.Q. Novel salicylic acid analogs induce a potent defense response in *Arabidopsis*. *Int. J. Mol. Sci.* **2019**, *20*, 3356. [CrossRef]
20. Majláth, I.; Szalai, G.; Soós, V.; Sebestyén, E.; Balázs, E.; Vanková, R.; Dobrev, P.I.; Tandori, J.; Janda, T. Effect of light on the gene expression and hormonal status of winter and spring wheat plants during cold hardening. *Phys. Plant.* **2012**, *145*, 296–314. [CrossRef]
21. Szalai, G.; Pap, M.; Janda, T. Light-induced frost tolerance differs in winter and spring wheat plants. *J. Plant Physiol.* **2009**, *166*, 1826–1831. [CrossRef] [PubMed]

22. Janda, T.; Szalai, G.; Tari, I.; Páldi, E. Hydroponic treatment with salicylic acid decreases the effects of chilling injury in maize (*Zea mays* L.) plants. *Planta* **1999**, *208*, 175–180. [CrossRef]
23. Scott, I.M.; Clarke, S.M.; Wood, J.E.; Mur, L.A.J. Salicylate Accumulation Inhibits Growth at Chilling Temperature in *Arabidopsis*. *Plant Physiol.* **2004**, *135*, 1040–1049. [CrossRef] [PubMed]
24. Pokotylo, I.; Kravets, V.; Ruelland, E. Salicylic acid binding proteins (SABPs): The hidden forefront of salicylic acid signalling. *Int. J. Mol. Sci.* **2019**, *20*, 4377. [CrossRef] [PubMed]
25. Westfall, C.S.; Xu, A.; Jez, J.M. Structural evolution of differential amino acid effector regulation in plant chorismate mutases. *J. Biol. Chem.* **2014**, *289*, 28619–28628. [CrossRef] [PubMed]
26. Vishwakarma, K.; Upadhyay, N.; Kumar, N.; Yadav, G.; Singh, J.; Mishra, R.K.; Kumar, V.; Verma, R.; Upadhyay, R.G.; Pandey, M.; et al. Abscisic Acid Signaling and Abiotic Stress Tolerance in Plants: A Review on Current Knowledge and Future Prospects. *Front. Plant Sci.* **2017**, *8*, 161. [CrossRef] [PubMed]
27. Xiong, L.; Zhu, J.-K. Regulation of abscisic acid biosynthesis. *Plant Physiol.* **2003**, *133*, 29–36. [CrossRef]
28. Soitamo, A.J.; Piippo, M.; Allahverdiyeva, Y.; Natalia Battchikova, N.; Aro, E.-M. Light has a specific role in modulating *Arabidopsis* gene expression at low temperature. *BMC Plant Biology* **2008**, *8*, 13. [CrossRef]
29. Janda, T.; Szalai, G.; Leskó, K.; Yordanova, R.; Apostol, S.; Popova, L. Factors contributing to the enhanced freezing tolerance in wheat during frost hardening in light. *Phytochemistry* **2007**, *68*, 1674–1682. [CrossRef]
30. Szalai, G.; Janda, T.; Páldi, E.; Szigeti, Z. Role of light in the development of post-chilling symptoms in maize. *J. Plant Physiol.* **1996**, *148*, 378–383. [CrossRef]
31. Janda, T.; Szalai, G.; Páldi, E. Chlorophyll fluorescence and anthocyanin content in chilled maize plants and after returned to a non-chilling temperature under various light conditions. *Biol. Plant.* **1996**, *38*, 625–627. [CrossRef]
32. Zagorchev, L.; Seal, C.E.; Kranner, I.; Odjakova, M. A Central Role for Thiols in Plant Tolerance to Abiotic Stress. *Int. J. Mol. Sci.* **2013**, *14*, 7405–7432. [CrossRef] [PubMed]
33. Toldi, D.; Gyugos, M.; Darkó, É.; Szalai, G.; Gulyás, Z.; Gierczik, K.; Székely, A.; Boldizsár, Á.; Galiba, G.; Müller, M.; et al. Light intensity and spectrum affect metabolism of glutathione and amino acids at transcriptional level. *PLoS ONE* **2019**, *14*, e0227271. [CrossRef] [PubMed]
34. Han, Y.; Chaouch, S.; Mhamdi, A.; Queval, G.; Zechmann, B.; Noctor, G. Functional analysis of *Arabidopsis* mutants points to novel roles for glutathione in coupling H_2O_2 to activation of salicylic acid accumulation and signalling. *Antioxid. Redox Signal.* **2013**, *18*, 2106–2121. [CrossRef] [PubMed]
35. Kellős, T.; Tímár, I.; Szilágyi, V.; Szalai, G.; Galiba, G.; Kocsy, G. Stress hormones and abiotic stresses have different effects on antioxidants in maize lines with different sensitivity. *Plant Biol.* **2008**, *10*, 563–572. [CrossRef]
36. Gondor, O.K.; Janda, T.; Soós, V.; Pál, M.; Majláth, I.; Adak, M.K.; Balázs, E.; Szalai, G. Salicylic acid induction of flavonoid biosynthesis pathways in wheat varies by treatment. *Front. Plant Sci.* **2016**, *7*, 1447. [CrossRef]
37. Meuwly, P.; Métraux, J.P. *Ortho*-anisic acid as internal standard for the simultaneous quantitation of salicylic acid and its putative biosynthetic precursors in cucumber leaves. *Anal. Biochem.* **1993**, *214*, 500–505. [CrossRef]
38. Pál, M.; Horváth, E.; Janda, T.; Páldi, E.; Szalai, G. Cadmium stimulates the accumulation of salicylic acid and its putative precursors in maize (*Zea mays* L.) plants. *Physiol. Plant.* **2005**, *125*, 356–364. [CrossRef]
39. Janda, T.; Lejmel, M.A.; Molnár, A.B.; Majláth, I.; Pál, M.; Nguyen, Q.T.; Nguyen, N.T.; Le, V.N.; Szalai, G. Interaction between elevated temperature and different types of Na-salicylate treatment in *Brachypodium dystachion*. *PLoS ONE* **2020**, *15*, e0227608. [CrossRef]
40. Gulyás, Z.; Boldizsár, Á.; Novák, A.; Szalai, G.; Pál, M.; Galiba, G.; Kocsy, G. Central role of the flowering repressor ZCCT2 in the redox control of freezing tolerance and the initial development of flower primordia in wheat. *BMC Plant Biol.* **2014**, *14*, 91. [CrossRef]
41. Gao, S.; Yan, R.; Cao, M.; Yang, W.; Wang, S.; Chen, F. Effects of copper on growth, antioxidant enzymes and phenylalanine ammonia-lyase activities in *Jatropha curcas* L. seedling. *Plant Soil Environ.* **2008**, *54*, 117–122. [CrossRef]
42. Kho, K.F.F.; Bolsman-Louwen, A.C.; Vuik, J.C.; Bennink, G.J.H. Anthocyanin synthesis in a white flowering mutant of *Petunia* hybrida—II. Accumulation of dihydroflavonol intermediates in white flowering mutants; uptake of intermediates in isolated corollas and conversion into anthocyanins. *Planta* **1977**, *135*, 109–118. [CrossRef] [PubMed]

International Journal of
Molecular Sciences

MDPI

Article

Improving Phenolic Total Content and Monoterpene in *Mentha* x *piperita* by Using Salicylic Acid or Methyl Jasmonate Combined with Rhizobacteria Inoculation

Lorena del Rosario Cappellari [1], Maricel Valeria Santoro [2], Axel Schmidt [2], Jonathan Gershenzon [2,*] and Erika Banchio [1,*]

1 INBIAS (CONICET-Universidad Nacional de Río Cuarto), Campus Universitario, 5800 Río Cuarto, Argentina; lcappellari@exa.unrc.edu.ar

2 Department of Biochemistry, Max Planck Institute for Chemical Ecology, Hans-Knöll-Str. 8, 07745 Jena, Germany; msantoro@ice.mpg.de (M.V.S.); aschmidt@ice.mpg.de (A.S.)

* Correspondence: gershenzon@ice.mpg.de (J.G.); ebanchio@exa.unrc.edu.ar (E.B.); Tel.: +54-358-4676-424 (J.G. & E.B.); Fax: +54-358-4676-232 (J.G. & E.B.)

Received: 9 November 2019; Accepted: 16 December 2019; Published: 19 December 2019

Abstract: The effects of plant inoculation with plant growth-promoting rhizobacteria (PGPR) and those resulting from the exogenous application of salicylic acid (SA) or methyl jasmonte (MeJA) on total phenolic content (TPC) and monoterpenes in *Mentha* x *piperita* plants were investigated. Although the PGPR inoculation response has been studied for many plant species, the combination of PGPR and exogenous phytohormones has not been investigated in aromatic plant species. The exogenous application of SA produced an increase in TPC that, in general, was of a similar level when applied alone as when combined with PGPR. This increase in TPC was correlated with an increase in the activity of the enzyme phenylalanine ammonia lyase (PAL). Also, the application of MeJA at different concentrations in combination with inoculation with PGPR produced an increase in TPC, which was more relevant at 4 mM, with a synergism effect being observed. With respect to the main monoterpene concentrations present in peppermint essential oil (EO), it was observed that SA or MeJA application produced a significant increase similar to that of the combination with rhizobacteria. However, when plants were exposed to 2 mM MeJA and inoculated, an important increase was produced in the concentration on menthol, pulegone, linalool, limonene, and menthone concentrations. Rhizobacteria inoculation, the treatment with SA and MeJA, and the combination of both were found to affect the amount of the main monoterpenes present in the EO of *M. piperita*. For this reason, the expressions of genes related to the biosynthesis of monoterpene were evaluated, with this expression being positively affected by MeJA application and PGPR inoculation, but was not modified by SA application. Our results demonstrate that MeJA or SA application combined with inoculation with PGPR constitutes an advantageous management practice for improving the production of secondary metabolites from *M. piperita*.

Keywords: salicylic acid; jasmonic acid; rhizobacteria; plant growth-promoting rhizobacteria (PGPR); mint; total phenolic content; monoterpene; menthol; pulegone

1. Introduction

Modern agriculture does not restrict itself to traditional food, forage, and fiber crops; rather, there is an increasing interest in crops that include species with secondary metabolites. These are present in all plant tissues at variable concentrations. They are low molecular weight compounds that are very important in plant ecology because they are responsible for the processes of adaptation of plants to their environment. Given their diverse biological and physio-chemical properties, secondary metabolites are of great interest as unique sources for pharmaceuticals, food additives, oils, waxes, perfumes, flavoring agents, dyes, and many other commercially important materials [1–3].

Peppermint (*Mentha* x *piperita* L.; family Labiatae) is an important, commonly-used flavoring agent world-wide. *M. piperita* plants contain ~3% volatile oils, consisting of >50 different compounds. The major EO components, which make up ~60% of its total oil volume, are limonene, linalool, menthone, menthol, and pulegone. Peppermint leaves are used for teas and flavoring foods and beverages, and its EOs are also used in chewing gum, candy, toothpaste, mouthwash, aromatherapy, pharmaceuticals, antimicrobial agents, and eco-friendly pesticides [4]. Not less important components of peppermint leaves are the phenolic compounds, including caffeic acid, rosmarinic acid, eriocitrin, and luteolin-7-*O*-glucoside [5,6], which represent about 20% of the dry weight. Seventy-five percent of these compounds can be extracted in an infusion [7–10]. The infusion of peppermint leaves is a common beverage with a refreshing flavor and a particular fragrance.

Many natural compounds extracted from plants have shown biological activities. There is a growing consumption of natural products with potential health benefits, and novel bioactive compounds are continually being discovered. Among others, plants are regarded as major sources of bioactive compounds with anticancer, antioxidant, antimicrobial, and anti-inflammatory effects [11,12]. Phytobioactive compounds commonly play important physiological roles in plants as secondary metabolites, which can be modified quantitatively and qualitatively through environmental changes as well as exposure to biotic and abiotic stress [13].

As knowledge has grown of the biological functionalities of secondary metabolites, so too has the search for new biotechnological alternatives to improve the production of economically important secondary metabolite compounds [2,3,14–16] with bioactive properties which can be used in agriculture as herbicides and pesticides, or as medicinal agents. Several strategies have been investigated to enhance the production of secondary metabolites from medicinal plants, including high-yielding cell line screening, media modification, elicitation, precursor feeding, large-scale cultivation systems, plant cell immobilization, hairy root culture, biotransformation, and others [17,18].

Jasmonic acid (JA), methyl jasmonate (MeJA), and salicylic acid (SA) are known as potent elicitors and plant defense hormones, which play significant roles in regulating plant defense responses against biotic and abiotic stress [19]. JA and its related compounds have long been known as transducers of elicitor signals for the production of plant secondary metabolites [20]. Exogenous application of JA signaling compounds stimulates the biosynthesis of secondary metabolites including a wide variety of plant secondary products such as terpenoids, flavonoids, alkaloids, and phenylpropanoids [21,22], while MeJA application has been reported to be both safe and inexpensive [23,24]. The JA signaling pathway is generally regarded as an integral signal for biosynthesis of many plant secondary products. Also, as many elicitors stimulate endogenous JA biosynthesis in plants, the JA signaling pathway is regarded as a transducer or mediator for elicitor signaling [25].

Salicylic acid (SA) is a well-known inducer of plant systematic acquired resistance (SAR) in plant–pathogen interaction, but it is not a universal inducer for the production of plant defensive metabolites. SA quickly accumulates at the infection site during pathogen attack and plant hypersensitive reaction, and it spreads to other parts of the plant to induce a wide range of defense responses, leading to the accumulation of plant secondary metabolites [26–28].

It is well known that a group of bacteria colonize the root systems of plants and can modulate plant growth, with beneficial effects on plant growth and crop yield and quality. Such bacteria, collectively termed "plant growth-promoting rhizobacteria" (PGPR), promote plant growth by enhancing the availability of nutrients, inducing metabolic activities by phytohormones and analogs, by shifting the phytohormonal balance, by inducing systemic resistance defense (ISR), or by reducing phytotoxic microbial communities [29–31]. Depending on the PGPR species, two or more of these growth-promoting mechanisms may be present [31–35].

Our previous studies showed that PGPR inoculation increased secondary metabolite production in various aromatic plant species [36–38]. The same beneficial bacteria inoculated in different plant species increased total EO yield (relative to controls), but not necessarily at the same level [37–39]. We observed that the effects of rhizobacteria on these secondary metabolites varied depending on the strain, suggesting that the rhizobacteria are recognized by the host plant in a strain-specific manner.

Thus, the elicitation of secondary metabolites appears to be a promising and innovative alternative. There are combinations of both biotic and abiotic forms of elicitors that stimulate metabolism but, while there are many studies of their potential use in improving secondary metabolism biosynthesis individually, few reports were found with the combination of both. Knowing that exogenous MeJA and SA applications, such as inoculation with PGPR, increase secondary metabolites, we inoculated the plants with PGPR and simultaneously applied phytohormones directly to learn how the combination of treatments affects total phenolic content (TPC) and the main monoterpenes of *Mentha* x *piperita*. New less aggressive biotechnological methods are needed to enhance the production of secondary metabolites from medicinal plant crops based on the use of beneficial microorganisms applied as biofertilizers.

2. Results

2.1. Total Phenolic Content (TPC)

Phenolic compounds are plant secondary metabolites that can be released under the influence of multiple biotic and abiotic stresses; thus, we sprayed inoculated plants with SA or JA at different concentrations in order to determine the effects of the combination of treatments in TPC. An increased accumulation of total phenols was observed in inoculated mint plants compared to untreated controls ($p < 0.05$) (Figure 1). TPC in leaves of WCS417r inoculated plants showed further increased accumulation of total phenolic compounds (347.98 mg/mg fresh weight); this phenolic accumulation was similar for the three strains evaluated. The TPC in plants sprayed with SA at 1 or 2 mM concentrations showed similar phenolic content to plants only inoculated (~400 mg/mg fw) (Figure 1, Supplementary Table S1). The main effect was observed when plants were treated with 2 mM in combination with GB03 increasing up to 60% compared to individual treatments (2 mM SA or PGPR; $p < 0.05$).The response of plants exposed to MeJA depended on the concentration applied (Figure 1); as the concentration increased, the effect was also greater, almost 2.5 times with 4 mM in relation to control plants. When plants were inoculated and treated with 1 or 2 mM MeJA, the TPC was comparable to that of plants treated only with MeJA regardless of the strain inoculated; when plants were treated with 4 mM+GB03 or WCS417, it increased up to 3 times (613.11–563.22 mg/mg fw; Supplementary Tables S1 and S2) compared to control, and was higher than MeJA alone or inoculated plants.

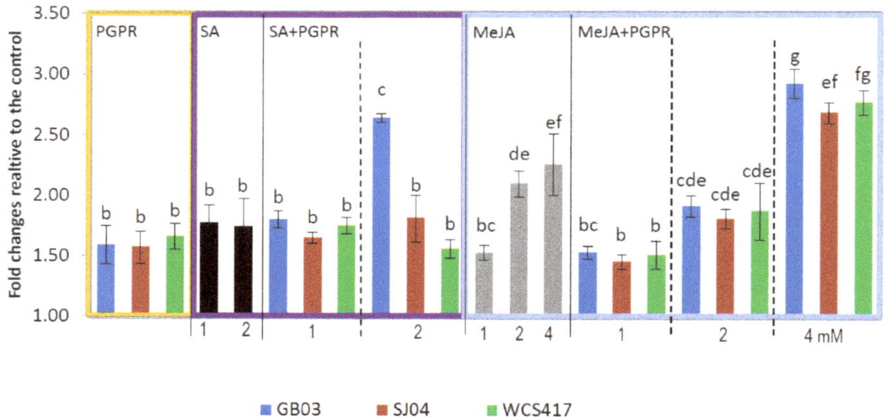

Figure 1. Inoculation with plant growth-promoting rhizobacteria (PGPR) and hormone treatments increase total phenolic compounds content in shoots of *M.* × *piperita* plants. Values are fold changes relative to the control. Different letters above bars for PGPR/*methyl jasmonte (MeJA)* + PGPR/salicylic acid (SA) + PGPR groups indicate significant differences according to the Tukey test ($p < 0.05$). The letter "a" indicates similar to the control. Three different ANOVA analyses were performed (i) PGPR (yellow box), (ii) SA–SA + PGPR (violet box), (iii) MeJA–MeJA + PGPR (blue box). TPC in control plants: for PGPR treatments = 209.84 ± 13.20 μg/g fresh weight, for SA treatments = 254.03 ± 35.26 μg/g fresh weight, for MeJA treatments = 192.43 ± 16.52 μg/g fresh weight. Native values are given in Supplementary Tables S1 and S2.

2.2. Phenylalanine Ammonia Lyase Activity (PAL)

Phenylalanine ammonia lyase activity was affected by inoculation and/or exogenous phytohormone application. Significantly higher PAL activity was observed in inoculated plants compared to untreated controls ($p < 0.05$; Figure 2). SA applications led to 10-fold increased activity regardless of the concentration applied ($p < 0.05$; Figure 2). However, when PGPR and 1 mM SA spraying were combined, the rise in PAL activity was similar to the increase seen in plants treated with SA alone ($p < 0.05$), while in plants inoculated with GB03 and treated with 2 mM, the activity was almost double that in 2 mM treated plants. Spraying with MeJA also produced an increase in PAL activity, which was higher at each higher concentration applied, being ~20-, ~30- and ~60-fold higher in plants treated with 2 and 4 mM, respectively, compared to control plants. When PGPR and MeJA spraying were combined, the effect on PAL activity was statistically similar to that of MeJA alone for the different concentrations evaluated, while with 1 mM were slightly higher.

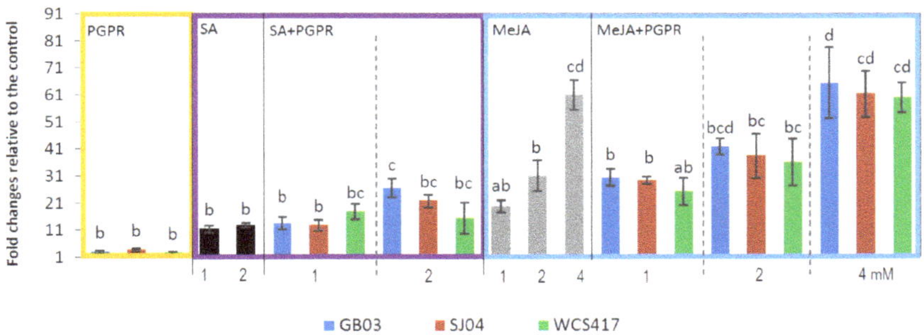

Figure 2. Inoculation with PGPR and hormone treatments modify phenylalanine ammonia lyase (PAL) activity in shoots of *M.* × *piperita* plants. Values are fold changes relative to the control. Different letters above bars for PGPR, MeJA + PGPR, SA + PGPR groups indicate significant differences according to the Tukey test ($p < 0.05$). The letter "a" indicates similar to the control. Three different ANOVA analyses were performed (i) PGPR (yellow box), (ii) SA–SA + PGPR (violet box), (iii) MeJA–MeJA + PGPR (blue box). PAL activity in control plants: for PGPR treatments = 4.62 ± 0.32 μg trans-cinnamic acid min^{-1} mg^{-1}protein, for SA treatments = 4.63 ± 0.30 μg trans-cinnamic acid/min × mg protein, for MeJA treatments = 4.60 ± 0.33 μg trans-cinnamic acid/min × mg protein. Native values are given in Supplementary Tables S1 and S2.

2.3. Quantification of Main EO Components

Since we showed in previous studies that PGPR and the exogenous application of SA and MeJA combined with inoculation increased EO content [40], we analyzed the response of the main EO components in plants treated with PGPR in addition to SA or MeJA. GC analysis of the yields of the major EO components ((+)-pulegone, (−)-menthone, (−)-menthol, limonene, and linalool) revealed great differences among plants treated with the phytohormone and control groups.

Plants only inoculated with the different strains produced an increase in all the major monoterpenes evaluated. The total amount of menthol increased (Figure 3) to 1.29 μg/g fresh weight in plants inoculated with GB03 in comparison with 0.20 μg/g fw in the control group ($p < 0.05$). The same trend was observed for pulegone (Figure 4), linalool, limonene, and menthone, where inoculation with GB03 generally produced the greater effect (Table 2).

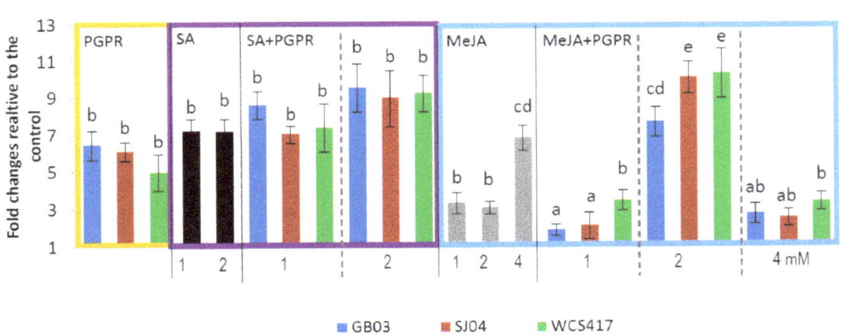

Figure 3. Inoculation with PGPR and hormone treatments increase menthol concentration in shoots of *M.* × *piperita* plants. Values are fold changes relative to the control. Different letters above bars for PGPR, MeJA + PGPR, SA + PGPR groups indicate significant differences according to the Tukey test ($p < 0.05$). The letter "a" indicates similar to the control. Three different ANOVA analyses were performed (i) PGPR (yellow box), (ii) SA–SA + PGPR (violet box), (iii) MeJA–MeJA + PGPR (blue box). Menthol concentrations in control plants: for PGPR treatments = 0.20 ± 0.02 μg/g fw, for SA treatments = 0.19 ± 0.02 μg/g fw, for MeJA treatments = 0.23 ± 0.07 μg/g fw.

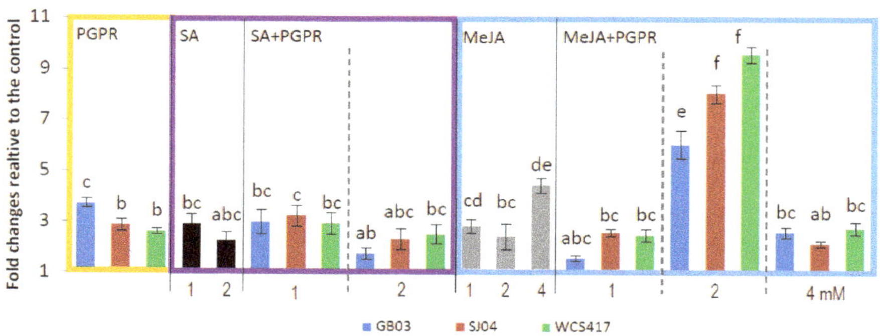

Figure 4. Inoculation with PGPR and hormone treatments increases pulegone concentration in shoots of *M.* × *piperita* plants. Values are fold changes relative to the control. Different letters above bars for PGPR, MeJA + PGPR, SA + PGPR groups indicate significant differences according to the Tukey test ($p < 0.05$). The letter "a" indicates similar to the control. Three different ANOVA analyses were performed (i) PGPR (yellow box), (ii) SA–SA + PGPR (violet box), (iii) MeJA–MeJA + PGPR (blue box). Pulegone concentrations in control plants: for PGPR treatments = 3.74 ± 0.16 μg/g fw, for SA treatments = 4.27 ± 0.27 μg/g fw, for MeJA treatments = 3.99 ± 0.28 μg/g fw.

The external application of SA modified the yield of the main monoterpene compounds of *M. piperita* (Table 1, Figures 3 and 4), but, in general, there was no difference between the application of 1 and 2 mM ($p > 0.05$). The combination of SA with PGPR did not show any significant difference with the application of SA alone for all the compounds evaluated.

Table 1. Effects of inoculation with PGPR strains and external application of SA on concentration of main shoot essential oil components of *M.* × *piperita* plants (mean ± SE). Means followed by the same letter in a given column are not significantly different according to the Tukey test ($p < 0.05$).

Treatments	Linalool (μg/g fw)	(−)-Limonene (μg/g fw)	(−)-Menthone (μg/g fw)
Control	0.15 ± 0.01 a	0.20 ± 0.02 a	0.50 ± 0.03 a
G	0.61 ± 0.07 b	0.76 ± 0.06 b	1.15± 0.05 bc
S	0.48 ± 0.08 b	0.73 ± 0.08 b	1.34 ± 0.15 c
W	0.52 ± 0.05 b	0.65 ± 0.07 b	0.99 ± 0.09 b
Control for SA	0.14 ± 0.02 a	0.19 ± 0.04 a	0.43 ± 0.09 a
1 mM SA	0.54 ± 0.07 b	1.09 ± 0.18 b	1.30 ± 0.09 b
2 mM SA	0.40 ± 0.07 ab	1.01 ± 0.20 b	1.40 ± 0.09 b
1 mM SA + G	0.41 ± 0.03 cd	0.91 ± 0.06 b	1.34 ± 0.07 b
1 mM SA + S	0.32 ± 0.03 ab	0.99 ± 0.18 b	1.71 ± 0.35 c
1 mM SA + W	0.40 ± 0.05 ab	0.66 ± 0.17 ab	1.45 ± 0.05 b
2 mM SA + G	0.34 ± 0.09 ab	1.14 ± 0.03 b	1.50 ± 0.22 b
2 mM SA + S	0.41 ± 0.05 ab	1.18 ± 0.08 b	1.53 ± 0.06 b
2 mM SA + W	0.22 ± 0.06 a	1.06 ± 0.17 b	1.31 ± 0.05 b

G—GB03; S—SJ04; W— WCS417; SA—Salicylic Acid.

Menthol content in plants sprayed with SA increased 7 times in relation to uninoculated plants (Figure 3), similar values to those in inoculated plants alone (Supplementary Table S1). The same effect was observed for pulegone: SA application increased pulegone approximately three-fold in comparison to the control (Figure 4), similar to the increase in PGPR-inoculated plants (Supplementary Table S1).

Linalool content (Table 1) increased significantly only with 1 mM SA application ($p < 0.05$).

Limonene content in plants treated with 1 or 2 mM showed the same effect, an increase of 4–5 times compared to controls, a similar effect to inoculated plants (Table 1).

Menthone yield showed the same trend as limonene. Plants treated with SA showed an increase compared to control plants ($p < 0.05$) (Table 1), but when the treatments were combined, menthone content was similar in plants treated only with SA and in those also inoculated ($p > 0.05$), with the exception of 1mM SA + S, which resulted in a yield increase of 30% with respect to 1 mM SA.

The response of plants treated with MeJA depended on the concentration applied. Plants treated with 1 or 2 mM MeJA increased menthol and pulegone content approximately three-fold in relation to the control, with the greater effect observed in plants treated with 4 mM MeJA (Figures 3 and 4), in which the amounts of menthol and pulegone increased 7 and 5 times, respectively (Figure 3, Supplementary Table S2). The effect of spraying with MeJA and inoculation increased menthol concentration only when 2 mM was applied, reaching 10-fold compared to control plants and 3 times compared to 2 mM-sprayed mint plants, while the combination of 1 or 4 mM and inoculation showed no yield effect compared to sprayed plants alone. A similar effect was observed for pulegone content (Figure 4), where the combination of inoculation with any of the strains evaluated and the application of 1 and 4 mM MeJA showed no significant difference with 1 and 4 mM sprayed plants ($p > 0.05$), while the combination with 2 mM increased pulegone content approximately 6-, 8- and 10-fold in relation to control plants in the inoculated strains, GB03, SJ04 and WCS417, respectively, with higher yields than those obtained with individual treatments (Figure 4; Supplementary Table S2).

Linalool, limonene and menthone content in plants treated with 1 and 2 mM MeJA was not modified ($p > 0.05$) in comparison to control plants, while 4 mM significantly increased the amount of linalool, limonene and menthone, 6-, 3-, and 4-fold, respectively (Table 2). When the inoculation was combined with MeJA, significant differences were observed only at 2 mM, increasing the content of linalool, limonene, and menthone approximately 5-fold in relation to 2 mM-treated plants (Table 2).

Table 2. Effects of direct inoculation with PGPR strains and external application of MeJA on concentration of main shoot essential oil components of *M. x piperita* plants (mean ± SE). Means followed by the same letter in a given column are not significantly different according to Fisher's LSD test ($p < 0.05$).

Treatments	Linalool (μg/g fw)	(−)-Limonene (μg/g fw)	(−)-Menthone (μg/g fw)
Control MeJA	0.15 ± 0.02 a	0.25 ± 0.02 a	0.35 ± 0.03 a
1 mM + MeJA	0.16 ± 0.02 a	0.25 ± 0.05 a	0.43 ± 0.08 a
2 mM + MeJA	0.22 ± 0.04 a	0.21 ± 0.04 a	0.47 ± 0.13 a
4 mM + MeJA	0.97 ± 0.12 c	0.83 ± 0.07 b	1.42 ± 0.14 b
1 mM MeJA + G	0.23 ± 0.04 a	0.40 ± 0.05 a	0.23 ± 0.03 a
1 mM MeJA + S	0.32 ± 0.07 a	0.25 ± 0.03 a	0.51 ± 0.13 a
1 mM MeJA + W	0.35 ± 0.07 ab	0.37 ± 0.04 a	0.44 ± 0.13 a
2 mM MeJA + G	0.79 ± 0.15 bc	1.09 ± 0.07 bc	1.75 ± 0.17 bc
2 mM MeJA + S	1.12 ± 0.17 c	1.23 ± 0.07 c	1.95 ± 0.17 bc
2 mM MeJA + W	0.84 ± 0.21 c	1.11 ± 0.10 bc	2.54 ± 0.45 c
4 mM MeJA + G	0.21 ± 0.08 a	0.22 ± 0.04 a	0.27 ± 0.07 a
4 mM MeJA + S	0.20 ± 0.02 a	0.23 ± 0.01a	0.20 ± 0.04 a
4 mM MeJA + W	0.23 ± 0.02 a	0.27 ± 0.09 a	0.42 ± 0.12 a

G—GB03; S—SJ04; W—WCS417; MeJA—Methyl Jasmonate.

2.4. PGPR Inoculation and External Phytohormones Application Induces Terpenoid Gene Expression in M. piperita

Both rhizobacteria inoculation and SA and MeJA treatment were found to affect the amount of the main monoterpenes present in the EO of *M. piperita*. We, therefore, evaluated qPCR gene expression of two enzymes involved in the biosynthetic pathway leading to the bioactive monoterpenes. Previous studies have established the biochemical pathway in mints that leads to the production of the most important monoterpenes [41]. We considered the early gene, limonene synthase (*Ls*), involved in the formation of limonene, which is responsible for the first dedicated step of monoterpene biosynthesis in mint species. Limonene synthase catalyzes the cyclization of geranyl diphosphate, the universal C_{10}

precursor of the monoterpenes, to (−)-4-*S*-limonene [41] and one of the end genes, *Pr*, coding for the enzyme pulegone reductase (*Pr*), which produced menthone and isomenthone at an average ratio of 2.5:1 [42].

Considering that not all the treatments in the present study produced an increase in the main EO compounds, we determined the gene expression of the most important treatments. *Ls*, coding for the enzyme that catalyzes the conversion of geranyl diphosphate to limonene, one of the simplest of all terpenoid cyclization reactions, was upregulated almost 2-fold in plants inoculated with SJ04 and WCS417($p < 0.05$). External application of 1 and 2 mM SA had no effect on the expression of *Ls* ($p > 0.05$; Figure 5). In contrast, the application of 2 and 4 mM MeJA upregulated *Ls* expression 3- and 4-fold, respectively ($p < 0.05$). When plants were inoculated and sprayed with 2 mM MeJA, they showed the same effect as with the application of 2 mM alone (Figure 5).

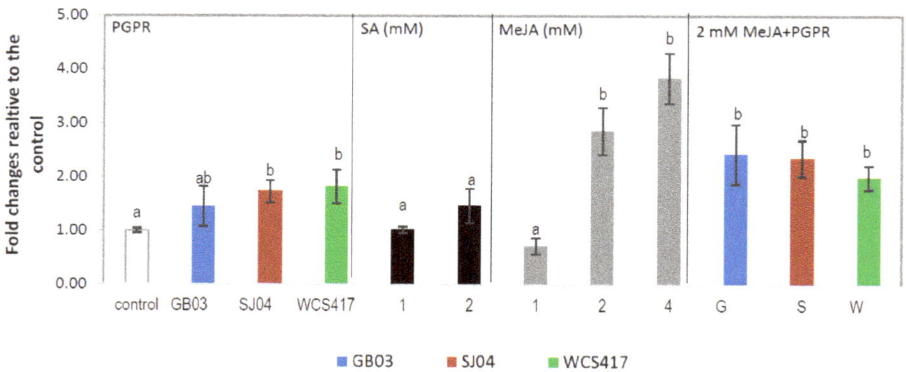

Figure 5. Limonene synthase gene expression of *M.* x *piperita* plants inoculated with PGPR and hormone treatments with respect to control leaves. Bars indicate the standard error over the mean of at least three biological replicates. Means followed by the same letter in a given column are not significantly different according to the Tukey test ($p < 0.05$).

Pr, which codes for pulegone reductase, the enzyme responsible for NADPH-dependent reduction of the conjugated double bond of terpenone to yield (−)-menthone and lesser amounts of (+)-isomenthone [42], was upregulated almost 2-fold by PGPR inoculation ($p < 0.05$; Figure 6). SA did not affect *Pr* expression ($p > 0.05$; Figure 6). MeJA external application produced the same effect as in *Ls* gene expression, which upregulated *Pr* by over 2-fold when it was applied at 2 and 4 mM, but these two values were not statistically different ($p > 0.05$). Moreover, this effect increased (to approximately 3.5-fold) when plants were sprayed with 2 mM MeJA and inoculated.

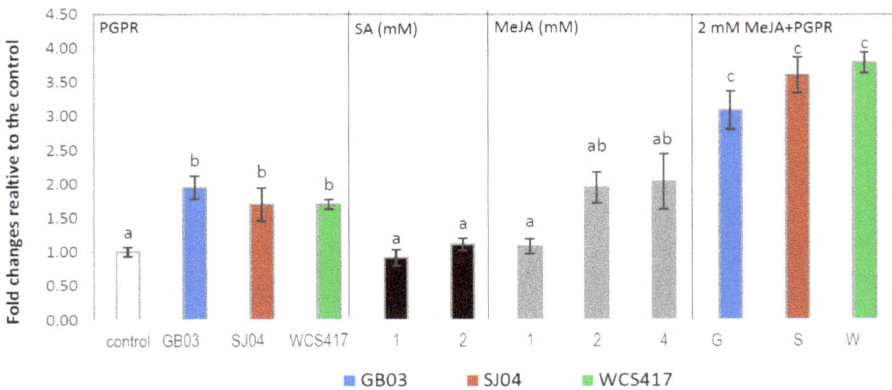

Figure 6. Inoculation with PGPR and hormone treatments increases pulegone reductase expression in shoots of *M.* x *piperita* plants. Values are fold changes relative to the control. Means followed by the same letter in a given column are not significantly different according to the Tukey test ($p < 0.05$). The letter "a" indicates similar to the control.

2.5. Principal Component Analysis

The PCA was performed to correlate the effects of inoculation with the three PGPR strains and spraying with SA or MeJA on the production of the main monoterpenes, TPC and PAL activity. This type of analysis provides a graph that facilitates the visualization and interpretation of the data set and the variables. Figure 7 shows the PCA correlating the effects of inoculation with the three PGPR strains combined with exogenous SA application. The variation in the data (78.2%) was explained by the first two principal components and gave a cophenetic correlation coefficient of 0.970. This plot shows that the strains, whichever was combined with 1 or 2 mM SA, were located in proximity to all the variables evaluated (blue circle): limonene content (LIM), linalool content (LIN), menthone (MTNE), pulegone (PUL), menthol content (MTO), TPC, and PAL. The control treatments SA and the strains alone are far from the variable evaluated (red circle), showing a low effect in relation to the other treatments. We observed a strong positive correlation (acute angle) between all monoterpene content and TPC content with the exception of PUL content (Figure 7).

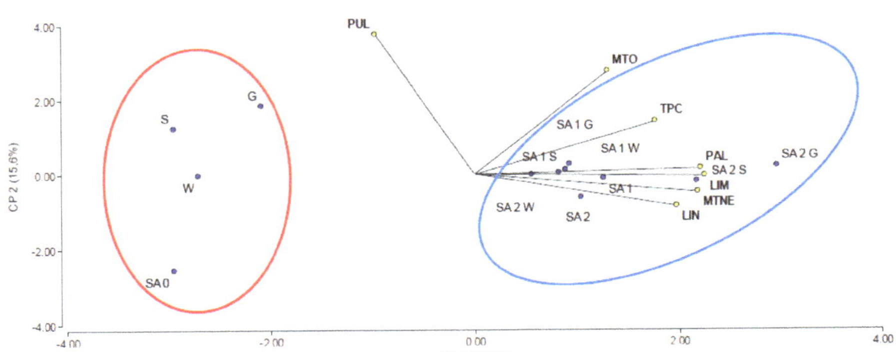

Figure 7. Principal component analysis illustrating relationships among PGR inoculation and SA external application on *M. x piperita* for TPC, PAL and major essential oil components: LIM (limonene content), LIN (linalool content), MTNE (menthone content), PUL (pulegone content), MTO (menthol content).

In relation to the PCA that correlates effects of inoculation with MeJA treatment, a plot defined by the first two principal components was sufficient for our purpose because it explained most (92.4%) of the variation in the data and gave a cophenetic correlation coefficient of 0.994 (Figure 8). In the two-dimensional coordinate system based on the first two principal components, it was possible to differentiate the strains combined with different MeJA concentrations. The inoculation and spraying with 1 and 4 mM MeJA were all together, far from the variable evaluated (red circle), while the three strains combined with 2 mM MeJA (blue circle) were located in proximity to the variables LIM, LIN, MTNE, PUL, and MTO. No one treatment was closer to TPC and PAL. In regard to associations among variables, there was a strong positive correlation (acute angle in Figure 8) between the limonene, menthone, menthol and pulegone content, as expected. Surprisingly, no associations (i.e., right angles in Figure 8) were observed between TPC and the major essential oil components.

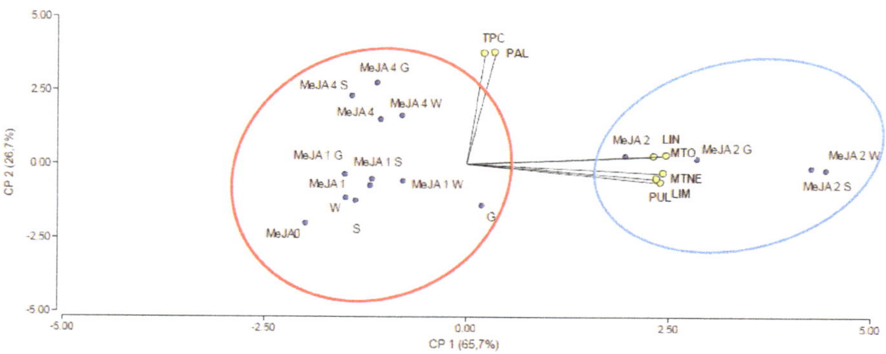

Figure 8. Principal component analysis illustrating relationships among strain inoculation and MeJA external application on *M. x piperita* for TPC, PAL, and major essential oil components: LIM (limonene content), LIN (linalool content), MTNE (menthone content), PUL (pulegone content), MTO (menthol content).

3. Discussion

Plants have been called chemical factories as they have the ability to fabricate important phytochemicals. Changes in secondary metabolites and the enhanced growth of host plants in interaction with different beneficial microbes have recently been studied [30,35].

M. piperita leaves contain high levels of polyphenolic compounds [5,6]. In the present study, we confirm the strong effect of rhizobacteria inoculation on TPC observed previously in *M. piperita* leaves [43]. The exogenous application of SA produces an approximately 1.5-fold increase in TPC compared to controls, whatever the concentration (1 or 2 mM) applied. In plants inoculated with GB03 and then sprayed with 2 mM SA, TPC increased to a greater extent than with the application of the phytohormone alone (60% higher). These results suggest that the combination produces a synergistic effect on the biosynthesis of phenolic compounds. A similar result was observed for MeJA treatments, where 2 and 4 mM produced an increase of TPC greater than in inoculated plants. Further, when plants were inoculated with GB03 or SJ04 and sprayed with 4 mM, the TPC increased 30% in relation to the individual treatment (MeJA or PGPR), suggesting also a synergism between 4 mM and PGPR inoculation, as observed for 2 mM SA.

The observed results match those of a previous study in which exogenous applications of JA and SA significantly induced the accumulation of phenolic compounds in a wide variety of plant species [44]. *M. piperita* suspension cultures exposed to JA and MeJA showed increased accumulation of rosmarinic acid (one of the main TPC of peppermint) [45]. The increase in TPC was also observed in suspension cultures supplemented with JA or SA of *Panax ginseng* root [46], of *Thevetia peruviana* [47], *Cucumis melo* [48], in plants of *Romaine lettuce* [49], in buckwheat [50], radish sprouts [51] and

Agastache rugosa treated only with JA [52]. Similarly, Figueroa Perez et al. [9] reported an increase in TPC of 65%, 35%, and 31% in peppermint treated with SA at 0.05, 0.10, and 0.50 mM, respectively. Similar results were reported in other aromatic plants, such as *Thymus vulgaris* treated with 1 and 2 mM SA [53], *Rosmarinus officinalis* [54], and *Achillea millefolium* [55].

The induction of phenolic accumulation by JA and SA is related with the stimulation of the phenylpropanoid pathway, increasing PAL activity [19,47] in agreement with our observation that PAL activity increases as the concentration of MeJA treatments increases and, for SA treatments, PAL activity increases whatever the concentration applied. Particularly with the combination of 2 mM SA and GB03, the increase of PAL activity was greater than the individual treatments, as observed in TPC content. However, with 4 mM MeJA and inoculation with GB03 or WCS417, PAL activity was similar to that observed for 4 mM treated plants. No synergism in the combination of treatments (4 mM + bacteria) was observed as had been seen in TPC production.

Kim et al. [49] observed an increase in transcript levels of phenylpropanoid biosynthetic genes after treatment with methyl jasmonate in cell cultures of *Agastache rugosa*. Liu et al. [56] found that 10 µmol JA treatment on pea leaves (*Pisum sativum*) led to a significant increase in the activities of plasma membrane NADPH oxidase and PAL. Jasmonate elicitation was also found to increase the production of phenylpropanoids and naphtodianthrones in *Hypericum perforatum* cell suspensions [57].

Rhizobacteria inoculation was also found to increase TPC and PAL activity in peppermint [43], as was also reported in *Piper betle* inoculated with *Serratia marcescens* and *Tagetes minuta* inoculated with WCS417r and *Azospirillum brasilense* Sp7 [38,58]. The inoculation of chickpea seeds with *P. fluorescens* and *P. aeruginosa*, singly or in combination, induced the synthesis of specific phenolic acids (gallic, ferulic, and chlorogenic) and increased total phenolic content at various stages of plant growth [59]. Salla et al. [60] showed that the inoculation of eucalyptus with *Streptomyces* increased total phenolic content in leaves. In addition, Panka et al. [61] reported that the presence of the endophyte fungus, *Neotyphodium lolii*, increased the content of phenolic compounds in the aerial part of three different genotypes of the perennial grass, *Lolium perenne*. Increased PAL activity was recorded in *P. fluorescens*-pretreated tomato plants challenged with pathogen compared to untreated control [62] and other plant species [63].

Regarding the concentrations of the main monterpene compounds, EO levels and composition in plants play several key roles in plant–environment interactions and plant–plant communication. Terpenoids are crucial components in plant defensive responses to abiotic and biotic stresses [64,65]. Our previous studies showed that PGPR inoculation increased pulegone, menthone, menthol, and linalool production in *M. piperita* [66] and in other aromatic plant species: *A. brasilense* inoculation increased the levels of ocimenone and tagetone by 71% and 66%, respectively, in *Tagetes minuta* [38]; in *Origanum majorana*, the main compounds, terpinen-4-ol, cis-sabinene hydrate, trans-sabinene hydrate, and a-terpineol were also increased by inoculation with *P. fluorescens* [67]; and greater amounts of terpineol and eugenol were reported for sweet basil inoculated plants [36].

In a previous study, we reported that *M. piperita* exposed to SA and MeJA increased the total EO yield. Particularly, external application of 1 and 2 mM SA increased the EO yield approximately two-fold, while the combination with inoculation led to a 3-fold increase compared to control plants [40]. Curiously, the main monoterpene in plants exposed to SA increased in greater proportions; particularly, menthol rose approximately 7-fold, and pulegone, linalool, limonene and menthone 3-fold in comparison with control plants. Moreover, in plants exposed to combined treatments (PGPR + SA), the levels of the main monoterpene were similar to those of exposure to SA alone. The genes involved in isoprenoid biosynthesis have been shown to be transcriptionally upregulated by SA in *Salvia miltiorhiza* and *Michelia chapensis* [68,69]. The expression of three prenyltransferases from the core terpenoid biosynthetic pathway has been shown to be upregulated by SA in different species as well [70–72]. Similarly, the levels of many isoprenoids, have been shown to be upregulated during drought or salt stress in parallel to increases in SA levels (see, for instance, [73,74]). In addition, holm oaks fumigated with SA showed higher monoterpene levels in leaves and enhanced volatile monoterpene emission [75]. Zhang et al. [76] reported that treatment of *Glycine max* with 1 mM SA induced transcription of a newly identified gene encoding a monoterpene synthase. Treatment of *Cistus creticus* subsp. Creticus with 5 mM SA increased the expression of two genes which encode for enzymes that catalyze the first reactions of methyl-erythritol-phosphate and mevalonate [77]. Xu et al. [78] reported that 8 days after exposure to SA, the medicinal herb, *Houttuynia cordata*, increased the accumulation of α-thujene, α-pinene, α-terpinene, β-pinene, β-myrcene, limonene, and β-ocimene. Although the complete mechanism of SA-mediated plant defense is still not completely understood, the central role of SA in plant defense is universally accepted [79]. Further research is needed to determine the effects of SA on the biosynthesis of particular terpenoids (especially in the committed steps leading to these particular terpenoids), and to establish the possible relationship between the biosynthetic pathways leading to different individual terpenoids.

In relation to the effects of MeJA on the total EO yield of peppermint, it was previously reported that it produced an increase depending on the concentration applied: 3-fold for 1 and 2 mM and 5-fold for 4 mM. The strongest effect was observed when plants were treated with 2 mM and inoculated with rhizobacteria, increasing approximately 8-fold in comparison with controls or inoculated plants, indicating that there is a synergism between PGPR and MeJA [40]. Regarding the amount of the main monoterpene evaluated, not all responded in the same way. Pulegone and menthol showed the same trend as total EOs reported, while limonene, linalool, and menthone did not differ significantly from control plants for exposure to 1 and 2 mM and approximately 4 times for 4 mM. In relation to the combination of treatments, in the present study, we observed a 10-fold increase in the amount of pulegone and menthone when combining 2 mM with inoculation with WCS417. This increase was 2-fold greater as shown in the total EO yield reported [40]. Wang and Wu [80] stated that MeJA is an effective inducer of the terpenoid in *Taxus* spp. and also induced rapid activation of PAL.

JA and its derivatives and precursors are involved in the induction of plant defense responses [81–85]. These hormones play a central role in the regulation of the biosynthesis of several secondary metabolites, such as alkaloids, terpenoids, phytoalexins, coumarins, anthocyanins, among others, by gene regulation, promoting an increase in the number of transcripts of the enzymes linked to the metabolic pathway of those compounds [86–88]. Similarly to the results obtained in the present study, Schmidt et al. [89] reported that the exogenous application of JA induced accumulation of monoterpenes and diterpenes in *Picea abies* stems. However, in this study, treatment with JA did not produce a significant increase in production of terpenoids in eucalyptus leaves, which contained abundant secretory cavities [90]. Likewise, both the emission of linalool and the content of this monoterpene in glandular trichomes of tomato plants increased after the application of JA 1 mM. Treatment with JA also increased the transcription levels of the *LeMTS1* gene, which codes for a linalool synthase located in trichomes of tomato plants [91].

Few studies have been reported on the effects of exogenous application of JAs on the production of secondary metabolites in aromatic and medicinal plants. Złotek et al. [92] reported that treatment with JA significantly increased the content of monoterpenes linalool, eugenol, and limonene in *Ocimum basilicum*. The results obtained in this work indicate that the biosynthesis of terpenoids in aromatic plants and other plant species can be induced by treatment with JA or its methylated derivative, MeJA.

The increase observed in monoterpene accumulation is correlated with an increase in the density of glandular trichomes [93,94]. Biochemical studies with isolated peltate glandular trichomes of peppermint have revealed that the secretory cells are responsible for the secretion of monoterpenes in the oil-storage space [95,96]. A previous study showed that exposure to SA and MeJA produced a 2-fold increase in the density of peltate trichomes in peppermint [40]; this explains the increase in content of the main monoterpenes observed in the present study. Moreover, peltate trichomes also serve as the site of monoterpene biosynthesis [96]. Thus, upregulation of limonene synthase *Ls* and *Pr* gene expression was observed. The biosynthesis of monoterpenes from primary metabolism requires a series of enzymatic steps. First, using geranyl diphosphate as substrate, *Ls* generates limonene and minor amounts of myrcene, alpha-pinene, and beta-pinene [96]. In the present study, an upregulation of *Ls* was observed that correlates with the increase observed in limonene content in plants exposed to 2 and 4 mM MeJA alone and 2 mM MeJA + PGPR. In contrast, no effect on *Ls* expression was observed in plants treated with SA, despite the 5-fold increase in limonene observed in those plants. A similar response was observed for pulegone reductase *Pr* gene expression; menthone and isomenthone (in a 2:1 to 10:1 ratio) were formed by the action of the NADPH-dependent *Pr* using pulegone as substrate [96]. The enzyme responsible for the synthesis of menthone was upregulated only in plants exposed to MeJA (4 mM) and in the combined treatments 2 mM MeJA + PGPR, while SA treatment did not affect *Pr* expression, despite the accumulation of menthone in SA-treated plants being increased in relation to control plants. The poor or the total lack of correlation between gene expression of *Pr* and *Ls* with the respective monoterpene content was probably because at the time that the qPCR were performed (7 days after phytohormone application) the increase in the biosynthesis of monoterpene may have already occurred, probably 48–72 h after phytormone application [97], and it is well known that the monoterpenes are stored in the glandular trichomes. On the other hand, protein stability may be increased due to post-translational modifications such as phosphorylation, acetylatylation, glycosylation, and also it could be that protein may be long lived and accumulates over time whereas mRNA turnover is quick [98].

In a previous study, we observed a significant increase in free jasmonic acid and the active form JA-Ile as well as SA in *M. piperita* plants inoculated with different rhizobacteria strains. The induction of SA and JA by rhizobacteria in the host plants suggests that plants may perceive these bacteria as a risk and thereby initiate a defensive response [40]. Elicitors such as SA and JA are considered as signal molecules that activate the signal-transduction cascade, leading to the activation and expression of genes related to the biosynthesis of secondary metabolites and playing a major role in the defense response [99]. It was reported that exogenous applications of MeJA result in the major reprogramming of genes, including genes that are known to be involved in plant stress responses [100,101], genes that induce defense by increasing activities of pathogenesis-related proteins, and genes that cause oxidative bursts, phytoalexin accumulation, lignification, and cell wall stiffening [102,103]. Furthermore, JA induces the expression of genes involved in biosynthesis, which leads to the accumulation of antimicrobial secondary metabolites, including alkaloids, terpenoids, flavonoids, anthraquinones, and glucosinolates in different plant species [104].

Total phenolic content has been related with stress tolerance, either through contributing with indirect photo protection or by participating directly as antioxidants [51,105]. The increase of phenolic compounds, which are considered the major antioxidant compounds in plants, resulted in a significant decrease in antioxidant capacity. Others organic compounds act as antioxidants like the monoterpenes, through the direct ROS scavenging pathway, and show a capacity for modulating the endogenous antioxidant system [106,107].

Interest in phenolic compounds has considerably increased in recent years because of their broad chemical spectrum and diverse biological properties [108]. In addition to their antioxidant properties, these compounds have been reported to be potential candidates in reducing cardiovascular diseases and anticarcinogenic activity, with antiallergenic, antiarthrogenic, antiinflammatory, antimicrobial, and antithrombotic effects [109]. Extracts of fruits, herbs, vegetables, cereals, and other plant materials rich in phenolics are increasingly of interest in the food industry, because they retard the oxidative degradation of lipids and thereby improve the quality and nutritional value of food [108].

Also, in view of the environmental, food-safety and health related issues associated with the application of chemical insecticides, growing emphasis is being laid on pest control through plant resources. In addition to the menthol and pulegone present in the essential oil of the Mentha species, which are the substances that give mints their characteristic aromas and flavors [4,110,111], they are also found to possess insecticidal, antiviral and fungicidal activities. Pulegone showed potent insecticidal activity against the mushroom scatopsid fly, *Scatopse* spp., in fumigant bioassay [112], and significant antibacterial activity is also observed for Mentha oil due to the presence of menthol [113]. These compounds form complexes with bacterial enzymes and protein and inhibit the growth of bacterial pathogens [114], causing the disruption of the plasma membrane, which increases its permeability and depolarizes its potential, finally leading to the death of the bacteria [115]. Menthol and limonene have also been reported as potential antifungal agents against plant pathogenic fungi [4].

4. Material and Methods

4.1. Plant Material, Bacterial Inoculation, and Treatments

Pseudomonas simiae WCS417r (formerly known as *P. fluorescens* WCS417r; [116]; *P. putida* SJ04) is a native fluorescent strain isolated from rhizospheric soil under a commercial crop of *Mentha* x *piperita* (San José) in Córdoba, Argentina, and demonstrated to have plant growth-promoting activity (GenBank KF312464.1) [117]; and *Bacillus amyloliquefaciens* GB03 (originally described as *Bacillus subtilis* GB03 [118]).

Bacteria were grown on LB medium [119] for routine use and maintained in nutrient broth with 15% glycerol at −80 °C for long-term storage.

Each bacterial culture was grown overnight at 30 °C with rotation at 120 rpm until reaching the exponential phase, washed twice in 0.9% NaCl with centrifugation (4300× *g*, 10 min, 4 °C), resuspended in sterile water, and adjusted to a final concentration of ~10^8 CFU/ mL for use as inoculum.

Plants were grown in plastic pots (diameter 12 cm, depth 22 cm) containing sterilized vermiculite. *M.* x *piperita* seedlings were planted (one per pot) in vermiculite and inoculated with 1000 μL bacterial suspension. Four experimental treatments were performed with the bacteria: sterile water (control), SJ04, WCS417r, and GB03.

4.2. Greenhouse Experiments

M. x *piperita* in vitro micropropagation was performed as described by Cappellari et al. [66]. On day 7 of culture, obtained by in vitro multiplication, they were transplanted directly into vermiculite in a greenhouse, and watered by a micro-irrigation system. All plants received Hoagland's nutrient medium (20 mL/pot) twice per week [38]. Plants were grown in a growth chamber under controlled conditions of light (16/8-h light/dark cycle), temperature (22 ± 2 °C), and relative humidity (~70%). Bacterial suspensions as described above were applied to experimental seedlings, and sterile water was applied to control seedlings. After 7 days of inoculation, plants were sprayed until run-off with 1, 2, or 4 mM methyl jasmonate solution (MeJA) (Sigma–Aldrich, St. Louis, MO, USA, 1% methanol in water, *v/v*) or 1 or 2 mM SA solution (1% ethanol in water, *v/v*). The plants were left to dry for 30–60 min. For the phytohormone control treatments, a solution of the solvent used was applied. After phytohormone or control treatments, plants were transferred to a climate chamber with the phytohormone treatments spatially separated from other treatments because MeJA is very volatile.

After 14 days of applied phytohormone treatments, plants were removed from pots. Experiments were replicated 3 times (10 pots per treatment; 1 plant per pot) and were performed under non-sterile conditions.

4.3. Determination of Total Phenolic Content

Total phenols were determined using Folin–Ciocalteu reagent [120]. Each plant extract (0.5 mL) or gallic acid (standard phenolic compound) was mixed with Folin–Ciocalteu reagent (0.5 mL, diluted with 8 mL distilled water) and aqueous Na_2CO_3 (1 mL, 1 M). After 1 h, the level of total phenols was determined by colorimetry at a wavelength of 760 nm. Total phenol values were expressed in terms of μg gallic acid (a common reference compound) equivalent per g plant dry weight [38].

4.4. Determination of PAL Enzyme Activity

100 mg mint leaves were homogenized with liquid nitrogen using a mortar and pestle containing appropriate buffer solution (50 mM potassium phosphate and 1 mM EDTA, pH 7.8) and 1% PVP (polyvinylpyrrolidone) and then filtered through a 0.20 mm nylon filter into a centrifuge tube. The

tissue extract was centrifuged at 12,000× g for 40 min at 4 °C. The supernatant to be used for enzyme activity determination was stored at 20 °C. Protein concentration was determined by the method described by Bradford [121].

PAL activity was assayed following the method described by Beaudoin–Eagan and Thorpe [122] by measuring the amount of trans-cinnamic acid formed at 290 nm. The reaction mixture consisted of 100 µL of enzyme extract, 900 µL 6 mM of L-phenylalanine, and 500 mM Tris HCl buffer solution (pH 8). The mixture was placed in a water bath at 37 °C for 70 min, and the reaction was stopped by the addition of 50 µL of 5 N HCl. Trans-cinnamic acid (1 mg/mL was used as standard and PAL activity was expressed as µg trans-cinnamic acid/min mg protein.

4.5. Extraction and Quantification of Main Monoterpene EO Components

Shoot samples were individually weighed and subjected to hydrodistillation in a Clevenger-like apparatus for 40 min. The volatile fraction was collected in dichloromethane, and β-pinene (1 µL in 50 µL ethanol) was added as an internal standard.

The major *M. piperita* EO components, which make up ~60% of total oil volume, are limonene, linalool, (−)-menthone, (−)-menthol, and (+)-pulegone. These compounds were quantified with respect to the standard added during the distillation procedure. The oil components were initially identified based on mass spectral and retention time data and confirmed by direct comparisons with commercial standards from Sigma–Aldrich Co. Flame ionization detector (FID) response factors for each compound generated essentially equivalent areas (differences <5%). Chemical analyses were performed as reported by Banchio et al. [36].

4.6. Total RNA Extraction and Quantitative Real-Time PCR

Total RNA from 50 mg lyophilized plant material was isolated using the Plant RNA Isolation Kit (Stratec, Berlin, BE, Germany), according to the manufacturer's instructions but including an additional DNA digestion step (RNase Free DNase set (Qiagen, Valencia, CA, USA). Using identical amounts of total RNA, template cDNA for subsequent PCR reactions was generated using Superscript™ III (Invitrogen, Karlsruhe, BW, Germany) according to the manufacturer's instructions. Quantitative real-time PCR was performed with SsoAdvanced Universal SYBR Green Supermix (BIO-RAD, Munich, Bavaria, Germany) and 10 pmol forward and 10 pmol reverse primer.

Relative RNA levels were calibrated and normalized with the level of housekeeping gene actin. Primer sequences for *Actin* (Act), *Limonene synthase* (Lim S), and *Pulegone reductase* (Pr) are shown in Table 3. PCR was performed using a CFX Connect Real-Time PCR system (BIO-RAD) according to the instruction manual. Transcript abundance was normalized to the transcript abundance of the actin.

Table 3. Primer sequences for RT-PCR.

Gene	Forward Primer Sequence (5′-3′)	Reverse Primer Sequence (5′-3′)
Act	GCTCCAAGGGCTGTGTTCC	TCTTTCTGTCCCATGCCAAC
Ls	TTGTGGCGAATTCTCTCGCT	GGCTTCTGAGCTGGTCACTT
Pr	GCATGGAGATCCCAGATGGC	AGTAGAGCCAGGAAGGATGGA

4.7. Statistical Analyses

Data were pooled and subjected to analysis of variance (ANOVA) followed by comparison of multiple treatment levels with controls using the Tukey test and principal component analysis (PCA). Control with solvent for MeJA and SA did not differ statistically with the control and is therefore not shown in the figures. For Figures 1–4, the statistics were performed on the native data, but the figures were made using "fold changes" in order to facilitate the interpretation of the results. Differences between means were considered significant for p values <0.05. The Infostat software program, v. 2008

(Group Infostat, Universidad Nacional de Córdoba, Córdoba, Argentina), was used for all statistical analyses. In the Supplementary Tables S1 and S2, the native data is shown.

5. Conclusions

Due to the multiple properties of secondary metabolites from *M. piperita*, monoterpene and phenolic compounds arouse the interest of the pharmaceutical and food industry as well as cosmetics producers. Elicitation of secondary metabolites appears to be a promising and innovative alternative; there are combinations of both biotic (PGPR) and abiotic (phytohormone) forms of elicitors that stimulate metabolism, and there have been many studies of their potential use individually for improving secondary metabolism biosynthesis, but no reports were found with the combination of both. This study revealed that peppermint plants treated with elicitors SA or JA and simultaneously inoculated with PGPR could enhance the production of phenolic compounds and monoterpenes. Considering the different concentrations of SA and MeJA evaluated, we suggest using a concentration of MeJA 2 mM for the external application on *M. piperita* 7 days before harvest. This is a cost-effective concentration, which increased the main secondary metabolite content, and taking into account the fact that a concentration of MeJA 4 mM is more expensive, not necessarily any more effective, and did not increase the main monoterpene by as much as the 2mM concentration. This is the first report demonstrating that inoculation with PGPR in combination with an external phytohormone increases phytochemical production in relation to each treatment alone. Results from this study will help improve secondary metabolite production for this crop.

Supplementary Materials: Supplementary materials can be found at http://www.mdpi.com/1422-0067/21/1/50/s1.

Author Contributions: L.d.R.C. and M.V.S. performed the experiments; E.B. designed the research and analyzed the data. E.B., A.S. and J.G. were involved in data interpretation. E.B., A.S. and J.G. wrote the manuscript. All authors have read and agreed to the published version of the manuscript.

Funding: This study was supported by grants from the Secretaría de Ciencia y Técnica de la Universidad Nacional de Río Cuarto, the Consejo Nacional de Investigaciones Científicas y Técnicas (CONICET), MinCyT Córdoba, the Agencia Nacional de Promoción Científica y Tecnológica (ANPCyT) (PICT 0636-14), financial support to E.B. from the Georg Forster Research Fellowship of the Alexander von Humboldt Foundation, and to M.V.S., A.S., and J.G. from the Max Planck Society. E.B. is a Career Member of CONICET. L.d.R.C. received fellowship from CONICET.

Acknowledgments: This study was supported by grants from the Secretaría de Ciencia y Técnica de la Universidad Nacional de Río Cuarto, the Consejo Nacional de Investigaciones Científicas y Técnicas (CONICET), MinCyT Córdoba, and the Agencia Nacional de Promoción Científica y Tecnológica (ANPCyT) PICT 0636-14, Argentina. E.B. obtained financial support from a Georg Forster-Research Fellowship of the Alexander von Humboldt Foundation. E.B. is Career Members of CONICET. L.d.R.C. received fellowships from CONICET- MinCyT. The authors are grateful to Joss Heywood, native speaker, for editorial assistance.

Conflicts of Interest: The authors declare no conflict of interest. The funders had no role in the design of the study; in the collection, analyses, or interpretation of data; in the writing of the manuscript, or in the decision to publish the results.

Abbreviations

EO	essential oil
PGPR	plant growth-promoting rhizobacteria
JA	jasmonic acid
MeJA	methyl jasmonate
SA	salicylic acid
TPC	total phenolic content
PAL	phenylalanine ammonia lyase

References

1. Akula, R.; Ravishankar, G.A. Influence of abiotic stress signals on secondary metabolites in plants. *Plant Signal. Behav.* **2011**, *6*, 1720–1731. [CrossRef] [PubMed]

2. Guerriero, G.; Berni, R.; Muñoz-Sanchez, J.A.; Apone, F.; Abdel-Salam, E.M.; Qahtan, A.A.; Alatar, A.A.; Cantini, C.; Cai, G.; Hausman, J.F.; et al. Production of plant secondary metabolites: Examples, tips and suggestions for biotechnologists. *Genes* **2018**, *9*, 309. [CrossRef] [PubMed]
3. Gonçalves, S.; Romano, A. Production of Plant Secondary Metabolites by Using Biotechnological Tools. In *Secondary Metabolites—Sources and Applications*; Vijayakumar, R., Raja, S.S.S., Eds.; IntechOpen: London, UK, 2018; pp. 81–99. [CrossRef]
4. Sing, P.; Pandey, A.K. Prospective of Essential Oils of the Genus Mentha as Biopesticides: A Review. *Front. Plant Sci.* **2018**. [CrossRef] [PubMed]
5. Dorman, H.J.; Koşar, M.; Başer, K.H.; Hiltunen, R. Phenolic profile and antioxidant evaluation of *Mentha* × *piperita* L. (peppermint) extracts. *Nat. Prod. Comun.* **2009**, *4*, 535–542. [CrossRef]
6. Farnad, N.; Heidari, R.; Aslanipour, B. Phenolic composition and comparison of antioxidant activity of alcoholic extracts of Peppermint (*Mentha piperita*). *J. Food Meas. Charact.* **2014**, *8*, 113–121. [CrossRef]
7. McKay, D.L.; Blumberg, J.B. A review of the bioactivity and potential health benefits of peppermint tea (*Mentha piperita* L.). *Phytother. Res.* **2006**, *20*, 619–633. [CrossRef]
8. Figueroa Pérez, M.G.; Rocha-Guzmán, N.E.; Mercado-Silva, E.; Loarca-Piña, G.; Reynoso Camacho, R. Effect of chemical elicitors on peppermint (*Mentha piperita*) plants and their impact on the metabolite profile and antioxidant capacity of resulting infusions. *Food Chem.* **2014**, *156*, 273–278. [CrossRef]
9. Figueroa-Pérez, M.G.; Rocha-Guzmán, N.E.; Pérez-Ramírez, I.F.; Mercado-Silva, E.; Reynoso-Camacho, R. Metabolite profile, antioxidant capacity, and inhibition of digestive enzymes in infusions of peppermint (*Mentha piperita*) grown under drought stress. *J. Agric. Food Chem.* **2014**, *62*, 12027–12033. [CrossRef]
10. Riachi, L.G.; De Maria, C.A.B. Peppermint antioxidants revisited. *Food Chem.* **2015**, *176*, 72–81. [CrossRef]
11. Fridlender, M.; Kapulnik, Y.; Koltai, H. Plant derived substances with anti-cancer activity: From folklore to practice. *Front. Plant Sci.* **2015**, *6*, 799. [CrossRef]
12. Yang, L.; Yang, C.; Li, C. Recent advances in biosynthesis of bioactive compounds in traditional Chinese medicinal plants. *Sci. Bull.* **2016**, *61*, 3–17. [CrossRef] [PubMed]
13. Verma, N.; Shukla, S. Impact of various factors responsible for fluctuation in plant secondary metabolites. *J. Appl. Res. Med. Aromat. Plants* **2015**, *2*, 105–113. [CrossRef]
14. Sangwan, N.S.; Farooqi, A.H.A.; Shabih, F.; Sangwan, R.S. Regulation of essential oil production in plants. *Plant Growth Regul.* **2001**, *34*, 3–21. [CrossRef]
15. Wheatley, R.E. The consequences of volatile organic compound mediated bacterial and fungal interactions. *Antonie Van Leeuwenhoek* **2002**, *81*, 357–364. [CrossRef]
16. Kai, M.; Haustein, M.; Molina, F.; Petri, A.; Scholz, B.; Piechulla, B. Bacterial volatiles and their action potential. *Appl. Microbiol. Biotechnol.* **2009**, *81*, 1001–1012. [CrossRef]
17. Rao, S.M.; Ravishankar, G.A. Plant cell cultures: Chemical factories of secondary metabolities. *Biotechnol. Adv.* **2002**, *20*, 101–153. [CrossRef]
18. Vanisree, M.; Lee, C.-Y.; Lo, S.-F.; Nalawade, S.M.; Lin, C.Y.; Tsay, H.-S. Studies on the production of some important secondary metabolites from medicinal plants by plant tissue cultures. *Bot. Bull. Acad. Sin.* **2004**, *45*, 1–22.
19. Thiruvengadam, M.; Rekha, K.; Chung, I.M. Induction of hairy roots by *Agrobacterium rhizogenes*-mediated transformation of spine gourd (*Momordica dioica* Roxb. ex. willd) for the assessment of phenolic compounds and biological activities. *Sci. Hort.* **2016**, *198*, 132–141. [CrossRef]
20. Farmer, E.E.; Alméras, E.; Krishnamurthy, V. Jasmonates and related oxylipins in plant responses to pathogenesis and herbivory. *Curr. Opin. Plant Biol.* **2003**, *6*, 372–378. [CrossRef]
21. Koo, A.J.K.; Howe, G.A. The wound hormone jasmonate. *Phytochemistry* **2009**, *70*, 1571–1580. [CrossRef]
22. Wasternack, C.; Strnad, M. Jasmonates are signals in the biosynthesis of secondary metabolites—Pathways, transcription factors and applied aspects—A brief review. *N. Biotechnol.* **2019**, *25*, 1–11. [CrossRef] [PubMed]
23. Ku, K.M.; Jeffery, E.H.; Juvik, J.A. Optimization of Methyl Jasmonate Application to Broccoli Florets to Enhance Health-promoting Phytochemical Content. *J. Sci. Food Agric.* **2014**, *94*, 2090–2096. [CrossRef] [PubMed]
24. Scognamiglio, J.; Jones, L.; Letizia, C.S.; Api, A.M. Fragrance material review on methyl jasmonate. *Food Chem. Toxicol.* **2012**, *50*, S572–S576. [CrossRef] [PubMed]
25. Zhao, J.; Davis, L.C.; Verpoorte, R. Elicitor signal transduction leading to production of plant secondary metabolites. *Biotechnol. Adv.* **2005**, *23*, 283–333. [CrossRef]

26. Taguchi, G.; Yazawa, T.; Hayashida, N.; Okazaki, M. Molecular cloning and heterologous expression of novel glucosyltransferases from tobacco cultured cells that have broad substrate specificity and are induced by salicylic acid and auxin. *Eur. J. Biochem.* **2001**, *268*, 4086–4094. [CrossRef]
27. Rivas-San Vicente, M.; Plasencia, J. Salicylic acid beyond: Its role in plant growth and development. *J. Exp. Bot.* **2011**, *62*, 3321–3338. [CrossRef]
28. Khan, N.; Bano, A.; Rahman, M.A.; Guo, J.; Kang, Z.; Babar, M.A. Comparative physiological and metabolic analysis reveals a complex mechanism involved in drought tolerance in chickpea (*Cicer arietinum* L.) induced by PGPR and PGRs. *Sci. Rep.* **2019**, *14*, 2097. [CrossRef]
29. Kloepper, J.W. Plant-Growth-Promoting Rhizobacteria as Biological Control Agents. In *Soil Microbial Ecology: Applications in Agricultural and Environmental Management*; Metting, F.B., Ed.; Marcel Dekker Inc.: New York, NY, USA, 1993; pp. 255–273.
30. Niranjan, R.S.; Shetty, H.S.; Reddy, M.S. Plant Growth Promoting Rhizobacteria: Potential Green Alternative for Plant Productivity. In *PGPR: Biocontrol and Biofertilization*; Siddiqui, Z.A., Ed.; Springer: Dordrecht, The Netherlands, 2006; pp. 197–216.
31. van Loon, L.C. Plant response to plant growth-promoting rhizobacteria. *Eur. J. Plant Pathol.* **2007**, *119*, 243–254. [CrossRef]
32. Pieterse, C.M.; Zamioudis, C.; Berendsen, R.L.; Weller, D.M.; van Wees, S.C.M.; Bakker, P.A. Induced systemic resistance by beneficial microbes. *Annu. Rev. Phytopathol.* **2014**, *52*, 347–375. [CrossRef]
33. Vessey, J.K. Plant growth promoting rhizobacteria as biofertilizers. *Plant Soil* **2003**, *255*, 571–586. [CrossRef]
34. Gupta, G.; Parihar, S.S.; Ahirwar, N.K.; Snehi, S.K.; Singh, V. Plant growth promoting rhizobacteria (PGPR): Current and future prospects for development of sustainable agriculture. *J. Microb. Biochem. Technol.* **2015**, *7*, 1013–1020. [CrossRef]
35. Backer, R.; Rokem, J.S.; Ilangumaran, G.; Lamont, J.; Praslickova, D.; Ricci, E.; Subramanian, S.; Smith, D.L. Plant growth-promoting rhizobacteria: Context, mechanisms of action, and roadmap to commercialization of biostimulants for sustainable agriculture. *Front. Plant Sci.* **2018**, *9*, 1473. [CrossRef] [PubMed]
36. Banchio, E.; Xie, X.; Zhang, H.; Paré, P.W. Soil bacteria elevate essential oil accumulation and emissions in sweet basil. *J. Agric. Food Chem.* **2009**, *5*, 653–657. [CrossRef] [PubMed]
37. Banchio, E.; Bogino, P.; Santoro, M.V.; Torres, L.; Zygadlo, J.; Giordano, W. Systemic induction of monoterpene biosynthesis in *Origanum x majoricum* by soil bacteria. *J. Agric. Food Chem.* **2010**, *58*, 650–654. [CrossRef] [PubMed]
38. Cappellari, L.; Santoro, M.V.; Nievas, F.; Giordano, W.; Banchio, E. Increase of secondary metabolite content in marigold by inoculation with plant growth-promoting rhizobacteria. *Appl. Soil Ecol.* **2013**, *70*, 16–22. [CrossRef]
39. Santoro, M.V.; Zygadlo, J.; Giordano, W.; Banchio, E. Volatile organic compounds from rhizobacteria increase biosynthesis of essential oils and growth parameters in peppermint (*Mentha piperita*). *Plant Physiol. Biochem.* **2011**, *49*, 1177–1182. [CrossRef]
40. Cappellari, L.; Santoro, V.M.; Schmidt, A.; Gershenzon, J.; Banchio, E. Induction of essential oil production in *Mentha* x *piperita* by plant growth promoting bacteria was correlated with an increase in jasmonate and salicylate levels and a higher density of glandular trichomes. *Plant Physiol. Biochem.* **2019**, *141*, 142–153. [CrossRef]
41. McConkey, M.E.; Gershenzon, J.; Croteau, R.B. Developmental regulation of monoterpene biosynthesis in the glandular trichomes of peppermint. *Plant Physiol.* **2000**, *122*, 215–224. [CrossRef]
42. Turner, G.W.; Gershenzon, J.; Nielson, E.E.; Froehlich, J.E.; Croteau, R.B. Limonene synthase, the enzyme responsible for monoterpene biosynthesis in peppermint, is localized to leucoplasts of oil gland secretory cells. *Plant Physiol.* **1999**, *120*, 879–886. [CrossRef]
43. Cappellari, L.R.; Chiappero, J.; Santoro, M.; Giordano, W.; Banchio, E. Inducing phenolic production and volatile organic compounds emission by inoculating *Mentha piperita* with plant growth-promoting rhizobacteria. *Sci. Hort.* **2017**, *220*, 193–198. [CrossRef]
44. Sudha, G.; Ravishankar, G.A. Involvement and interaction of various signaling compounds on the plant metabolic events during defense response, resistance to stress factors, formation of secondary metabolites and their molecular aspects. *Plant Cell Tissue Organ. Cult.* **2002**, *71*, 181–212. [CrossRef]

45. Krzyzanowska, J.; Czubacka, A.; Pecio, L.; Przybys, M.; Doroszewska, T.; Stochmal, A.; Oleszek, W. The effects of jasmonic acid and methyl jasmonate on rosmarinic acid production in *Mentha × piperita* cell suspension cultures. *Plant Cell Tiss. Organ Cult.* **2012**, *108*, 73–81. [CrossRef]
46. Ali, M.B.; Hahn, E.J.; Paek, K.Y. Methyl jasmonate and salicylic acid induced oxidative stress and accumulation of phenolics in *Panax ginseng* bioreactor root suspension cultures. *Molecules* **2007**, *12*, 607–621. [CrossRef] [PubMed]
47. Mendoza, D.; Cuaspud, O.; Arias, J.P.; Ruiz, O.; Arias, M. Effect of salicylic acid and methyl jasmonate in the production of phenolic compounds in plant cell suspension cultures of *Thevetia peruviana*. *Biotechnol. Rep.* **2018**, *3*, e00273. [CrossRef] [PubMed]
48. Nafie, E.; Hathout, T.; Mokadem, A.S.A. Jasmonic acid elicits oxidative defense and detoxification systems in *Cucumis melo* L. cells. *Braz. J. Plant Physiol.* **2011**, *23*, 161–174. [CrossRef]
49. Kim, H.J.; Fonseca, J.M.; Choi, J.H.; Kubota, C. Effect of methyl jasmonate on phenolic compounds and carotenoids of romaine lettuce (*Lactuca sativa* L.). *J. Agric. Food Chem.* **2007**, *12*, 5510366. [CrossRef]
50. Park, C.H.; Yeo, H.J.; Park, Y.E.; Chun, S.W.; Chung, Y.S.; Lee, S.Y.; Park, S.U. Influence of chitosan, salicylic acid and jasmonic acid on phenylpropanoid accumulation in germinated buckwheat (*Fagopyrum esculentum* Moench). *Foods* **2019**, *6*, 153. [CrossRef]
51. Chen, H.; Seguin, P.; Archambault, A.; Constan, L.; Jabaji, S. Gene expression and isoflavone concentrations in soybean sprouts treated with chitosan. *Crop Sci.* **2009**, *49*, 224–236. [CrossRef]
52. Kim, Y.B.; Kim, J.K.; Uddin, M.R.; Xu, H.; Park, W.T.; Tuan, P.A.; Li, X.; Chung, E.; Lee, J.-H.; Park, S.U. Metabolomics analysis and biosynthesis of rosmarinic acid in *Agastache rugosa* Kuntze treated with methyl jasmonate. *PLoS ONE* **2013**, *8*, e64199. [CrossRef]
53. Khalil, N.; Fekry, M.; Bishr, M.; El-Zalabani, S.; Salama, O. Foliar spraying of salicylic acid induced accumulation of phenolics, increased radical scavenging activity and modified the composition of the essential oil of water stressed *Thymus vulgaris* L. *Plant Physiol. Biochem.* **2018**, *123*, 65–74. [CrossRef]
54. El-Esawi, M.A.; Elansary, H.O.; El-Shanhorey, N.A.; Abdel-Hamid, A.M.E.; Ali, H.M.; Elshikh, M.S. Salicylic acid-regulated antioxidant mechanisms and gene expression enhance rosemary performance under saline conditions. *Front. Physiol.* **2017**, *8*, 716. [CrossRef] [PubMed]
55. Gorni, P.H.; Pacheco, A.C. Growth promotion and elicitor activity of salicylic acid in *Achillea millefolium* L. *Afr. J. Biotechnol.* **2016**, *15*, 657–665. [CrossRef]
56. Liu, Y.; Pan, Q.-H.; Yang, H.-R.; Liu, Y.-Y.; Huang, W.-D. Relationship between H_2O_2 and jasmonic acid in pea leaf wounding response. *Russ. J. Plant Physl.* **2008**, *55*, 765. [CrossRef]
57. Gadzovska, S.; Maury, S.; Delaunay, A. Jasmonic acid elicitation of *Hypericum perforatum* L. cell suspensions and effects on the production of phenylpropanoids and naphtodianthrones. *Plant Cell Tissue Organ. Cult.* **2007**, *89*, 1–13. [CrossRef]
58. Lavania, M.; Chauhan, P.S.; Chauhan, S.V.S.; Singh, H.B.; Nautiyal, C.H. Induction of plant defense enzymes and phenolics by treatment with plant growth promoting rhizobacteria *Serratia marcescens* NBRI1213. *Curr. Microbiol.* **2006**, *52*, 363–368. [CrossRef]
59. Singh, U.P.; Sarma, B.K.; Singh, D.P. Effect of plant growth-promoting rhizobacteria and culture filtrate of *Sclerotium rolfsii* on phenolic and salicylic acid contents in Chickpea (*Cicer arietinum*). *Curr. Microbiol.* **2003**, *46*, 131–140. [CrossRef]
60. Salla, T.D.; Ramos, T.; Astarita, L.V.; Santarém, E.R. *Streptomyces* rhizobacteria modulate the secondary metabolism of *Eucalyptus* plants. *Plant Physiol. Biochem.* **2014**, *85*, 14–20. [CrossRef]
61. Panka, D.; Piesik, D.; Jeske, M.; Baturo-Ciesniewska, A. Production of phenolics and the emission of volatile organic compounds by perennial ryegrass (*Lolium perenne* L.)/*Neotyphodium lolii* association as a response to infection by *Fusarium poae*. *J. Plant Physiol.* **2013**, *170*, 1010–1019. [CrossRef]
62. Vanitha, S.C.; Umesha, S. *Pseudomonas fluorescens* mediated systemic resistance in tomato is driven through an elevated synthesis of defense enzymes. *Biol. Plant* **2011**, *55*, 317. [CrossRef]
63. Ramamoorthy, V.; Samiyappan, R. Induction of defenserelatedgenes in *Pseudomonas fluorescens* treated chilli plants in response to infection by *Colletotrichum capsici*. *J. Mycol. Plant Pathol.* **2001**, *31*, 146–155.
64. Unsicker, S.B.; Kunert, G.; Gershenzon, J. Protective perfumes: The role of vegetative volatiles in plant defense against herbivores. *Curr. Opin. Plant Biol.* **2009**, *1*, 479–485. [CrossRef] [PubMed]
65. Vickers, C.E.; Gershenzon, J.; Lardau, M.T.; Loreto, F. A unified mechanism of action for volatile isoprenoids in plant abiotic stress. *Nat. Chem. Biol.* **2009**, *5*, 283–291. [CrossRef] [PubMed]

66. Cappellari, L.; Santoro, M.; Reinoso, H.; Travaglia, C.; Giordano, W.; Banchio, E. Anatomical, morphological, and phytochemical effects of inoculation with plant growth promoting rhizobacteria on peppermint (*Mentha piperita*). *J. Chem. Ecol.* **2015**, *41*, 149–158. [CrossRef] [PubMed]
67. Banchio, E.; Bogino, P.; Zygadlo, J.; Giordano, W. Plant growth promoting rhizobacteria improve growth and essential oil yield in *Origanum majorana* L. *Biochem. Syst. Ecol.* **2008**, *36*, 766–771. [CrossRef]
68. Yan, X.; Zhang, L.; Wang, J. Molecular characterization and expression of 1-deoxy- d-xylulose 5-phosphate reductoisomerase (DXR) gene from *Salvia miltiorrhiza*. *Acta Physiol. Plant* **2009**, *31*, 1015. [CrossRef]
69. Cao, X.Y.; Li, C.G.; Miao, Q.; Zheng, Z.J.; Jiang, J.H. Molecular cloning and expression analysis of a leaf-specific expressing 3-hydroxy-3-methylglutaryl-CoA (HMG-CoA) reductase gene from *Michelia chapensis* Dandy. *J. Med. Plants Res.* **2011**, *5*, 3868–3875. [CrossRef]
70. Cao, X.Y.; Yin, T.; Miao, Q.; Li, C.G.; Ju, X.Y.; Sun, Y. Molecular characterization and expression analysis of a gene encoding for farnesyl diphosphate synthase from *Euphorbia pekinensis* Rupr. *Mol. Biol. Rep.* **2012**, *39*, 1487–1492. [CrossRef]
71. Shabani, L.; Ehsanpour, A.A.; Asghari, G.; Emami, J. Glycyrrhizin production by in vitro cultured *glycyrrhiza glabra* elicited by methyl jasmonate and salicylic acid. *Russ. J. Plant Physiol.* **2009**, *56*, 621–626. [CrossRef]
72. Kai, M.; Crespo, E.; Cristescu, S.M.; Harren, F.J.M.; Francke, W.; Piechulla, B. *Serratia odorifera*: Analysis of volatile emission and biological impact of volatile compounds on *Arabidopsis thaliana*. *Appl. Microbiol. Biotechnol.* **2010**, *88*, 965–976. [CrossRef]
73. Munne-Bosch, S.; Peñuelas, J. Photo- and antioxidative protection, and a role for salicylic acid during drought and recovery in field-grown *Phillyrea angustifolia* plants. *Planta* **2003**, *217*, 758–766. [CrossRef]
74. Eraslan, F.; Inal, A.; Gunes, A.; Alpaslan, M. Impact of exogenous salicylic acid on the growth, antioxidant activity and physiology of carrot plants subjected to combined salinity and boron toxicity. *Sci. Hort.* **2007**, *113*, 120–128. [CrossRef]
75. Peñuelas, J.; Llusià, J.; Filella, I. Methyl salicylate fumigation increases monoterpene emission rates. *Biol. Plant.* **2007**, *51*, 372–376. [CrossRef]
76. Zhang, M.; Liu, J.; Li, K.; Yu, D. Identification and characterization of a novel monoterpene synthase from soybean restricted to neryldiphosphate precursor. *PLoS ONE* **2013**, *4*, e75972. [CrossRef]
77. Pateraki, I.; Kanellis, A.K. Stress and developmental responses of terpenoid biosynthetic genes in *Cistus creticus* subsp. Creticus. *Plant Cell Rep.* **2010**, *29*, 629–641. [CrossRef]
78. Xu, Y.W.; Lv, S.S.; Zhao, D.; Chen, J.W.; Yang, W.T.; Wu, W. Effects of salicylic acid on monoterpene production and antioxidant systems in *Houttuynia cordata*. *Afr. J. Biotechnol.* **2012**, *11*, 1364–1372.
79. Anand, A.; Uppalapati, S.R.; Ryu, C.M.; Allen, S.N.; Kang, L.; Tang, Y.; Mysore, K.S. Salicylic acid and systemic acquired resistance play a role in attenuating crown gall disease caused by *Agrobacterium tumefaciens*. *Plant Physiol.* **2008**, *146*, 703–715. [CrossRef]
80. Wang, J.W.; Wu, J.Y. Nitric oxide is involved in methyl jasmonate-induced defense responses and secondary metabolism activities of *Taxus* cells. *Plant Cell Physiol.* **2005**, *46*, 923–930. [CrossRef]
81. Karban, R.; Baldwin, I.T. *Induced Responses to Herbivory*; University Chicago Press: Chicago, IL, USA, 1997; pp. 33–38.
82. Hu, P.; Zhou, W.; Cheng, Z.; Fan, M.; Wang, L.; Xie, D. JAV1 controls jasmonate-regulated plant defense. *Mol. Cell* **2013**, *50*, 504–515. [CrossRef]
83. Kautz, S.; Trisel, J.A.; Ballhorn, D.J. Jasmonic acid enhances plant cyanogenesis and resistance to herbivory in lima bean. *J. Chem. Ecol.* **2014**, *40*, 1186–1996. [CrossRef]
84. Tian, D.; Peiffer, M.; De Moraes, C.M.; Felton, G.W. Roles of ethylene and jasmonic acid in systemic induced defense in tomato (*Solanum lycopersicum*) against *Helicoverpa zea*. *Planta* **2014**, *239*, 577–589. [CrossRef]
85. Yan, C.; Xie, D. Jasmonate in plant defence: Sentinel or double agent? *Plant Biotech. J.* **2015**, *13*, 1233–1240. [CrossRef] [PubMed]
86. Howe, G.A. Cyclopentenone signals for plant defense: Remodeling the jasmonic acid response. *Proc. Natl. Acad. Sci. USA* **2001**, *98*, 12317–12319. [CrossRef] [PubMed]
87. Maes, L.; Van Nieuwerburgh, F.C.; Zhang, Y.; Reed, D.W.; Pollier, J.; Vande Casteele, S.R.; Inzé, D.; Covello, P.S.; Deforce, D.L.; Goossens, A. Dissection of the phytohormonal regulation of trichome formation and biosynthesis of the antimalarial compound artemisinin in *Artemisia annua* plants. *New Phytol.* **2011**, *189*, 176–189. [CrossRef] [PubMed]

88. Wasternack, C.; Song, S. Jasmonates: Biosynthesis, metabolism, and signaling by proteins activating and repressing transciption. *J. Exp. Bot.* **2016**, *68*, 1303–1321. [CrossRef] [PubMed]
89. Schmidt, A.; Nagel, R.; Krekling, T.; Christiansen, E.; Gershenzon, J.; Krokene, P. Induction of isoprenyl diphosphate synthases, plant hormones and defense signalling genes correlates with traumatic resin duct formation in Norway spruce (*Picea abies*). *Plant Mol. Biol.* **2011**, *77*, 577–590. [CrossRef]
90. Henery, M.L.; Wallis, I.R.; Stone, C.; Foley, W.J. Methyl jasmonate does not induce changes in *Eucalyptus grandis* leaves that alter the effect of constitutive defense on larvae of a specialist herbivore. *Oecologia* **2008**, *56*, 847–859. [CrossRef]
91. Rodríguez-Saona, C.; Crafts-Brandner, S.J.; Paré, P.W.; Henneberry, T.J. Exogenous methyl jasmonate induces volatile emissions in cotton plants. *J. Chem. Ecol.* **2001**, *27*, 679–695. [CrossRef]
92. Złotek, U.; Michalak-Majewska, M.; Szymanowska, U. Effect of jasmonic acid elicitation on the yield, chemical composition, and antioxidant and anti-inflammatory properties of essential oil of lettuce leaf basil (*Ocimum basilicum* L.). *Food Chem.* **2016**, *213*, 1–7. [CrossRef]
93. Lange, B.M.; Mahmoud, S.S.; Wildung, M.R.; Turner, G.W.; Davis, E.M.; Lange, I.; Baker, R.C.; Boydston, R.A.; Croteau, R.B. Improving peppermint essential oil yield and composition by metabolic engineering. *Proc. Natl. Acad. Sci. USA* **2011**, *108*, 16944–16949. [CrossRef]
94. Lange, B.M.; Turner, G.W. Terpenoid biosynthesis in trichomes-current status and future opportunities. *Plant Biotechnol. J.* **2013**, *11*, 2–22. [CrossRef]
95. Gershenzon, J.; McCaskill, D.; Rajaonarivony, J.I.M.; Mihaliak, C.; Karp, F.; Croteau, R. Isolation of secretory cells from plant glandular trichomes and their use in biosynthetic studies of monoterpenes and other gland products. *Anal. Biochem.* **1992**, *200*, 130–138. [CrossRef]
96. McCaskill, D.; Gershenzon, J.; Croteau, R. Morphology and monoterpene biosynthetic capabilities of secretory cell clusters isolated from glandular trichomes of peppermint (*Mentha piperita* L.). *Planta* **1992**, *187*, 445–454. [CrossRef] [PubMed]
97. Boughton, A.J.; Hoover, K.; Felton, G.W. Methyl jasmonate application induces increased densities of glandular trichomes on tomate, *Lycopersicon esculentum*. *J. Chem. Ecol.* **2005**, *31*, 2211–2216. [CrossRef] [PubMed]
98. Greenbaum, D.; Colangelo, C.; Williams, K. Comparing protein abundance and mRNA expression levels on a genomic scale. *Genome Biol.* **2003**, *4*, 117. [CrossRef]
99. van der Fits, L.; Memelink, J. ORCA3, a jasmonate-responsive transcriptional regulator of plant primary and secondary metabolism. *Science* **2000**, *289*, 295–297. [CrossRef]
100. Schenk, P.M.; Kazan, K.; Wilson, I.; Anderson, J.P.; Richmond, T.; Somerville, S.C.; Manners, J.M. Coordinated plant defense responses in Arabidopsis revealed by microarray analysis. *Proc. Natl. Acad. Sci. USA* **2000**, *97*, 11655–11660. [CrossRef]
101. Reymond, P.; Bodenhausen, N.; Van Poecke, R.M.P.; Krishnamurthy, V.; Dicke, M.; Farmer, E.E. A conserved transcript pattern in response to a specialist and a generalist herbivore. *Plant Cell* **2000**, *16*, 3132–3147. [CrossRef]
102. Durrant, W.E.; Dong, X. Systemic acquired resistance. *Annu. Rev. Phytopathol.* **2004**, *42*, 185–209. [CrossRef]
103. Guo, X.; Stotz, H.U. Defense against *Sclerotinia sclerotiorum* in *Arabidopsis* is dependent on jasmonic acid, salicylic acid, and ethylene signaling. *Mol. Plant Microbe Interact.* **2007**, *20*, 1384–1395. [CrossRef]
104. Memelink, J.; Verpoorte, R.; Kijne, J.W. ORCAnization of jasmonate-responsive gene expression in alkaloid metabolism. *Trends Plant Sci.* **2001**, *6*, 212–219. [CrossRef]
105. Agati, G.; Tattini, M. Multiple functional roles of flavonoids in photoprotection. *New Phytol.* **2010**, *186*, 786–793. [CrossRef] [PubMed]
106. González-Burgos, E.; Gómez-Serranillos, M.P. Terpene compounds in nature: A review of their potential antioxidant activity. *Curr. Med. Chem.* **2012**, *19*, 5319–5341. [CrossRef] [PubMed]
107. Janda, T.; Gondor, O.K.; Yordanova, R.; Szalai, G.; Pál, M. Salicylic acid and photosynthesis: Signalling and effects. *Acta Physiol. Plant* **2014**, *36*, 2537–2546. [CrossRef]
108. Lattanzio, V. Phenolic Compounds: Introduction. In *Natural Products*; Ramawat, K.G., Me'rillon, J.M., Eds.; Springer: Berlin/Heidelberg, Germany, 2013.
109. Huxley, R.R.; Neil, H.A. The relation between dietary flavonol intake and coronary heart disease mortality: A meta-analysis of prospective cohort studies. *Eur. J. Clin. Nutr.* **2003**, *57*, 904–908. [CrossRef] [PubMed]

110. Lubbe, A.; Verpoorte, R. Cultivation of medicinal and aromatic plants for specialty industrial materials. *Ind. Crop. Prod.* **2011**, *34*, 785–801. [CrossRef]
111. Salehi, B.; Stojanović-Radić, Z.; Matejić, J.; Sharopov, F.; Antolak, H.; Kręgiel, D.; Sen, S.; Sharifi-Rad, M.; Acharya, K.; Sharifi-Rad, R.; et al. Plants of Genus Mentha: From Farm to Food Factory. *Plants* **2018**, *7*, 70. [CrossRef]
112. Basbagci, G.; Erler, F. Evaluation of some essential oils and their major components against mushroom scatopsid flies as fumigants. *Fresenius Environ. Bull.* **2013**, *22*, 3173–3181.
113. Dorman, H.J.D.; Deans, S.G. Antibacterial agents from plants: Antibacterial activity of plant volatile oils. *J. Appl. Microbiol.* **2000**, *88*, 308–316. [CrossRef]
114. Rhouma, A.; Daoud, H.B.; Ghanmi, S.; Salah, H.B.; Romdhane, M.; Demak, M. Antimicrobial activities of leaf extracts of *Pistacia* and *Schinus* species against some plant pathogenic fungi and bacteria. *J. Plant Pathol.* **2009**, *91*, 339–345.
115. Xu, J.; Zhou, F.; Ji, B.P.; Pei, R.S.; Xu, N. The antibacterial mechanism of carvacrol and thymol against *Escherichia coli*. *Lett. Appl. Microbiol.* **2008**, *47*, 174–179. [CrossRef]
116. Berendsen, R.L.; van Verk, M.C.; Stringlis, I.A. Unearthing the genomes of plant-beneficial Pseudomonas model strains WCS358, WCS374 and WCS417. *BMC Genom.* **2015**, *16*, 539. [CrossRef] [PubMed]
117. Santoro, V.M.; Bogino, P.C.; Nocelli, N.; Cappellari, L.; Giordano, W.F.; Banchio, E. Analysis of plant growth-promoting effects of fluorescent Pseudomonas strains isolated from *Mentha piperita* rhizosphere and effects of their volatile organic compounds on essential oil composition. *Front. Microb.* **2016**, *7*, 1085. [CrossRef] [PubMed]
118. Choi, S.K.; Jeong, H.; Kloepper, J.W.; Ryu, C.M. Genome sequence of *Bacillus amyloliquefaciens* GB03, an active ingredient of the first commercial biological control product. *Genome Announc.* **2014**, *2*, e01092-14. [CrossRef] [PubMed]
119. Bertani, G. Studies on lysogenesis. The mode of phage liberation by lysogenic *Escherichia coli*. *J. Bacteriol.* **1951**, *62*, 293–300.
120. Singleton, V.L.; Rossi, J.A. Colorimetry of total phenolics with phosphomolybdic-phosphotungstic acid reagents. *Am. J. Enol. Vitic.* **1965**, *16*, 144–158.
121. Bradford, M.A. Rapid and sensitive method for the quantification of microgram quantities of protein utilizing the principle of protein dye binding. *Anal. Biochem.* **1976**, *72*, 248–254. [CrossRef]
122. Beaudoin-Eagan, L.D.; Thorpe, T.A. Tyrosine and phenylalanine ammonia-lyase activities during shoot initiation in tobacco callus cultures. *Plant Physiol.* **1985**, *78*, 438–441. [CrossRef]

International Journal of
Molecular Sciences

MDPI

Article

Novel Salicylic Acid Analogs Induce a Potent Defense Response in Arabidopsis

Ian Arthur Palmer [1], Huan Chen [1,2], Jian Chen [1,2], Ming Chang [1,2], Min Li [1], Fengquan Liu [2,*] and Zheng Qing Fu [1,*]

1 Department of Biological Sciences, University of South Carolina, Columbia, SC 29208, USA
2 Institute of Plant Protection, Jiangsu Academy of Agricultural Sciences, Jiangsu Key Laboratory for Food Quality and Safety-State Key Laboratory Cultivation Base of Ministry of Science and Technology, Nanjing 210014, China
* Correspondence: fqliu20011@sina.com (F.L.); zfu@mailbox.sc.edu (Z.F.)

Received: 11 June 2019; Accepted: 5 July 2019; Published: 8 July 2019

Abstract: The master regulator of salicylic acid (SA)-mediated plant defense, NPR1 (NONEXPRESSER OF PR GENES 1) and its paralogs NPR3 and NPR4, act as SA receptors. After the perception of a pathogen, plant cells produce SA in the chloroplast. In the presence of SA, NPR1 protein is reduced from oligomers to monomers, and translocated into the nucleus. There, NPR1 binds to TGA, TCP, and WRKY transcription factors to induce expression of plant defense genes. A list of compounds structurally similar to SA was generated using ChemMine Tools and its Clustering Toolbox. Several of these analogs can induce SA-mediated defense and inhibit growth of *Pseudomonas syringae* in *Arabidopsis*. These analogs, when sprayed on *Arabidopsis*, can induce the accumulation of the master regulator of plant defense NPR1. In a yeast two-hybrid system, these analogs can strengthen the interactions among NPR proteins. We demonstrated that these analogs can induce the expression of the defense marker gene *PR1*. Furthermore, we hypothesized that these SA analogs could be potent tools against the citrus greening pathogen *Candidatus liberibacter* spp. In fact, our results suggest that the SA analogs we tested using *Arabidopsis* may also be effective for inducing a defense response in citrus. Several SA analogs consistently strengthened the interactions between citrus NPR1 and NPR3 proteins in a yeast two-hybrid system. In future assays, we plan to test whether these analogs avoid degradation by SA hydroxylases from plant pathogens. In future assays, we plan to test whether these analogs avoid degradation by SA hydroxylases from plant pathogens.

Keywords: salicylic acid; analogs; NPR1; NPR3; NPR4; PR1; *Pseudomonas syringae*

1. Introduction

Plant immunity can be described as consisting of four phases, known as the zig-zag model [1]. First, pathogen-associated molecular patterns (PAMPs) are recognized by pattern recognition receptors (PRRs) on the plant cell's surface. Pathogen-associated molecular patterns are evolutionarily conserved molecules associated with pathogens such as flagellin, EF-Tu, and chitin [2–4]. Pathogen-associated molecular pattern recognition results in PAMP-triggered immunity (PTI). PAMP-triggered immunity consists of an increase in cytosolic Ca^{2+} [5], oxidative burst [6], MAPK activation [7], ethylene production [8], stomatal closure, transcriptional reprogramming, accumulation of the plant defense hormone salicylic acid (SA) [9], and callose deposition [10]. This response is basal disease resistance against pathogens that can halt colonization. During the second phase of the zig-zag model, pathogens deliver effectors into plant cells to interfere with PTI, resulting in effector-triggered susceptibility (ETS). Plants have evolved rsistance (R) proteins capable of specifically recognizing secreted effectors, resulting in effector-triggered immunity (ETI), as phase three. Resistance proteins are nucleotide-binding leucine-rich repeat (NB-LRR) proteins that can respond to effectors from plant pathogens [11,12].

Resistance proteins usually recognize effectors indirectly. They may act as accessory recognition proteins that detect effector modification of the effector's true virulence targets or decoys that mimic the effector's targets [13]. In phase four, pathogens either lose effector genes or acquire additional effector genes that can continue to suppress ETI and PTI. The loss of recognized effectors or the gain of novel effectors, causes selective pressure on the host to evolve new R proteins, resulting in ETI [1].

SA acts as a major plant hormone, playing a regulatory role in various physiological processes. These processes are diverse, and include seed germination, storage, fruit maturity, regulation of flower development, sex differentiation, stomatal movement, and photoperiod [14]. SA is necessary to induce a defense response against pathogens [14] and exogenous application of SA is sufficient to induce a defense response [15]. Transgenic *Arabidopsis* plants expressing *NahG* from *Pseudomonas putida*, encoding a SA hydroxylase enzyme which degrades SA into catechol, are rendered more susceptible to a variety of pathogens [16].

SA is synthesized in the chloroplast after pathogen perception. In *Arabidopsis*, SA biosynthesis is produced primarily through the isochorismate pathway, in which chorismate is converted into isochorismate by ICS1 (isochorismate synthase 1) and then isochorismate is presumably converted into SA by an unidentified IPL (isochorismate pyruvate lyase) [17]. Isochorismate synthase 1 is localized in the plastid and is responsible for the majority of SA accumulation in response to the presence of hemibiotrophic and biotrophic pathogens [18,19]. The *Arabidopsis ics1* mutants are unable to accumulate SA, and consequently, *ics1* mutants are more susceptible to pathogen infection. SA can induce a potent systemic immune response known as systemic acquired resistance (SAR) [20].

The NPR1 and NPR1 paralogs NPR3 and NPR4 bind SA and function as SA receptors [21–23]. NPR1 functions as a transcriptional co-activator. Upon pathogen infection, NPR1 oligomers in the cytosol are reduced into monomers and then NPR1 monomers enter the nucleus and interact with TGA and TCP transcription factor to activate the expression of *PR* genes, which encode small proteins that may have antimicrobial properties [24–26]. Induction of the expression of *PR1* is directly correlated with an increase of SA levels [27]. The NPR1 paralogs NPR3 and NPR4 act as adaptor proteins for Cullin 3 E3 ubiquitin ligase, leading to the ubiquitination and degradation of NPR1, dependent on SA concentration—a high level of SA disrupts the interaction between NPR1 and NPR4, while promoting the interaction between NPR1 and NPR3; this creates a biphasic pattern of NPR1 level and defense response [28]. Both NPR3 and NPR4 are also known to form homo- and heterodimers and the formation of NPR3 and NPR4 homo- and heterodimers is strengthened by the presence of SA [21].

In addition to inducing a local defense response, SA promotes systemic acquired resistance (SAR) after an invading pathogen is recognized [29,30]. Systemic acquired resistance protects the plant against further pathogen colonization by causing a systemic defense reaction including the production of pathogenesis related (PR) proteins, phytoalexins, and the strengthening of cell walls. SA is also responsible for regulating these later responses to pathogenic invasion [31], and application of SA is sufficient to induce plant defense including SAR [32].

The SA-mediated plant defense pathway can be activated by exogenous application of SA, 2,6-dichloroisonicotinic acid (INA), or benzothiadiazole (BTH) (Figure 1) [33,34]. Additionally, some synthetic compounds have been used in the past to elicit a defense response, protecting crops from disease. These synthetic compounds include 3-allyloxy-1,2-benzisothiazole-1,1-dioxide (probenazole, PBZ), applied to *Oryza sativa* to prevent rice blast caused by *Magnaporthea grisea* [35]; the previously mentioned INA on *Cucumis sativus* and *Nicotiana tabacum* to prevent anthracnose (caused by *Colletotrichum lagenarium*) and tobacco mosaic virus infection, respectively [15,36]; N-cyanomethyl-2-chloroisonicotinamide (NCI) on *O. sativa* to induce defense against *Pyricularia oryzae*, an anamorph of *M. oryzae*, and many others [37,38].

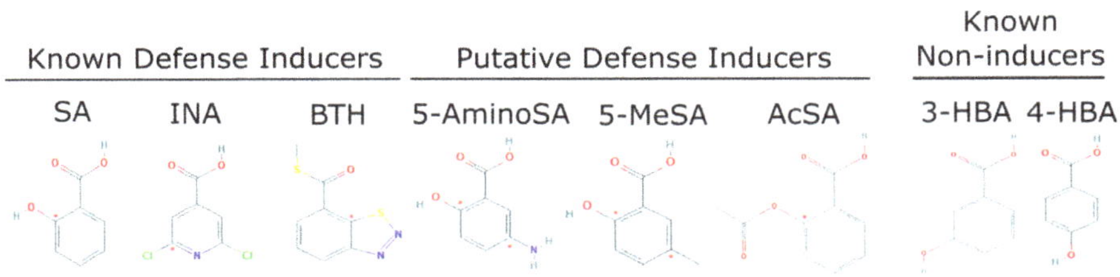

Figure 1. Comparison of known defense inducers and known non-inducers. Known inducers often have substitutions on carbon two and/or carbon five of the aromatic ring. Non-inducers have substitutions on carbon three or four. Substitutions on carbons two or five are indicated by a red asterisk. SA is salicylic acid, INA is 2,6-dichloroisonicotinic acid, BTH is benzothiadiazole, 5-AminoSA is 5-aminosalicylic acid, 5-MeSA is 5-methylsalicylic acid, AcSA is acetylsalicylic acid, 3-HBA is 3-hydroxybenzoic acid, and 4-HBA is 4-hydroxybenzoic acid.

Unsurprisingly, due to the necessity of SA for defense induction, pathogens have evolved enzymes capable of degrading this key phytohormone. Bacterial members of the genera *Pseudomonas, Bacillus, Agrobacterium, Rhizobium, Sinorhizobium, Ralstonia*, and *Burkholderia* have genes encoding SA hydroxylases capable of metabolizing SA into less active forms [39]. SA hydroxylases function typically by binding SA and NADH or NADPH, then binding molecular oxygen. The resulting products are catechol, H_2O, and CO_2 [40]. Ectopically expressing the bacterial SA hydroxylase gene, *NahG*, from *Pseudomonas putida* in *Arabidopsis* suppresses the defense response against both bacterial and fungal pathogens and abolishes SA accumulation after pathogen infection [41].

Here, we present the results of a screen of SA analogs. We demonstrate that by applying several of these analogs to *Arabidopsis* Col-0 plants, the accumulation of the master regulator of SA-mediated plant defense, NPR1, can be induced. We show that the application of these SA analogs results in the induction of *PR1* expression. We demonstrate that these SA analogs can strengthen the protein–protein interactions between NPR1 paralogs NPR3 and NPR4 in a yeast two-hybrid system. We demonstrate that these analogs are effective in inhibiting bacterial growth, causing increased resistance against pathogen infection. We also demonstrate that a similar group of SA analogs that are functional in *Arabidopsis* are also capable of strengthening the interactions between NPR1 and NPR3 homologs in *Citrus sinensis*.

2. Results

2.1. ChemMine Results

The simplified molecular-input line-entry system (SMILES) string for SA, c1ccc(c(c1)C(=O)O)O, was used as input for ChemMine Tools. This online suite of tools allows for comparing pairwise structural similarities between compounds and provides ultra-fast structure similarity search algorithms. ChemMine Tools also contains a Clustering Toolbox to group the mined chemicals based on systematic structure and predicted activity [42]. This suite of tools was used to find the 50 most similar compounds to SA, compiled into an excel workbook. Candidate chemical compounds were then sorted by logP value and eliminated from the list based on predicted logP value (Table 1).

Of the list of 50 most similar compounds to SA, we selected compounds that had substitutions on the second or fifth carbon of the six-carbon ring. We hypothesized that substitutions made on the second or fifth carbon may be key to developing novel SA analogs that are functional but may resist degradation by bacterial pathogens (Figure 1), based on a comparison of the molecular structures of known defense inducers, SA, INA, and BTH, compared with known non-inducers, 3-hydroxybenzoic acid (3-HBA) and 4-hydroxybenzoic acid (4-HBA). The complete list of SA analogs tested in this work and in Figure 1 can be found in Table 2. In this paper, we focused on sodium salicylate (NaSA) as a positive control.

Sodium salicylate is a water-soluble form of SA that dissociates, forming SA in solution. Ethyl salicylate (EtSA), acetylsalicylate (ACSA), 5-methylsalicylic acid (5-MeSA), 5-aminosalicylic acid (5-amino-SA) or mesalamine, 5-fluoro-2-hydroxybenzoic acid (5-F-2-HBA), and 5-iodosalicylic acid (5-I-SA), 2HTPA (2-hydroxyterephthalic acid), 2,4-dihydroxybenzoic acid (2, 4DHBA), 2,5-dihydroxybenzoic acid (2,5 DHBA), and 4HBA (4-hydroxyl-benzoic acid) as a negative control.

Table 1. List of mined SA analogs sorted by logP value.

Acid	Name	Molecular_Weight	LogP
3469	2,5-dihydroxybenzoic acid	1.53×10^2	6.67×10^{-1}
9338	2,6-dihydroxybenzoic acid	1.53×10^2	6.67×10^{-1}
55251260	Lithium 2,5-dihydroxybenzoate	1.60×10^2	6.67×10^{-1}
1491	2,4-Dihydroxybenzoic acid	1.53×10^2	6.67×10^{-1}
23663423	Monosodium 2,4-dihydroxybenzoate	1.76×10^2	6.67×10^{-1}
3418	Fosfosal	2.17×10^2	1.1109
11812	2-Hydroxyisophthalic acid	1.80×10^2	1.3557
97257	2-Hydroxyterephthalic acid	1.80×10^2	1.3557
6998	Salicylaldehyde	1.22×10^2	1.4218
67658	5-Fluorosalicylic acid	1.55×10^2	1.4986
54675839	2,5-Dihydroxybenzoate	1.52×10^2	1.5033
54712708	2,4-Dihydroxybenzoate	1.52×10^2	1.5033
53629521	62TEY51RR1	3.64×10^2	1.6432
16682734	Bismuth subsalicylate	3.63×10^2	1.8035
8388	5-Iodosalicylic acid	2.63×10^2	1.9641
72874	2-Hydroxy-4-iodobenzoic acid	2.63×10^2	1.9641
4133	Methyl salicylate	1.52×10^2	2.0602
8375	2′-Hydroxyacetophenone	1.36×10^2	2.1286
6738	3-Methylsalicylic acid	1.51×10^2	2.1672
6973	5-Methylsalicylic acid	1.51×10^2	2.1672
5788	4-methylsalicylic acid	1.51×10^2	2.1672
11279	2-hydroxy-6-methylbenzoic acid	1.51×10^2	2.1672
164578	4-Trifluoromethylsalicylic acid	2.05×10^2	2.3783
8631	3,5-Diiodosalicylic acid	3.89×10^2	2.4457
8365	Ethyl salicylate	1.66×10^2	2.767
54683201	Copper disalicylate	3.38×10^2	2.9625
54684589	Magnesium salicylate	2.99×10^2	2.965
64738	Magnesium salicylate	2.99×10^2	2.965
1.02E+08	Magan	2.99×10^2	2.965
517068	Calcium salicylate	3.14×10^2	2.965
54684600	Calcium disalicylate	3.14×10^2	2.965
1.32E+08	Magnesium salicylate	3.17×10^2	3.1257
201887	2-Hydroxy-3-isopropylbenzoic acid	1.79×10^2	3.5808
5282387	Magnesium salicylate tetrahydrate	3.71×10^2	3.6078
54708862	Magnesium salicylate tetrahydrate	3.71×10^2	3.6078
133124	Whitfield's ointment	2.58×10^2	3.7803
6873	Isobutyl salicylate	1.94×10^2	4.1806
16330	Butyl salicylate	1.94×10^2	4.1806

Table 1. *Cont.*

Acid	Name	Molecular_Weight	LogP
50216	Prenyl salicylate	2.06×10^2	4.276
16299	Amyl salicylate	2.08×10^2	4.8874
6437473	trans-2-Hexenyl salicylate	2.20×10^2	4.9828
5371102	cis-3-Hexenyl salicylate	2.20×10^2	4.9828
103379	Benzoic acid, 2-hydroxy-, (3Z)-3-hexenyl ester	2.20×10^2	4.9828
6021887	3-Hexenyl salicylate	2.20×10^2	4.9828
22629	Hexylsalicylate	2.22×10^2	5.5942
153705	3-Hexylsalicylic acid	2.21×10^2	5.7012
196549	Tcp (antiseptic)	5.56×10^2	6.2422

Table 2. List of Tested SA Analogs with Chemical Structures.

ID	Name	Abbv.	Structure	Formula	Mol. Weight
1	Sodium Salicylate	NaSA	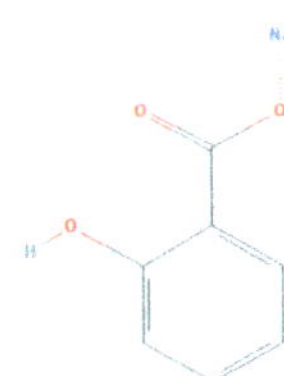	$C_7H_5NaO_3$	160.104 g/mol
2	4-Hydroxybenzoic Acid	4-HBA	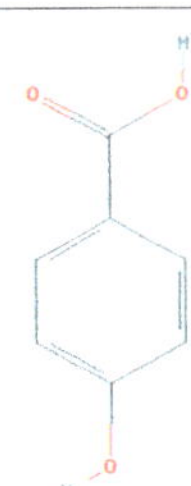	$C_7H_6O_3$	138.122 g/mol
3	Acetylsalicylic Acid	AcSA	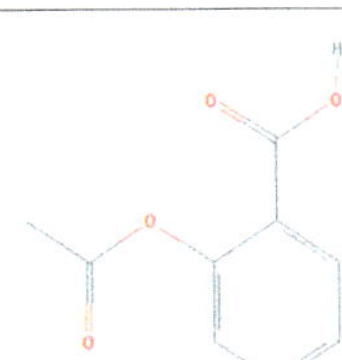	$C_9H_8O_4$	180.159 g/mol
4	5-Aminosalicylic Acid	5-AminoSA	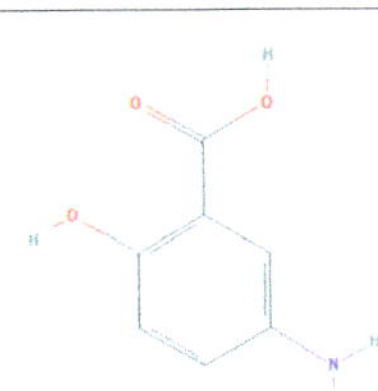	$C_7H_7NO_3$	183.137 g/mol

Table 2. *Cont.*

ID	Name	Abbv.	Structure	Formula	Mol. Weight
5	Ethyl Salicylate	EtSA	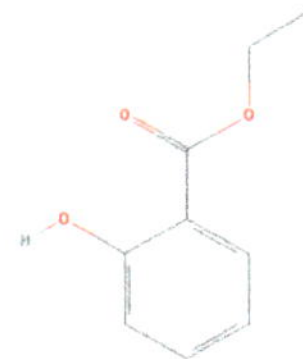	$C_9H_{10}O_3$	166.167 g/mol
6	2,5-Dihydroxybenzoic Acid	2,5-DHBA	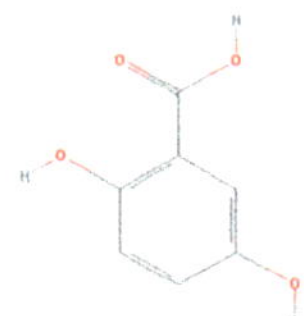	$C_7H_6O_4$	154.121 g/mol
7	5-Methylsalicylic Acid	5-MeSA	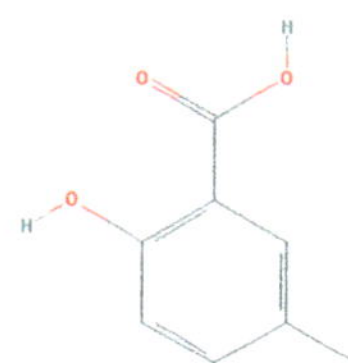	$C_8H_8O_3$	152.149 g/mol
8	5-Iodosalicylic Acid	5-I-SA	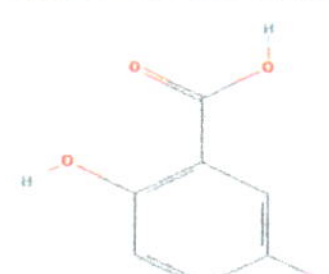	$C_7H_5IO_3$	264.018 g/mol
9	5-Fluoro-2-Hydroxybenzoic Acid	5-F-2-HBA	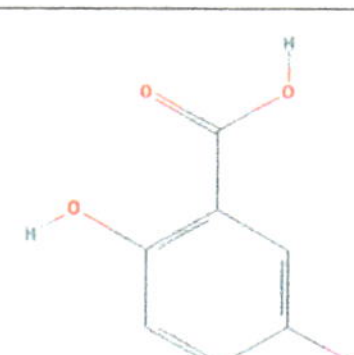	$C_7H_5FO_3$	156.112 g/mol
10	2,4-Dihydroxybenzoic Acid	2,4-DHBA	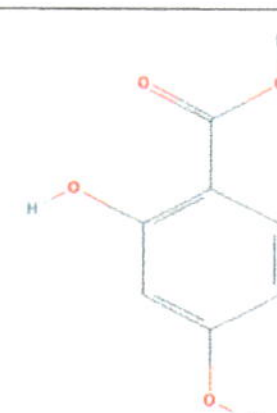	$C_7H_6O_4$	154.121 g/mol

Table 2. *Cont.*

ID	Name	Abbv.	Structure	Formula	Mol. Weight
11	2-Hydroxyterephthalic Acid	2-HTPA	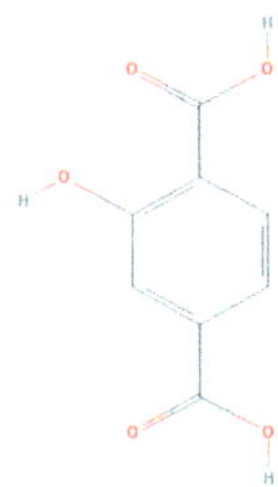	$C_8H_6O_5$	182.131 g/mol

List of tested SA analogs, including their chemical formulas, abbreviations, molecular weights, and structures [43].

2.2. *Several Putative SA Analogs Increased the Strength of Interactions among NPR3/4 in Y2H*

Due to the critical role that NPR1 paralogs NPR3 and NPR4 play in SA-mediated defense, we hypothesized that active SA analogs would increase the strength of the interactions among these proteins in a yeast two-hybrid system. Because the interaction between NPR1 and NPR3 is strengthened in response to SA and the interaction between NPR1 and NPR4 is disrupted by SA, we chose to examine the effects of SA analogs on the NPR3 and NPR4 interactions, which are strengthened by the presence of SA [21]. By examining the interactions between NPR1 paralogs instead of NPR1 itself, we hoped to remove some ambiguity from our Y2H results, resulting from the SA analogs both strengthening and disrupting interactions between NPR1 and its paralogs in Y2H. Indeed, we observed that several SA analogs cause an increase in the number of yeast colonies that survive on quadruple dropout media. The number of surviving colonies treated with SA analogs can be compared to the number that grow when treated with sodium salicylate, appearing when diluted to OD_{600} 0.01. As shown in Figure 2, we observed that 5-meSA and 5-F-2HBA can strengthen the interaction between NPR3 with NPR3 and NPR4 with NPR4. 5-I-SA can strengthen the interaction between NPR3 and NPR3. AcSA can strengthen the interaction between both NPR3/4 homo- and heterodimers.

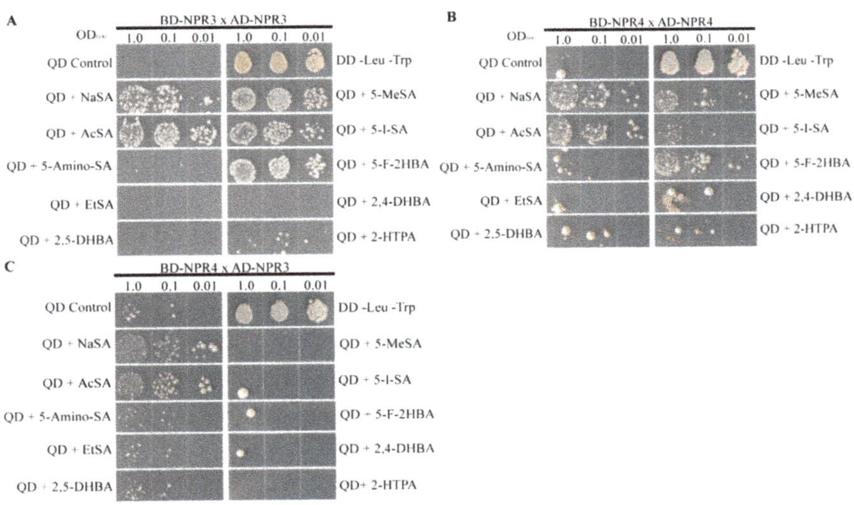

Figure 2. Several SA analogs consistently strengthened the interactions among NPR proteins in a Y2H system.

(**A**) Interaction between NPR3 and NPR3; (**B**) interaction between NPR4 and NPR4; (**C**) interaction between NPR4 and NPR3. Yeast strains were incubated for 24 h in double dropout liquid media before being washed in sterile deionized water, diluted, and plated on quadruple dropout agar media with or without 200 μM NaSA or SA analogs. Plates were incubated at 30 °C for 72 h. QD is quadruple dropout –Leu–Trp–His–Ade. DD is double dropout –Leu–Trp. The assay was repeated three times with similar results.

2.3. *Several SA Analogs Induced NPR1 Accumulation*

Next, to determine whether these SA analogs could induce the accumulation of NPR1, we treated wild-type *Arabidopsis* with a 1 mM spray of SA analogs or SA, and compared the NPR1 protein levels, using untreated plants as a negative control. Previous research has shown that exogenous application of SA is sufficient to elicit a defense response, including the accumulation of NPR1. As shown in Figure 3, we found that AcSA, 5-I-SA, 5-F-2HBA, and 5-MeSA can induce NPR1 accumulation. 4-HBA and non-treated plants were included as negative controls. NaSA was included as a positive control.

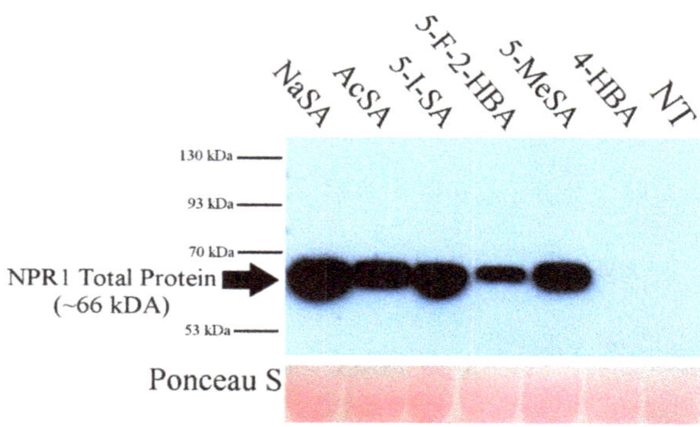

Figure 3. SA analog treatment induces accumulation of NPR1. Three-week-old *Arabidopsis thaliana* were sprayed with 1 mM NaSA or SA analogs. Samples were collected 6 hpi. Composite samples were taken consisting of one leaf each of a similar size and age from four plants, and 100 μg of protein was electrophoresed per sample. The membrane was incubated with anti-NPR1 antibody overnight at 4 °C. NT is non-treated. The assay was repeated three times with similar results.

2.4. *Several SA Analogs Inhibited Bacterial Growth*

After observing that SA analogs could induce the accumulation of NPR1 in planta, we were curious whether treatment with SA analogs could inhibit the growth of plant bacterial pathogens. We observed that all but one SA analog, 2,5-DHBA, could reduce the number of CFUs per leaf disc by at least one order of magnitude, when compared with non-treated plants (Figure 4). Additionally, we observed no significant difference between the number of bacteria found in the SA-analog-treated plants and the SA-treated plants, again with the exception of 2,5-DHBA.

2.5. *SA Analogs that Induced NPR1 Accumulation Were Inducers of* PR1 *Expression*

After observing that almost all SA analogs could inhibit the growth of *P. syringae* and that several analogs were potent inducers of NPR1 accumulation, we hypothesized that an increase in NPR1 protein must trigger the expression of *PR1*, a gene encoding a small peptide which is known to inhibit the growth of bacterial pathogens. We sprayed Col-0 *Arabidopsis* with 1 mM SA or SA analogs, then collected leaf samples for RT-qPCR after 24 h. We observed that AcSA induces the highest level of *PR1* accumulation, even higher than the same concentration of NaSA. we observed that several other SA analogs could induce *PR1* expression, but at lower levels than NaSA or AcSA (Figure 5).

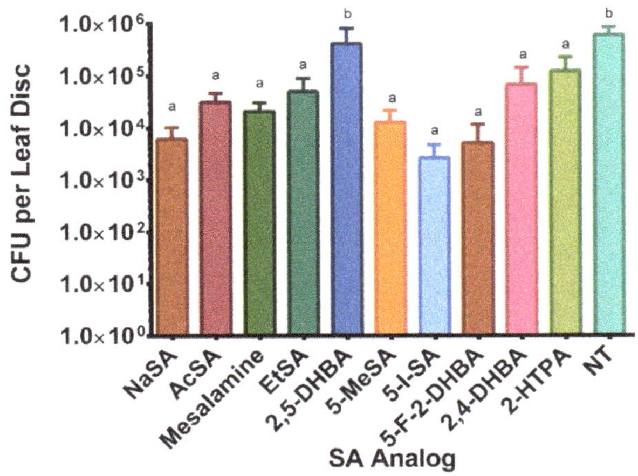

Figure 4. SA analog treatment reduced the number of bacteria present in leaves of treated plants. Three-week old *A. thaliana* Col-0 were sprayed with 1 mM NaSA or SA analogs. After 24 h, two leaves each from three plants per treatment were infiltrated with *Pseudomonas syringae* pv. maculicula ES4326 at OD600 0.001 in 10 mM MgSO4. After 72 h, 2 discs were sampled from each leaf. Dunnett's multiple comparison test was used to generate groups of statistical significance. $p \leq 0.05$. NT is non-treated. The assay was performed twice with similar results.

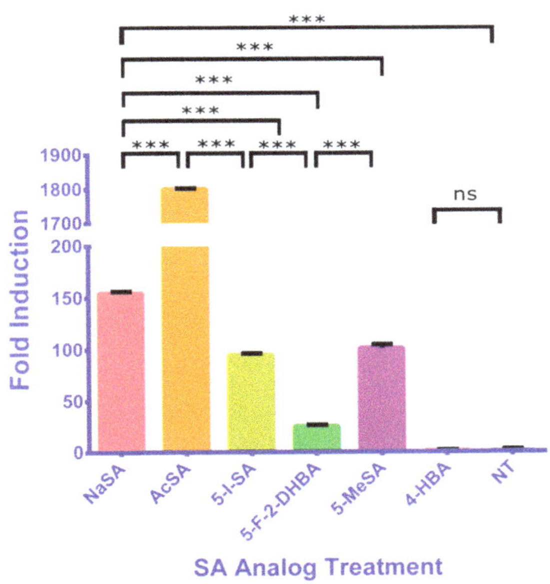

Figure 5. Relative normalized *PR1* expression 24 h after 1 mM SA analog spray. Composite samples were made from five biological replicates. Samples were assayed using three technical replicates. Expression levels were calculated using the double-delta Ct method. Error bars represent standard error of measurement. Expression levels of *NPR1* were normalized to the expression levels of *Ubiquitin 5* (*UBQ5*) (ΔCt ANOVA $p < 0.0001$; Student's *t*-test *** $p < 0.0001$; ns is no significance).

2.6. The Interaction Between CsNPR1 and CsNPR3 Was Strengthened by Several SA Analogs

We hypothesized that these SA analogs could be potent tools against the citrus greening pathogen, *Candidatus liberibacter* spp., which is known to produce an SA hydroxylase enzyme that functions to suppress plant defense [39]. We cloned the NPR1 and NPR3 homologs from *Citrus sinensis* Valencia and tested whether the SA analogs could also strengthen the interaction among citrus NPR proteins using Y2H (Figure 6). The NPR homologs have been identified previously in citrus [44]. The citrus NPR1 homolog CtNH1 (Citrus NPR1 homolog 1) has been previously shown to induce *PR* gene expression in *Citrus maxima* and confer resistance to the bacterial pathogen *Xanthomonas axonopodis* pv. *citri* [44]. We observed that NaSA, AcSA, 5-MeSA, 5-we-SA, 5-F-2-HBA, and 2-HTPA all can strengthen the interaction among citrus NPR proteins in our Y2H system. This finding is significant, because it suggests that the SA analogs we tested using *Arabidopsis* may also be effective for inducing a defense response in citrus. If these SA analogs are active in citrus, then we speculate that they may be candidates for fighting the citrus greening pathogen, because they may not be able to be degraded by the pathogen's SA hydroxylase enzyme.

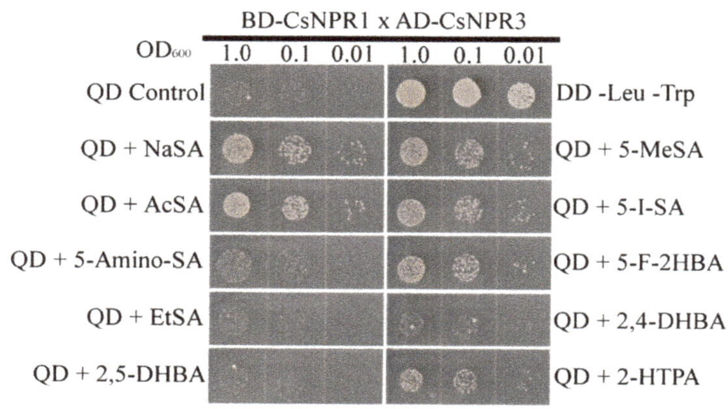

Figure 6. Several SA analogs consistently strengthen the interactions between citrus NPR1 and NPR3 proteins in a Y2H system. Yeast strains were incubated for 24 h in double dropout liquid media before being washed in sterile deionized water, diluted, and plated on quadruple dropout agar media with or without 200 μM SA or SA analogs. Plates were incubated at 30 °C for 72 h. QD is quadruple dropout –Leu–Trp–His–Ade. DD is double dropout–Leu–Trp.

3. Discussion

In this study, we found acetylsalicylate, 5-methylsalicylic acid, 5-fluoro-2-hydroxybenzoic acid, and 5-iodosalicylic acid to be reliable inducers of plant defense. The data we have presented here suggest that these SA analogs would be worthy candidates for use against bacterial pathogens. Their ability to invoke a defense response from *Arabidopsis* and confer bacterial resistance are traits that warrant further investigation.

Previous research suggested that acetylsalicylate was effective against tobacco mosaic virus in tobacco [45]; however, there is little research into its use against bacterial pathogens. Acetylsalicylate's ability to induce defense is not entirely surprising when one considers that acetylsalicylic acid and SA also share a function in mammals. The ability for acetylsalicylate to induce a higher level of PR1 accumulation and *PR1* expression may be due to an increase in membrane permeability of that compound in relation to sodium salicylate. A compound's polar surface area can be used as a measure of that compound's H-bonding potential, and therefore, its membrane penetration potential [46]. Acetylsalicylate has a slightly higher polar surface area at 63.6 Å^2 than sodium salicylate which is 60.4 Å^2 [47], which could make it slightly more bioavailable to the treated plant's cells.

5-Fluoro-2-hydroxybenzoic acid and 5-iodosalicylic acid are likely inducers of plant defense because of their structural similarity to SA. Usually, the chemical interaction between a protein and a small molecule is dictated by electrostatic forces—H-bonding and Van der Waals forces—but halogen atoms can also generate intermolecular forces capable of stabilizing a protein complex that are similar to H-bonding in both strength and directionality [48]. This realization has enabled researchers to develop new halogen-substituted ligands that are more membrane permeable and have a longer biological half-life by avoiding the normal catabolic processes that normally degrade the drug [48]. For these reasons, 5-F-2HBA and 5-we-SA would make great candidates for use against pathogens that produce SA hydroxylase enzymes.

Previous research has suggested that SA acts directly on *P. syringae*, acting as an anti-microbial agent and reducing biofilm formation [49]. In addition to their ability to induce plant defense, these SA analogs may also act directly to inhibit the growth of *P. syringae* and other pathogens. As demonstrated in Figure 4, all SA analogs, except for 2,5-DHBA, reduced the bioburden of *P. syringae* after treatment; however, more research is needed to demonstrate the direct action of these SA analogs on pathogen growth and biofilm formation.

Our research demonstrates that 5-methylsalicylic acid can induce NPR1 accumulation, *PR1* expression, inhibit pathogen growth, and promote the interaction between NPR proteins. 5-MeSA differs from methyl salicylate (MeSA), which has a methyl group appended to the carboxyl group on carbon 1 of the aromatic ring, rather than the methyl substitution on carbon 5. Unlike methyl salicylate which is a volatile, wintergreen-scented compound that is a liquid at room temperature, 5-MeSA is a white, odorless compound that is solid at room temperature. 5-MeSA's use as a defense inducer warrants further research, because it is similar enough in structure to SA, but may be able to avoid degradation by bacterial SA hydroxylases due to the methyl group substitution on carbon 5.

Currently, INA and BTH are widely used as active salicylic acid analogs. However, both of these active analogs have major drawbacks. Because of its toxic side effects, INA has never been commercialized [38]. BTH has been commercialized as Bion® or Actigard® but is very expensive because the complex structure is costly to synthesize. Our newly identified salicylic acid analogs could potentially be used to replace INA and BTH for controlling plant diseases.

4. Materials and Methods

4.1. Yeast Two-Hybrid (Y2H) Assays

Yeast strains were mated in yeast extract, peptone, dextrose, adenine (YPDA) media for 48 h at 30 °C. Diploid yeast strains were plated on double dropout selective media. Colonies were selected, then grown for 48 h in liquid double dropout media at 30 °C. The resulting liquid culture was serially diluted to an OD600 value of 1.0, 0.1, and 0.01, then plated on quadruple synthetic dropout media with and without SA or SA analogs and incubated at 30 °C for 72 h. CsNPR1 and CsNPR3 were cloned from *Citrus sinensis* Valencia into pDONR® 207 using the Gateway BP reaction. The Gateway LR reaction was used to generate pGADT7 and pGBKT7 yeast expression vectors containing CsNPR1 or CsNPR3. These vectors were transformed into yeast strains Y187 or AH109, respectively, then the yeast strains were mated and plated on synthetic quadruple dropout (QD) media with and without SA or SA analogs like the previously conducted Y2H assays. The yeast strains expressing *Arabidopsis* NPR1, NPR3, and NPR4 were described in previous works [21]. QD agar lacking SA or SA analogs was used a negative control for Y2H, because the NPR protein interactions were previously described using the same Y2H system.

4.2. SA Analog Spray Treatment

SA analogs were diluted in 50 mL sterile purified water to a final concentration of 1 mM. The SA analog solutions were sprayed using a Preval® Sprayer. The *Arabidopsis* leaves were sprayed from multiple angles until the leaves were visibly wet to ensure complete coverage. Between applications, the Preval® Sprayer was washed, and 15 mL of sterile purified water was sprayed through to ensure no cross contamination of SA analogs.

4.3. Immunoblotting

Three-week-old *Arabidopsis thaliana* plants were sprayed with 1 mM SA or SA analogs as above. Samples were collected 6 h after treatment for assaying NPR1 accumulation. Composite samples were taken consisting of one leaf each of a similar size and age from four plants. Leaves were frozen in liquid nitrogen, then ground using a metal bead by crushing for 2 min at 1200 RPM. Protein was extracted using 1× protein extraction buffer (50 mM Tris-HCL, pH 7.5, 150 mM NaCl, 5 mM EDTA, 0.1% Triton X-100, 0.2% IGEPAL CA-630) with 1× protease inhibitor cocktail (Millipore Sigma, Burlington, MA, USA), 10 mM DTT, 1 mM PMSF, and 10 mM MG115. Samples were centrifuged at 15,000× *g* for 30 min at 4 °C and the supernatant removed to a new tube. The centrifugation was repeated twice. The protein concentration was determined by mixing 5 µL of protein sample with 200 µL of 5× Bradford reagent (Bio-Rad, Hercules, CA, USA) in a spectrophotometer cuvette and filling to 1 mL with sterile deionized water. The samples were analyzed for absorbance at 595 nm. Protein concentration was determined by comparing the absorbance to a standard curve. One-hundred micrograms of protein were boiled for 10 min in 1× Laemmli sample buffer (2% *w/v* SDS, 10% Glycerol, 60 mM Tris-HCL pH 6.8, 0.01% bromophenol blue, 0.2% 2-mercaptoethanol), then samples were electrophoresed for 1 h at 120 V. Protein was transferred to a nitrocellulose membrane by transferring for 1 h at 100 V. The membrane was incubated in 5% non-fat milk for 1 h at room temperature, then incubated with anti-NPR1 antibody (Agrisera, Vännäs, Sweden) overnight at 4 °C. The membrane was washed three times for ten minutes in 1× PBST (0.1% Tween20), then secondary antibody was added at a ratio of 1:5000 and incubated at room temperature for 2 h. The membrane was washed as above, then incubated in Bio-Rad ECL substrate for 5 min at room temperature. X-ray film was used to capture the resulting chemiluminescence.

4.4. RT-qPCR

Three-week-old *A. thaliana* were sprayed with 1 mM SA or SA analogs as above, and samples were collected after 24 h. Composite samples were collected consisting of one leaf from ten biological replicates. Each leaf was of a similar size and age. Samples were frozen immediately in liquid nitrogen and crushed using a Genogrinder at 1200 RPM for 2 min. The RNA was extracted using RNAzol® RT from Millipore Sigma per the manufacturer's instructions. The RNA concentration and purity were quantified spectroscopically by measuring absorbance at 260 and 280 nm. qScript™ cDNA SuperMix from QuantaBio was used to generate cDNA from 1 µg of the extracted RNA according to the manufacturer's instructions. PerfeCTa SYBR® Green SuperMix from QuantaBio was used to perform qPCR per the manufacturer's instructions. Relative expression levels were calculated using the double-delta Ct method. The assays were performed with ten biological replicates and six technical replicates. The primers used are listed in the table below (Table 3).

Table 3. List of RT-qPCR Primer Sequences.

Name	Sequence
UBQ5 forward RT	TCTCCGTGGTGGTGCTAAG
UBQ5 reverse RT	GAACCTTTCCAGATCCATCG
PR1 forward RT	GCAACTGCAGACTCATACAC
PR1 reverse RT	GTTGTAGTTAGCCTTCTCGC

4.5. Preparation of SA Analog Solutions

SA and AcSA were dissolved in sterile, de-ionized water, then diluted to the necessary concentrations. All other SA analogs were diluted in 100 µL of 100% ethanol or DMSO, then diluted to their respective concentrations with sterile, de-ionized water as needed. All solutions were filter-sterilized using a 0.2 µm syringe filter prior to use.

Author Contributions: Conceptualization, I.A.P., F.L., and Z.Q.F.; Investigation, I.A.P., H.C., J.C., M.C., M.L.; Resources, Z.Q.F.; Writing—Original Draft, I.A.P.; Writing—Review and Editing, I.A.P., M.L., F.L., and Z.Q.F.; Supervision, Z.Q.F.; Funding Acquisition, Z.Q.F.

Funding: This research was funded by the National Science Foundation (Early-Concept Grants for Exploratory Research (EAGER) grant 1464527 and grant number IOS-1758994 to Z.Q.F.) and an ASPIRE I grant from the University of South Carolina to Z.Q.F.

Conflicts of Interest: The authors declare no conflict of interest. The funders had no role in the design of the study; in the collection, analyses, or interpretation of data; in the writing of the manuscript, or in the decision to publish the results.

References

1. Jones, J.D.G.; Dangl, J.L. The plant immune system. *Nature* **2006**, *444*, 323–329. [CrossRef] [PubMed]
2. Gómez-Gómez, L.; Boller, T. Flagellin perception: A paradigm for innate immunity. *Trends Plant Sci.* **2002**, *7*, 251–256. [CrossRef]
3. Zipfel, C.; Kunze, G.; Chinchilla, D.; Caniard, A.; Jones, J.D.; Boller, T.; Felix, G. Perception of the bacterial pamp ef-tu by the receptor efr restricts agrobacterium-mediated transformation. *Cell* **2006**, *125*, 749–760. [CrossRef] [PubMed]
4. Eckardt, N.A. Chitin signaling in plants: Insights into the perception of fungal pathogens and rhizobacterial symbionts. *Am. Soc. Plant. Biol.* **2008**, *20*, 241–243. [CrossRef] [PubMed]
5. Boller, T.; Felix, G. A renaissance of elicitors: Perception of microbe-associated molecular patterns and danger signals by pattern-recognition receptors. *Annu. Rev. Plant Biol.* **2009**, *60*, 379–406. [CrossRef] [PubMed]
6. Lamb, C.J.; Lawton, M.A.; Dron, M.; Dixon, R.A. Signals and transduction mechanisms for activation of plant defenses against microbial attack. *Cell* **1989**, *56*, 215–224. [CrossRef]
7. Boudsocq, M.; Willmann, M.R.; McCormack, M.; Lee, H.; Shan, L.; He, P.; Bush, J.; Cheng, S.-H.; Sheen, J. Differential innate immune signalling via ca 2+ sensor protein kinases. *Nature* **2010**, *464*, 418. [CrossRef] [PubMed]
8. Tintor, N.; Ross, A.; Kanehara, K.; Yamada, K.; Fan, L.; Kemmerling, B.; Nürnberger, T.; Tsuda, K.; Saijo, Y. Layered pattern receptor signaling via ethylene and endogenous elicitor peptides during arabidopsis immunity to bacterial infection. *Proc. Natl. Acad. Sci. USA* **2013**, *110*, 6211–6216. [CrossRef] [PubMed]
9. Mishina, T.E.; Zeier, J. Pathogen-associated molecular pattern recognition rather than development of tissue necrosis contributes to bacterial induction of systemic acquired resistance in arabidopsis. *Plant J.* **2007**, *50*, 500–513. [CrossRef] [PubMed]
10. Luna, E.; Pastor, V.; Robert, J.; Flors, V.; Mauch-Mani, B.; Ton, J. Callose deposition: A multifaceted plant defense response. *Mol. Plant-Microbe Interact.* **2011**, *24*, 183–193. [CrossRef] [PubMed]
11. Elmore, J.M.; Lin, Z.-J.D.; Coaker, G. Plant nb-lrr signaling: Upstreams and downstreams. *Curr. Opin. Plant Biol.* **2011**, *14*, 365–371. [CrossRef]
12. Wu, L.; Chen, H.; Curtis, C.; Fu, Z.Q. Go in for the kill: How plants deploy effector-triggered immunity to combat pathogens. *Virulence* **2014**, *5*, 710–721. [CrossRef] [PubMed]
13. Van der Hoorn, R.A.; Kamoun, S. From guard to decoy: A new model for perception of plant pathogen effectors. *Plant Cell* **2008**, *20*, 2009–2017. [CrossRef]
14. Raskin, I. Role of salicylic acid in plants. *Annu. Rev. Plant Physiol. Plant Mol. Biol.* **1992**, 439–462. [CrossRef]
15. Ward, E.R.; Uknes, S.J.; Williams, S.C.; Dincher, S.S.; Wiederhold, D.L.; Alexander, D.C.; Ahl-Goy, P.; Métraux, J.-P.; Ryals, J.A. Coordinate gene activity in response to agents that induce systemic acquired resistance. *Plant Cell* **1991**, *3*, 1085–1094. [CrossRef] [PubMed]

16. Delaney, T.; Uknes, S.; Vernooij, B.; Friedrich, L.; Weymann, K.; Negrotto, D.; Gaffney, T.; Gut-Rella, M.; Kessmann, H.; Ward, E.; et al. A central role of salicylic acid in plant disease resistance. *Science* **1994**, *266*, 1247. [CrossRef] [PubMed]
17. Wildermuth, M.C.; Dewdney, J.; Wu, G.; Ausubel, F.M. Isochorismate synthase is required to synthesize salicylic acid for plant defence. *Nature* **2001**, *414*, 562. [CrossRef] [PubMed]
18. Strawn, M.A.; Marr, S.K.; Inoue, K.; Inada, N.; Zubieta, C.; Wildermuth, M.C. Arabidopsis isochorismate synthase functional in pathogen-induced salicylate biosynthesis exhibits properties consistent with a role in diverse stress responses. *J. Biol. Chem.* **2007**, *282*, 5919–5933. [CrossRef] [PubMed]
19. Fragnière, C.; Serrano, M.; Abou-Mansour, E.; Métraux, J.; L'Haridon, F. Salicylic acid and its location in response to biotic and abiotic stress. *FEBS Lett.* **2011**, *585*, 1847–1852. [CrossRef]
20. Vlot, A.C.; Dempsey, D.A.; Klessig, D.F. Salicylic acid, a multifaceted hormone to combat disease. *Annu. Rev. Phytopathol.* **2009**, *47*, 177–206. [CrossRef]
21. Fu, Z.Q.; Yan, S.; Saleh, A.; Wang, W.; Ruble, J.; Oka, N.; Mohan, R.; Spoel, S.H.; Tada, Y.; Zheng, N.; et al. Npr3 and npr4 are receptors for the immune signal salicylic acid in plants. *Nature* **2012**, *486*, 228–232. [CrossRef] [PubMed]
22. Wu, Y.; Zhang, D.; Chu, J.Y.; Boyle, P.; Wang, Y.; Brindle, I.D. The arabidopsis npr1 protein is a receptor for the plant defense hormore salicylic acid. *Cell Rep.* **2012**, *1*, 639–647. [CrossRef] [PubMed]
23. Ding, Y.; Sun, T.; Ao, K.; Peng, Y.; Zhang, Y.; Li, X.; Zhang, Y. Opposite roles of salicylic acid receptors npr1 and npr3/npr4 in transcriptional regulation of plant immunity. *Cell* **2018**, *173*, 1454–1467. [CrossRef] [PubMed]
24. Mou, Z.; Fan, W.; Dong, X. Inducers of plant systemic acquired resistance regulate npr1 function through redox changes. *Cell* **2003**, *113*, 935–944. [CrossRef]
25. Fu, Z.Q.; Dong, X. Systemic acquired resistance: Turning local infection into global defense. *Annu. Rev. Plant Biol.* **2013**, *64*, 839–863. [CrossRef] [PubMed]
26. Li, M.; Chen, H.; Chen, J.; Chang, M.; Palmer, I.A.; Gassmann, W.; Liu, F.; Fu, Z.Q. Tcp transcription factors interact with npr1 and contribute redundantly to systemic acquired resistance. *Front. Plant Sci.* **2018**, *9*, 1153. [CrossRef] [PubMed]
27. Malamy, J.; Carr, J.P.; Klessig, D.F.; Raskin, I. Salicylic acid: A likely endogenous signal in the resistance response of tobacco to viral infection. *Science* **1990**, *250*, 1002–1004. [CrossRef]
28. Moreau, M.; Tian, M.; Klessig, D.F. Salicylic acid binds npr3 and npr4 to regulate npr1-dependent defense responses. *Cell Res.* **2012**, *22*, 1631–1633. [CrossRef]
29. An, C.; Mou, Z. Salicylic acid and its function in plant immunity. *J. Integr. Plant Biol.* **2011**, *53*, 412–428. [CrossRef]
30. Palmer, I.A.; Shang, Z.; Fu, Z.Q. Salicylic acid-mediated plant defense: Recent developments, missing links, and future outlook. *Front. Biol.* **2017**, *12*, 1–13. [CrossRef]
31. Lu, H.; Greenberg, J.T.; Holuigue, L. Editorial: Salicylic acid signaling networks. *Front. Plant Sci.* **2016**, *7*, 238. [CrossRef] [PubMed]
32. Anand, A.; Uppalapati, S.R.; Ryu, C.-M.; Allen, S.N.; Kang, L.; Tang, Y.; Mysore, K.S. Salicylic acid and systemic acquired resistance play a role in attenuating crown gall disease caused by agrobacterium tumefaciens. *Plant Physiol.* **2008**, *146*, 703–715. [CrossRef] [PubMed]
33. Vernooij, B.; Friedrich, L.; Goy, P.A.; Staub, T.; Kessmann, H.; Ryals, J. 2, 6-dichloroisonicotinic acid-induced resistance to pathogens without the accumulation of salicylic acid. *Mol. Plant Microbe Interact.* **1995**, *8*, 228–234. [CrossRef]
34. Leslie, F.; Kay, L.; Wilhelm, R.; Peter, M.; Nicole, S.; Gut, R.M.; Beatrice, M.; Sandra, D.; Theodor, S.; Scott, U.; et al. A benzothiadiazole derivative induces systemic acquired resistance in tobacco. *Plant J.* **1996**, *10*, 61–70.
35. Watanabe, T. Effects of probenazole (oryzemate [registered name]) on each stage of rice blast fungus (*Pyricularia oryzae Cavara*). *J. Pestic. Sci.* **1977**, *2*, 395–404. [CrossRef]
36. Métraux, J.; Ahlgoy, P.; Staub, T.; Speich, J.; Steinemann, A.; Ryals, J.; Ward, E. Induced systemic resistance in cucumber in response to 2, 6-dichloro-isonicotinic acid and pathogens. In *Advances in Molecular Genetics of Plant-Microbe Interactions*; Springer: Berlin, Germany, 1991; Volume 1, pp. 432–439.
37. Yoshida, H.; Konishi, K.; Koike, K.; Nakagawa, T.; Sekido, S.; Yamaguchi, I. Effect of n-cyanomethyl-2-chloroisonicotinamide for control of rice blast. *J. Pestic. Sci.* **1990**, *15*, 413–417. [CrossRef]
38. Bektas, Y.; Eulgem, T. Synthetic plant defense elicitors. *Front. Plant Sci.* **2015**, *5*, 804. [CrossRef]

39. Li, J.; Pang, Z.; Trivedi, P.; Zhou, X.; Ying, X.; Jia, H.; Wang, N. 'Candidatus liberibacter asiaticus' encodes a functional salicylic acid (sa) hydroxylase that degrades sa to suppress plant defenses. *Mol. Plant Microbe Interact.* **2017**, *30*, 620–630. [CrossRef]
40. You, I.-S.; Murray, R.; Jollie, D.; Gunsalus, I. Purification and characterization of salicylate hydroxylase from pseudomonas putida ppg7. *Biochem. Biophys. Res. Commun.* **1990**, *169*, 1049–1054. [CrossRef]
41. Lawton, K.; Weymann, K.; Friedrich, L.; Vernooij, B.; Uknes, S.; Ryals, J. Systemic acquired resistance in arabidopsis requires salicylic acid but not ethylene. *Mol. Plant Microbe Interact.* **1995**, *8*, 863–870. [CrossRef]
42. Backman, T.W.; Cao, Y.; Girke, T. Chemmine tools: An online service for analyzing and clustering small molecules. *Nucleic Acids Res.* **2011**, *39*, W486–W491. [CrossRef] [PubMed]
43. Kim, S.; Chen, J.; Cheng, T.; Gindulyte, A.; He, J.; He, S.; Li, Q.; Shoemaker, B.A.; Thiessen, P.A.; Yu, B. Pubchem 2019 update: Improved access to chemical data. *Nucleic Acids Res.* **2018**, *47*, D1102–D1109. [CrossRef] [PubMed]
44. Chen, X.; Barnaby, J.Y.; Sreedharan, A.; Huang, X.; Orbović, V.; Grosser, J.W.; Wang, N.; Dong, X.; Song, W.-Y. Over-expression of the citrus gene ctnh1 confers resistance to bacterial canker disease. *Physiol. Mol. Plant Pathol.* **2013**, *84*, 115–122. [CrossRef]
45. White, R.F. Acetylsalicylic acid (aspirin) induces resistance to tobacco mosaic virus in tobacco. *Virology* **1979**, *99*, 410–412. [CrossRef]
46. Van de Waterbeemd, H.; Camenisch, G.; Folkers, G.; Chretien, J.R.; Raevsky, O.A. Estimation of blood-brain barrier crossing of drugs using molecular size and shape, and h-bonding descriptors. *J. Drug Target.* **1998**, *6*, 151–165. [CrossRef]
47. Kim, S.; Thiessen, P.A.; Bolton, E.E.; Chen, J.; Fu, G.; Gindulyte, A.; Han, L.; He, J.; He, S.; Shoemaker, B.A.; et al. Pubchem substance and compound databases. *Nucleic Acids Res* **2016**, *44*, D1202–D1213. [CrossRef]
48. Parisini, E.; Metrangolo, P.; Pilati, T.; Resnati, G.; Terraneo, G. Halogen bonding in halocarbon–protein complexes: A structural survey. *Chem. Soc. Rev.* **2011**, *40*, 2267–2278. [CrossRef]
49. Wilson, D.C.; Kempthorne, C.J.; Carella, P.; Liscombe, D.K.; Cameron, R.K. Age-related resistance in *Arabidopsis thaliana* involves the mads-domain transcription factor short vegetative phase and direct action of salicylic acid on *Pseudomonas syringae*. *Mol. Plant Microbe Interact.* **2017**, *30*, 919–929. [CrossRef]

MDPI
St. Alban-Anlage 66
4052 Basel
Switzerland
Tel. +41 61 683 77 34
Fax +41 61 302 89 18
www.mdpi.com

International Journal of Molecular Sciences Editorial Office
E-mail: ijms@mdpi.com
www.mdpi.com/journal/ijms

www.ingramcontent.com/pod-product-compliance
Lightning Source LLC
Chambersburg PA
CBHW061501210926
51850CB00008B/1787
9783039289813